THE ERA OF NANOTECHNOLOGY

Emergence and Essentials

THE ERA OF NANOTECHNOLOGY

Emergence and Essentials

Edited by
Cherry Bhargava, PhD
Pardeep Kumar Sharma, PhD
Amit Sachdeva, PhD

AAP APPLE ACADEMIC PRESS

First edition published 2022

Apple Academic Press Inc.
1265 Goldenrod Circle, NE,
Palm Bay, FL 32905 USA

4164 Lakeshore Road, Burlington,
ON, L7L 1A4 Canada

CRC Press
6000 Broken Sound Parkway NW,
Suite 300, Boca Raton, FL 33487-2742 USA

4 Park Square, Milton Park,
Abingdon, Oxon, OX14 4RN UK

© 2022 Apple Academic Press, Inc.

Apple Academic Press exclusively co-publishes with CRC Press, an imprint of Taylor & Francis Group, LLC

Library and Archives Canada Cataloguing in Publication

Title: The era of nanotechnology : emergence and essentials / edited by Cherry Bhargava, PhD, Pardeep Kumar Sharma, PhD, Amit Sachdeva, PhD.

Names: Bhargava, Cherry, 1982- editor. | Sharma, Pardeep Kumar, editor. | Sachdeva, Amit, editor.

Description: First edition. | Includes bibliographical references and index.

Identifiers: Canadiana (print) 20210203048 | Canadiana (ebook) 20210203080 | ISBN 9781771889858 (hardcover) | ISBN 9781774639337 (softcover) | ISBN 9781003160588 (ebook)

Subjects: LCSH: Nanotechnology.

Classification: LCC T174.7 .E73 2022 | DDC 620/.5—dc23

Library of Congress Cataloging-in-Publication Data

Names: Bhargava, Cherry, 1982- editor. | Sharma, Pardeep Kumar, editor. | Sachdeva, Pardeep, editor.

Title: The era of nanotechnology : emergence and essentials / edited by Cherry Bhargava, PhD, Pardeep Kumar Sharma, PhD, Amit Sachdeva, PhD.

Description: First edition. | Palm Bay, FL, USA : Apple Academic Press, 2022. | Includes bibliographical references and index. | Summary: "This book, The Era of Nanotechnology: Emergence and Essentials, presents a broad overview of the field of nanotechnology, focusing on key essentials, and delivers examples of applications in various fields. This book targets the future of nanotechnology in various industries and disciplines. It offers basic to advanced level study of the emerging developing and growing nanotechnology field by highlighting the key fundamentals and the application of advanced nanotechnology in real-life applications. The book looks at nanotechnology applications in a variety of fields, including health care, pharmaceutical sciences and drug delivery, nanomedicine, renewable energy, and more. The chapters offer some realistic examples and the latest research in the field of nanoscience and nanotechnology. With chapters written by internationally recognized experts that describe developments in the field of nanotechnology and nanostructured materials, this volume will provide a valuable resource for all involved in the study related to nanotechnology"-- Provided by publisher.

Identifiers: LCCN 2021019216 (print) | LCCN 2021019217 (ebook) | ISBN 9781771889858 (hbk) | ISBN 9781774639337 (pbk) | ISBN 9781003160588 (ebk)

Subjects: LCSH: Nanotechnology--Industrial applications.

Classification: LCC T174.7 .E73 2022 (print) | LCC T174.7 (ebook) | DDC 620/.5--dc23

LC record available at https://lccn.loc.gov/2021019216

LC ebook record available at https://lccn.loc.gov/2021019217

ISBN: 978-1-77188-985-8 (hbk)
ISBN: 978-1-77463-933-7 (pbk)
ISBN: 978-1-00316-058-8 (ebk)

About the Editors

Cherry Bhargava, PhD
Associate Professor, Department of Computer Science and Engineering, Symbiosis Institute of Technology, Pune, Maharashtra-412115

Cherry Bhargava, PhD, is working as an Associate Professor, Department of Computer Science and Engineering, Symbiosis Institute of Technology, A Constituent of Symbiosis International (Deemed University), Pune, Maharashtra . She has more than 16 years of teaching and research experience. She holds a PhD (ECE) from IKG Punjab Technical University, State Government University, Punjab; an MTech (VLSI Design and CAD) from Thapar University, and a BTech (EIE) from Kurukshetra University. She is GATE qualified with All India Rank of 428. She has authored about 50 technical research papers in SCI and Scopus-indexed journals and national/international conference proceedings. She is also an active reviewer and editorial member of numerous prominent SCI and Scopus indexed journals. She has 19 books to her credit. She has registered six copyrights and filed 21 Indian patents. Four Australian innovation patents are granted to her. She is a recipient of various national and international awards for being an outstanding faculty in engineering and an excellent researcher. Her research area is nanotechnology and artificial intelligence.

Pardeep Kumar Sharma, PhD
Associate Professor, Lovely Professional University, Punjab, India

Pardeep Kumar Sharma, PhD, is working as an Associate Professor at Lovely Professional University, Punjab, India. He has more than 13 years of teaching experience in the field of applied chemistry, experimental analysis, design of experiments, and reliability prediction. He has a PhD from Lovely Professional University and an MSc (Applied Chemistry) from Guru Nanak Dev

University, Amritsar, India. He has authored about 20 research papers in SCI and Scopus-indexed quality journals and national/international conference proceedings. He has six books to his credit in the field of nanotechnology and artificial intelligence. He has filed 18 Indian patents and registered two copyrights. Four Australian innovation patents have been granted to him. He is a recipient of various national and international awards. He is an active reviewer of various indexed journals.

Amit Sachdeva, PhD
Associate Professor, Lovely Professional University, Jalandhar, Punjab, India

Amit Sachdeva, PhD, is an Associate Professor at Lovely Professional University, Jalandhar. Punjab, India. He has teaching experience of more than seven years, and his field of specialization is materials technology. Dr. Sachdeva has authored around 20 technical research papers in SCI and Scopus-indexed quality journals and at national and international conferences. Dr. Sachdeva is also an editorial member of various scientific indexed journals and is a lifetime member of the International Association of Engineers and the Institute for Engineering Research and Publication. Dr. Sachdeva has received a Young Scientist Award from the University of Malaya at the International Conference on Frontiers of Polymers and Advanced Materials 2019 held at Penang Island, Malaysia. He also chaired a session and was also selected as a judge for evaluating poster sessions. He has participated in many international conferences and has also been part of the organizing committees for six international conferences. Dr. Sachdeva also coordinates all the FDPs along with academia-industry interface in charge of Lovely Professional University.

Contents

Contributors .... *ix*

Abbreviations .... *xiii*

Preface .... *xvii*

1. **Nanostructures in Health Care** 1

    Jayprakash Gupta, Lipika, Vivek Gupta, Ankit Kumar Yadav,
    Sachin Kumar Singh, and Pardeep Kumar Sharma

2. **Impact of Nanotechnology in Pharmaceutical Sciences** 31

    Mousmee Sharma and Parteek Prasher

3. **Impact of Nanotechnology on Human Life** 51

    Priyanka, Shipra, and Virinder Kumar Singla

4. **Potential Applications of Nanomedicines in the
    Treatment of Diabetic Neuropathy** 63

    Ankita Sood, Pankaj Prashar, Anamika Gautam, Bimlesh Kumar, Indu Melkani,
    Sakshi Panchal, Sachin Kumar Singh, Monica Gulati, Narendra Kumar Pandey,
    Pardeep Kumar Sharma, Linu Dash, Anupriya, and
    Varimadugu Bhanukirankumar Reddy

5. **An Overview of Preparation, Characterization, and
    Application of Aquasomes** 97

    Sanusha Santhosh, Narendra Kumar Pandey, Sachin Kumar Singh,
    Bimlesh Kumar, Monica Gulati, and Hardeep

6. **Self-Emulsifying Drug Delivery Systems: A Strategy to Improve
    the Bioavailability of Hydrophobic Drugs** 129

    Ankita Banerjee, Rajesh Kumar, Monica Gulati, Sachin Kumar Singh,
    Kamal Dua, Gurvinder Singh, Pardeep Kumar, and Ankit Sharma

7. **Solid Lipid Nanoparticles: An Overview of Production
    Techniques and Applications** 165

    Deepika Sharma, Rajesh Kumar, Rashi Mathur, Deep Shikha Sharma,
    Mangesh Pradeep Kulkarni, Chandan Bhogendra Jha, Gurvinder Singh, and
    Pardeep Kumar

8. **Evaluation of Polymer Electrolytes for Application in DSSC** 199

    Shivani Arora Abrol and Cherry Bhargava

9. Performance of Nano-Structured Thermal Spray
    Coatings in the Renewable Energy Sector.. 215
    Gaurav Prashar and Hitesh Vasudev

10. Molecular Dynamics Simulation for Finding Variation in Young's
    Modulus of Defective Single-Walled Carbon Nanotubes 227
    Manish Dhawan and Sumit Sharma

11. Design and Development of Efficient Semiconductor-Based
    Surface Acoustic Wave Gas Sensor: A Systematic Review..................... 245
    Payal Patial

Index... 257

Contributors

Shivani Arora Abrol
School of Electronics and Electrical Engineering, Lovely Professional University, Phagwara – 144411, Punjab, India

Anupriya
School of Pharmaceutical Sciences, Lovely Professional University, Phagwara – 144411, Punjab, India

Ankita Banerjee
School of Pharmaceutical Sciences, Lovely Professional University, Phagwara, Punjab, India

Cherry Bhargava
Department of Computer Science and Engineering, Symbiosis Institute of Technology, Pune, Maharashtra-412115

Linu Dash
School of Pharmaceutical Sciences, Lovely Professional University, Phagwara – 144411, Punjab, India

Manish Dhawan
Associate Professor, Department of Mechanical Engineering, Lovely Professional, University, Phagwara – 144411, Punjab, India, E-mail: mds_78@rediffmail.com

Kamal Dua
Discipline of Pharmacy, Graduate School of Health, University of Technology, Sydney, Australia

Anamika Gautam
School of Pharmaceutical Sciences, Lovely Professional University, Phagwara – 144411, Punjab, India

Monica Gulati
School of Pharmaceutical Sciences, Lovely Professional University, Phagwara – 144411, Punjab, India

Jayprakash Gupta
Student, School of Pharmaceutical Sciences, Lovely Professional University, Phagwara – 144411, Punjab, India

Vivek Gupta
Associate Professor, School of Pharmaceutical Sciences, Lovely Professional University, Phagwara – 144411, Punjab, India

Hardeep
School of Pharmaceutical Sciences, Lovely Professional University, Phagwara – 144411, Punjab, India

Chandan Bhogendra Jha
Division of Cyclotron and Radiopharmaceutical Sciences, Institute of Nuclear Medicine and Allied Science, DRDO, New Delhi, India

Mangesh Pradeep Kulkarni
School of Pharmaceutical Sciences, Lovely Professional University, Phagwara, Punjab, India

Bimlesh Kumar
School of Pharmaceutical Sciences, Lovely Professional University, Phagwara – 144411, Punjab, India, Tel.: +919888720835, Fax: +91 1824501900;
E-mails: Bimlesh.12474@pu.co.in; bimlesh1pham@gmail.com

Pardeep Kumar
School of Pharmaceutical Sciences, Lovely Professional University, Phagwara, Punjab, India

Rajesh Kumar
School of Pharmaceutical Sciences, Lovely Professional University, Phagwara, Punjab, India,
E-mail: rajksach09@gmail.com

Lipika
Student, School of Pharmaceutical Sciences, Lovely Professional University, Phagwara – 144411,
Punjab, India

Rashi Mathur
Division of Cyclotron and Radiopharmaceutical Sciences, Institute of Nuclear Medicine and Allied
Science, DRDO, New Delhi, India

Indu Melkani
School of Pharmaceutical Sciences, Lovely Professional University, Phagwara – 144411, Punjab, India

Sakshi Panchal
School of Pharmaceutical Sciences, Lovely Professional University, Phagwara – 144411, Punjab, India

Narendra Kumar Pandey
School of Pharmaceutical Sciences, Lovely Professional University, Phagwara – 144411, Punjab, India,
E-mail: herenarendra4u@gmail.com

Payal Patial
Department of Electronics and Communication Engineering, Chandigarh University, Gharuan,
Kharar, Mohali – 140413, Punjab, India, E-mail: payal.patial@gmail.com

Gaurav Prashar
School of Mechanical Engineering, Lovely Professional University, Phagwara – 144411, Punjab, India

Pankaj Prashar
School of Pharmaceutical Sciences, Lovely Professional University, Phagwara – 144411, Punjab, India

Parteek Prasher
Department of Chemistry, Uttaranchal University, Arcadia Grant, Dehradun – 248007, Uttarakhand,
India; Department of Chemistry, University of Petroleum and Energy Studies, Energy Acres,
Dehradun – 248007, Uttarakhand, India

Priyanka
COEM, Punjabi University Neighborhood Campus, Rampura Phul, Punjab, India

Varimadugu Bhanukirankumar Reddy
School of Pharmaceutical Sciences, Lovely Professional University, Phagwara – 144411, Punjab, India

Sanusha Santhosh
School of Pharmaceutical Sciences, Lovely Professional University, Phagwara – 144411, Punjab, India

Ankit Sharma
School of Pharmaceutical Sciences, Lovely Professional University, Phagwara, Punjab, India

Deep Shikha Sharma
School of Pharmaceutical Sciences, Lovely Professional University, Phagwara, Punjab, India

Deepika Sharma
School of Pharmaceutical Sciences, Lovely Professional University, Phagwara, Punjab, India

Mousmee Sharma
Department of Chemistry, Uttaranchal University, Arcadia Grant, Dehradun – 248007, Uttarakhand,
India, E-mail: mousmee.sharma90@gmail.com

Pardeep Kumar Sharma
Associate Professor, School of Pharmaceutical Sciences, Lovely Professional University,
Phagwara – 144411, Punjab, India, E-mail: pardeep.kumar1@lpu.co.in

Sumit Sharma
Assistant Professor, Department of Mechanical Engineering, Dr. B. R. Ambedkar National
Institute of Technology, Jalandhar, Punjab, India

Shipra
COEM, Punjabi University Neighborhood Campus, Rampura Phul, Punjab, India,
E-mail: shipracoem@pbi.ac.in

Gurvinder Singh
School of Pharmaceutical Sciences, Lovely Professional University, Phagwara, Punjab, India

Sachin Kumar Singh
Associate Professor, School of Pharmaceutical Sciences, Lovely Professional University,
Phagwara – 144411, Punjab, India

Virinder Kumar Singla
COEM, Punjabi University Neighborhood Campus, Rampura Phul, Punjab, India

Ankita Sood
School of Pharmaceutical Sciences, Lovely Professional University, Phagwara – 144411, Punjab, India

Hitesh Vasudev
School of Mechanical Engineering, Lovely Professional University, Phagwara – 144411, Punjab, India,
E-mail: hiteshvasudev@yahoo.in

Ankit Kumar Yadav
Research Scholar, School of Pharmaceutical Sciences, Lovely Professional University,
Phagwara – 144411, Punjab, India

Abbreviations

ACNC	alginate gel encapsulated chitosan-coated nanocore
AFM	atomic force microscopy
Ag-FSE	furosemide-silver complex
ALA	α-lipoic acid
ALC	acetyl-L-carnitine
APCs	antigen-presenting cell
ASTM	American Society for Testing Materials
BA-SLN	Baicalin SLNs
BA-sol	Baicalin solution
BBB	blood-brain barrier
BSA	bovine serum albumin
$CaHPO_4$	calcium hydrogen phosphate
CeO_2	cerium oxide
C_{max}	maximum concentration
CMI	cell-mediated immunity
CMOS	complementary metal-oxide semiconductor
CNC	chitosan-coated nanocores
CNTs	carbon nanotubes
COMPASS	condensed phase optimized molecular potential for atomistic simulation studies
CS	chitosan
DLS	dynamic light scattering
DMMP	dimethyl methyl phosphonate
DPN	diabetic neuropathy
DSC	differential scanning calorimetry
EE	entrapment efficiency
ENMs	engineered nanomaterials
EPR	enhanced-permeability-and-retention
FDA	Food and Drug Administration
FEM	finite element method
FF	fill factor
F-SLN	fluorescent SLNs
FTIR	Fourier transform infrared
GABA	dopamine and g-amino butyrate

HA	hyaluronic acid
HCC	hepatocellular carcinoma
HPH	high-pressure homogenization
HVOF	high-velocity oxygen fuel
IDT	inter digitized transducers
IVIVC	*in vitro-in vivo* correlation
LBDDS	lipid-based drug delivery system
LCST	lower critical solution temperature
LFCS	lipid formulation classification system
LPE	liquid phase epitaxy
MBE	molecular beam epitaxy
MC	Monte Carlo
MD	molecular dynamics
MDR	multi-drug-resistant
MDT	mean residence time
MLBLs	mannose-like binding lectins
MM	molecular mechanics
MPI	magnetic particle imaging
MRI	magnetic resonance imaging
NCD	nanobot control design
NeP	neuropathic pain
NIH	National Health Institutes
NIR	near-infrared locale
NLCs	nanostructured lipid carriers
NPs	nanoparticles
NSCLC	non-little cell lung disease
NTI	National Technology Initiative
OVA	ovalbumin
PAMAM	poly(amidoamine)
PAN	polyacrylonitrile
Pb	Plumbu
PBS	phosphate buffer saline
PCS	photon correlation spectroscopy
PDI	polydispersity index
PEG	polyethylene glycol
PLGA	poly(lactic-co-glycolic acid)
PMEM	poly methyl ether methacrylate
PMMA	polymethyl methacrylate
pSLNs	paramagnetic solid lipid nanoparticles

PVA	polyvinyl acetate
PXRD	powder X-ray diffraction
QD-FRET	QD-mediated Forster resonance energy transfer
RDF	radial distribution function
RES	reticuloendothelial system
RESS	rapid expansion of the supercritical solution
RF	radio frequency
RNA	ribonucleic acid
RNAi	RNA interference
ROS	reactive oxygen species
SAB	surface additive bonding
SAS	supper critical anti-solvent
SAW	surface acoustic wave
SDEDDS	self-double nano-emulsifying systems
SE	self-emulsifying
SEDDS	self-emulsifying drug delivery system
SEM	scanning electron microscope
SLNs	solid lipid nanoparticles
SLS	selective layer sintering
SMEDDS	self-micro emulsifying drug delivery systems
SNA	spherical nucleic acid
SNEDDS	self-nano emulsifying drug delivery systems
SNRIs	serotonin and norepinephrine reuptake
SPIONs	superparamagnetic iron oxide nanoparticles
STP	serratiopeptidase
SW	Stone-Wales
SWCNTs	single-walled carbon nanotubes
TEM	transmission electron microscopy
TPGS	tocopherol poly(ethylene glycol)succinate
tSLNs	targeted solid lipid nanoparticles

Preface

Nanotechnology is the study and application of extremely small things and can be used multidisciplinary. Nanotechnology is defined as the technology that deals with particles having dimensions in the range 1–100 nm. All the objects in this size range represent the nanoworld. In order to realize the nanoworld scientist around the globe adopts either a top-down approach or bottom-up approach. In the top-down approach, the large-sized particle is crushed down to dimensions equivalent to the nanoscale range. In the bottom-up approach, particles or devices are created by joining a single atom or molecules together via cohesive/adhesive forces. Major utilizer of such small dimension devices are biotechnologists, physicist, or electronics. It started with microelectronics, and now we have moved up to nanoelectronics. The nanoparticles (NPs) are available in nature before the advent of the human race. It is available in the form of a nano-dimensioned particle in the atmosphere, proteins, DNA, RNA, and cells in the human body.

Today, technology has gone to such a level that we can visualize individual atoms using scanning tunneling microscopy. IBM has written its name using individual atoms of Fe_3O_4. So individual atoms can also be manipulated; they can be moved to left-right or in an upward-downward direction also. Now LEDs are being replaced by Quantum dot technology. In the coming future, QD technology will rule the television industry based on its picture quality and performance. Quantum dots are a particle of nanoscale dimension having a range of 4–10 nm and are made of 10–20 atoms in total. These particles have gained so much popularity in recent times as the loss of energy from these particles is almost zero. Nanoparticles are basically known for their high surface area to volume ratio or aspect ratio. It has advanced and more efficient methods of drug delivery, disease diagnosis, and therapy. Therefore, the study of nanotechnology involves a multidisciplinary approach where various domains combine together, leading to a combined application.

This book, "The Era of Nanotechnology: Emergence and Essentials," presents a broad scoop of the entire field of nanotechnology, focusing on key essentials, and delivers examples of applications in various fields. This book targets the future of nanotechnology in various industries and disciplines. This book offers basic to advance the level study of the emerging developing and growing nanotechnology field by highlighting the key fundamentals and the application of advanced nanotechnology in real life applications.

CHAPTER 1

Nanostructures in Health Care

JAYPRAKASH GUPTA,[1] LIPIKA,[1] VIVEK GUPTA,[2] ANKIT KUMAR YADAV,[3] SACHIN KUMAR SINGH,[2] and PARDEEP KUMAR SHARMA[2]

[1]*Student, School of Pharmaceutical Sciences,*
Lovely Professional University, Phagwara – 144411, Punjab, India

[2]*Associate Professor, School of Pharmaceutical Sciences,*
Lovely Professional University, Phagwara – 144411, Punjab, India,
E-mail: pardeep.kumar1@lpu.co.in (P. K. Sharma)

[3]*Research Scholar, School of Pharmaceutical Sciences,*
Lovely Professional University, Phagwara – 144411, Punjab, India

ABSTRACT

With the advancement in the technology in the field of medical sciences, which promotes the detection and treatment of various diseases, nanotechnology has gained the attention due to its size which ranges in nanometer and makes it suitable to reach at the target site and showing the action on that particular target site only. Nanotechnology involves the preparation of various nanostructures such as nanorobots, nanosomes, dendrimers, nanovesicles, nanoemulsion, etc., which entraps the drug inside them and thus prevention the drug degradation too. From all the nanostructures, nanobots are the emerging nanoelectromechanical device which is used for the imaging, detection, and the treatment of the diseases. Nanobots are the smart nano-sized robotic system, designed to perform the specific task with higher degree of accuracy and precision. With the help of this approach, the targeted delivery of highly toxic anti-cancer agents is achieved without affecting the normal tissues. A microprocessor is fitted inside the nanobots which help to direct and to control the motion of nanobot inside the human body, thus making it suitable for clinical use. The current chapter features on the nanobots, its applications in target drug delivery, treatment of various life-threatening diseases such as cancer and the future of nanobots.

1.1   INTRODUCTION

The global health care industry has shown the significant progression in terms of diagnosis and the treatment of the various life-threatening diseases [1–4]. From the invention of a vaccine to the inventions of highly superior machines such as MRI (magnetic resonance imaging), they came too far, and they are moving towards the nanotechnology. The implementation of nanotechnology in health care system will lead to a new era in the history of medicines [1, 3, 5, 6]. All the available current technologies provide the data from the outside of the body, but what will happen if we could gather the data from the inside of the body itself [6, 7]. The data which will be getting from the inside of the body will be more accurate and very specific which will help in the detection process [8]. More accurate the diagnosis, more effective the treatment will be. So, what does nanotechnology means? Nowadays, it has become most interesting a rear of research in which scientists are developing the nano-sized structures [8, 9]. By combining all the nanostructures together, there is formation of a fully automatic robotic controlled system takes place and that system is called "Nanobots" [10, 11]. Nanobots are helpful in the early diagnosis of the diseases and their effective treatment too, making the health care system better by decreasing the both morbidity and mortality rate globally [8, 12]. This advance technique will leave all the conventional techniques behind in terms of diagnosis, investigation, and the efficacious treatment with zero or minimum chances of failure [9, 13]. Soon in health care system, all the procedures will be robotically controlled for the higher accuracy and few are already implemented for the same [14]. The Scientists and researchers are currently working on the technique which will be curing and defending from inside of the body instead of treating it from outside, and here the medical nanobots comes in [15]. This technology has more advantages over the conventional or existing techniques such as:

- No or minimum tissue injury;
- Very less recovery time [15];
- After the treatment, less care is required;
- Very quick response to any change occurring inside the body;
- Internal assessment of the progression of disease and the treatment process [7, 11, 15].

In addition, also the treatment process could be started before the conditions get worsen. Apart from the above-listed advantages; it has some additional features which make them suitable for the use such as (Figure 1.1) [15, 16]:

- It stores the previous data and the process followed for the treatment of the disease and thus helps to decide pattern to be followed for the treatment [15, 17].
- Nanobots are controlled externally and targeted to the specific site, which smoothens the treatment process [18, 19].
- Delivery of various payloads such as proteins, drugs, etc., to the specific location.
- Easily get disassembled and excreted out of the body after performing the given task [14, 19].

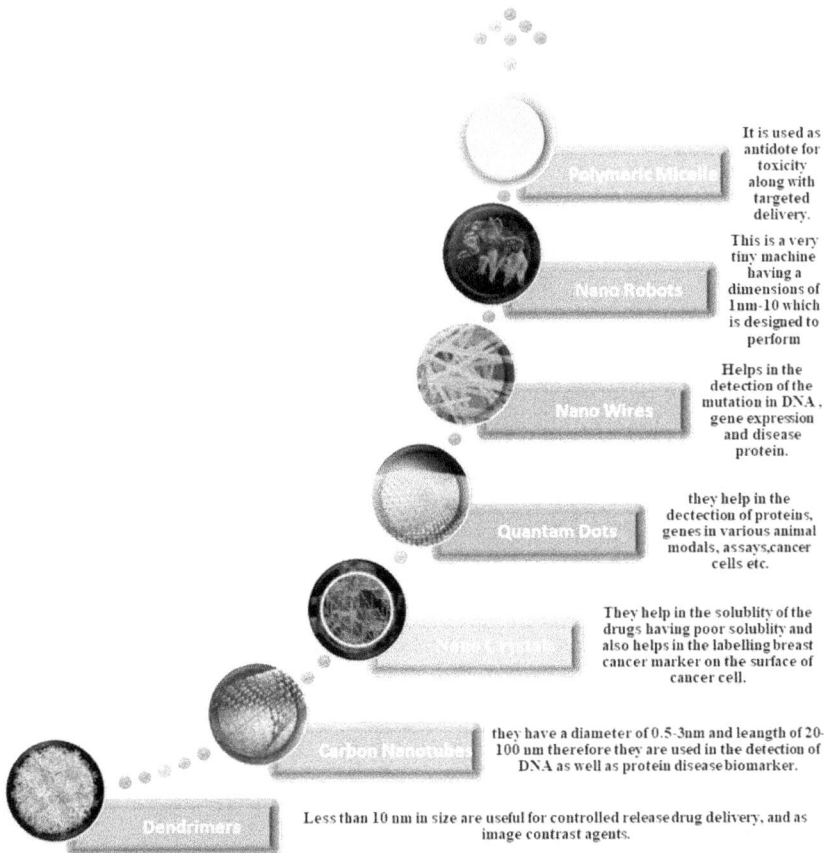

It is used as antidote for toxicity along with targeted delivery.

Polymeric Micell

This is a very tiny machine having a dimensions of 1nm-10 which is designed to perform

Nano Robots

Helps in the detection of the mutation in DNA, gene expression and disease protein.

Nano Wires

they help in the dectection of proteins, genes in various animal modals, assays,cancer cells etc.

Quantam Dots

They help in the solublity of the drugs having poor solublity and also helps in the labelling breast cancer marker on the surface of cancer cell.

Nano Crystal

they have a diameter of 0.5-3nm and leangth of 20-100 nm therefore they are used in the detection of DNA as well as protein disease biomarker.

Carbon Nanotubes

Less than 10 nm in size are useful for controlled release drug delivery, and as image contrast agents.

Dendrimers

FIGURE 1.1   Different types of nanostructures used in health care system for medical applications.

Conventional techniques having the several limitations which enable the use of nanobots. Few limitations are enlisted below:

- Harm to the surrounding tissues and delay healing process [19, 20].
- Treatment process is painful even if anesthesia is given; it only helps in the tolerance of pain but for a short period of time [20].
- Most delicate surgeries such as coronary revascularization, pancreatectomy, eye surgery, esophagectomy, etc., do not have 100% success rate [20–22].
- Whenever a patient is treated by any invasive method, the life of the patient depends on the surgeons/physicians, a single mistake could lead to the death of the patient [20].

1.2   GOALS OF NANOTECHNOLOGY

In the past few decades, nanotechnology has gained the attention and has shown the better progression in the health care system; helps to understand the fundamentals of nature; advances the health care system and education; also responsible for the growth of the various industries and directly manage the economy of a country, environment, and global market size. A recent study has concluded that in the next 10–15 years, this technology will cover the market of over $1 trillion globally per annum, which is a huge market size:

1. **Manufacturing:** Now we are moving towards particle having smaller size to manufacture a compatible object and this could only be achieved by nanotechnology which involves the fabrication and use of nano-sized materials. These nano-sized materials are too much efficient in performing their task. This technology provides understanding between the development and utilization of the nanoengineered tools. Materials with high performance rate required chemical and physical properties, stability, and functions could not be produced, but nanomaterials made it successful. The fabrication and use of nano-materials will reach at the top and will make a great impact on global market by contributing $340 billion annually [23–26].

2. **Electronics:** Nanotechnology majorly contributes in the production of semiconductors as well integrated circuits and it has been projected

that the market size will rise up to $300 billion all across the globe within 10–12 years, due to the implementation of nanotechnology which will provide a new direction in the field of electronics and communication technology [25, 27].

3. **Healthcare System:** Nanotechnology will ease the process of detection and treatment and helps in the faster processing which will increase the life expectancy, improved facilities and extended physical capabilities of human [28–30].

4. **Pharmacy:** More than 50% of the production will be dependent upon nanotechnology and will contribute for more than $180 billion annually in upcoming years [31].

5. **Chemical Plants:** There are various catalysts which are nano-structured and are used for refining and processing of petrol and oils in chemical and petroleum industries. It has been estimated that it will grow to $100 billion from $30 billion with the rate of 10% annually [32].

6. **Transportation:** This technology modifies and eases the transpor-tation facility by providing faster, safer, and lighter vehicles [33]. The vehicles which are made by using the nano-materials are most reliable than the other one. These vehicles are more durable and cos-effective and could be used by any of the transportation way such as roads, railway tracks, bridges, pipelines, runways, etc. In the next 10 years, the transportation facility will cover the market size of more than $70 billion per annum which make a great impact [34, 35].

7. **Sustainability:** Nanotechnology also helps to improve the yields of crops by enabling the various facilities such as water filtration, desalination, and solar energy which creates a pollution free clean environment [36]. It minimizes the extra materials requirements and increases the solar energy supply. Nanotechnology makes a great impact on agricultural industry and it could grow to a larger extend to cover a market size of $100 billion per annum with a significant rate of 10% and also have less carbon dioxide and monoxide emission (Figure 1.2) [34, 37].

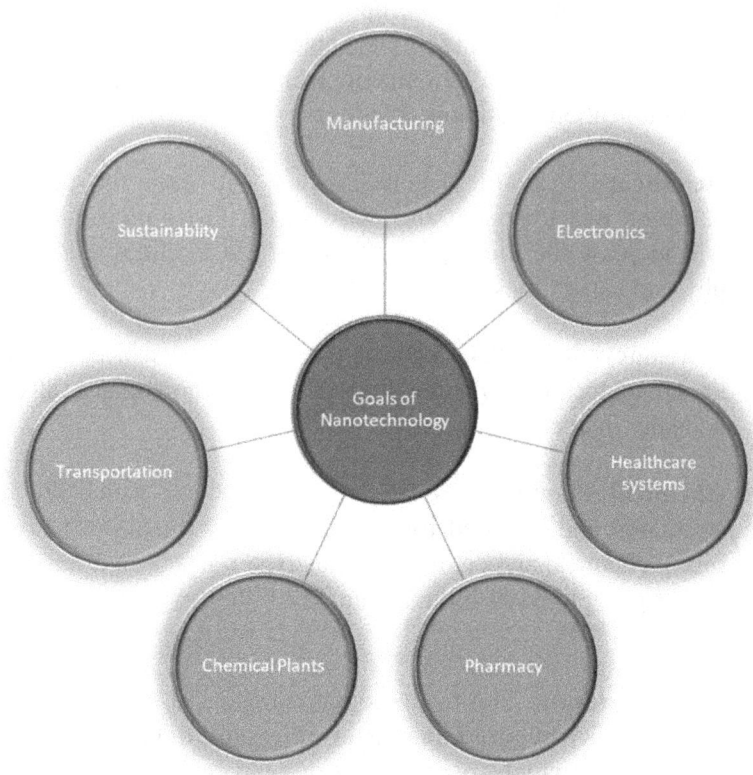

FIGURE 1.2   Different goals of nanotechnology in different fields.

1.3 NANOBOTS

Nanobots offer the unique features in terms of diagnosis and treatment in medical field [38]. These are the basically nano-sized robotic controlled system (size ranging from 1 to 100 nm) manipulated at atomic scale [9, 39]. Due to its smaller size and reliability, nanobots could be used for the variety of applications including drug delivery, diagnosis, surgery, etc., which can lead to technological development and innovative medical products [40, 41]. Initially, robots used to have the solar cells or the batteries inside them for the functioning, but it is too large for nanobot [42]. Then the researchers thought on this problem and finally came to a solution where they have used a thin layer of radioactive material and fixed it inside a nanobot and seen that the nanobot was automatically getting the fuel by the particles which were releasing by the decaying atoms [43–45]. After that it was concluded that this technique can be efficient for the fabrication of a self-driven nano-sized robotic system which will work independently and will never require fuel cell replacement as they won't have any type of cell or power system [46, 47].

Basically, nanobots are the microscopic assembly made up from various nanoscale parts which ranges from 1–100 nm [46, 48]. All the nanoscale parts equipped together to form a working robotic system ranges from 0.5–3 microns in diameter [46, 49]. Nanobots with 3 microns diameter are the largest one used as blood borne medical nanobots, because they have to pass out from blood capillaries and 3 microns in the passage requirement [50, 51]. In reality, they could perform the assigned task at molecular, atomic, and cellular level in both industrial and medical field at their level best [52]. Nanobots as nanomedicine are so smaller particle that they can easily travel inside the human body. The outer structure of nanobot is diamondoid because of inert properties of diamond and higher strength, a carbon allotrope, which makes it suitable for the use. Super smooth surfaces will lessen the likelihood of triggering the body's immune system, allowing the nanobots to go about their business unimpeded [53–55]. Nanomachines are largely in the research and development phase and some of them are already tested. The most important application of the nanomachine will be in health care system where they will be using for the detection of cancerous cells and to destroy them [56]. They might be used for the environmental purpose too such as detection of highly toxic chemicals and measuring their level in the environment [47].

Various life-threatening diseases such as cancer, cardiovascular diseases, and neurodegenerative disorders are continuously progression through a greater extent and causing increase in mortality rate around the globe [40, 57]. To limit these, nanotechnology came into existence. In past few decades, development of the various nanocarriers for the delivery of therapeutics at the target site has shown remarkable progress [58, 59]. These multicomponent nanobots offer high accuracy and are more reliable than the other existing systems [43, 47]. In the early 1970s, there was the concept of antibody-conjugates, polymer-conjugates, and the liposomes and were competing techniques for the drug delivery but out of these only one could have been emerging as an 'antidote' for targeted delivery of therapeutics [60]. As we all know, each has one positive point and one negative and the same was with all the above-mentioned techniques [61]. Antibody conjugates having potential in drug targeting but it can be immunogenic, whereas liposomes having high drug loading capacity but it could delivery total amount of drug at a time or could release very slowly and are very prone to capture by reticuloendothelial system (RES) [49, 62, 63]. An ideal delivery system is the combinations of benefits of two or more technologies simultaneously. Nanotechnology could bring the unexpected revolution in health care system which will provide a new direction in the detection and treatment of diseases for the establishment of a better and efficacious health care system [64]. Nanotechnology is playing a very crucial role and contributing for health care system, but this is initial phase only. Nanobots has the property to navigate itself as blood-borne devices, they could help for the early detection of the various diseases and treatment processes with site specific delivery of therapeutics [55, 65, 66]. Nanobots are highly efficient in performing their task such as early detection of cancer and engulfment of those cancerous cells to limit the spreading the cancer in the body. Nanobots as drug carrier system help to retain the drug for longer duration inside the body and responsible for the slow release of the drug to maintain its effect for longer period of time. It is more beneficial for the delivery of anti-cancer agents for the maintenance of dosage regimen into the systemic circulation to achieve desired pharmacokinetic parameters for the treatment of cancer by chemotherapy [67–70]. It also helps to avoid the interventions with the surrounding healthy tissues and thus minimizes the chances of adverse effects. When we talk about the detection process, nanobot consists of chemical-based nanosensors which are programmed in such a way that they can easily detect the fluctuation in the levels of nitric oxide synthase in case of cerebral aneurysm development at an early stage [39]. Nanobot integrated with chemical nanosensors, or any radio frequency (RF) wireless communication, is used for the determination of such changes in the body responsible for the progression of any disease. It easily identifies the chemical signals inside the body and thus helping in target identification, especially in cerebral treatment [32, 34]. The deposition of amyloid β-protein in the brain is responsible for causing Alzheimer's disease [21]. Generally, when Alzheimer's disease is detected in a patient, he/she is treated with suitable immunotherapy options and also delivered with various neurotransmitters such as dopamine and g-amino butyrate (GABA). But in the case of nanobots, they can directly deliver the myelin basic protein at the signaling site, avoiding the deposition of amyloid protein and preventing them from the attack of the body's self-immune system leading to better impulse generation and conduction in the neurons and tissues [40]. The installation of medical nanobots in the human body will ease the treatment process for the fast elimination of disease and also will help in painless recovery from any kind of physical injury or trauma [71]. They can also deliver the genes into the DNA for the correction of genetic defects or the treatment of any genetic disorder, which makes a great impact on health care system. In addition, the medical nanobots could be used for the enhancement of human's natural capacity. However, these are the electro-mechanical devices which are made outside of the body with most accurate procedure in various regulated nano-factories. Therefore, nanobots are not allowed neither for self-replication process inside the body nor any kind of repair inside the human body, they directly get eliminated out of the subject [72].

1.3.1 PROPERTIES OF MEDICAL NANOBOTS

1. The size of a typical medical nanobot ranging from 0.5 to 3 micron consisting of various parts of sizes 1 to 100 nm, which makes them suitable for the use as 3 microns size of nanobot is maximum and they will easily travel in blood capillaries without blocking the blood flow [73].

2. A medical nanobot is having two spaces, one is outer, and other is inner. The outer part of nanobot is diamondoid due to its inert properties and will be in contact with body's fluid, whereas the inner part of nanobot is closed and won't come in contact with body's chemical fluid unless or until there is any need of chemical analysis [74].

3. The outer surface of nanobot is always kept flawless and smooth and made up of diamond, chemically inert and having very low bioactivity, to prevent itself from the attack of body's immune system by decreasing leukocytes activity.

4. Communication with the medical nanobots is established by the generation of a specific type of acoustic carrier signal which at the wave frequencies of 1–100 MHz inside the body [75]. After the deploy of nanobots in the body, doctors or surgeons will give a command to perform a specific task, that command will be received by the nanobot in the form of acoustic signal and nanobot will decode that command and perform the assigned task [43, 76, 77].

5. After the completion of the task by the medical nanobots, they will be disposed out from the human body to prevent the interference with the normal functionalization of the body tissue and cells [78, 79].

6. If unluckily, the induced nanobot is got trapped by the phagocytic cells, then by inducing exocytosis of the phagocytic cell, it will come out of that or will inhibit the fusion and metabolism by the phago-cytic cell. Thus, coming out of that and assuring the safety of itself [80, 81].

7. Medical nanobots can be recovered after the completion of assigned task by the elimination of nanobots via body's excretory system or it could be destroyed by the scavenger systems [82, 83].

The replication of medical nanobots should inhibit inside the human body. It should be prepared in the regulated nano-factories with accuracy and precision. If the replication of any bacteria or virus inside the body could harm human and responsible for causing life-threatening diseases, then the same with nanobots is unimaginable [80–82, 84].

1.4 COMPONENTS OF MEDICAL NANOBOT

The dimension of medical nanobot varies and depends upon the work they have to perform. Still, it's impossible to say the exact structure of nanobot that how it would look like [85, 86]. They are intended to reach art their target site via traveling into the blood stream which decides the size and shape of nanobot. The nanobots which are non-blood borne and tissue traversing could be larger in size, i.e., as larger as 50–100 microns, whereas the nanobots for the bronchial or alimentary traveling might be even larger than the other ones [8, 87–89]. Each medical nanobot are designed specifi-cally of specific shape and size to accomplish their intended task [81, 90, 91]. The basic components of medical nanobot are:

1. **Skeleton:** It refers to the exterior of the nanobot which made up of allotrope of carbon. Carbon is the principle element contributing 90% of exterior structure of nanobot which mainly involves diamond or diamondoid/fullerene nanocomponents. Other elements which are used in the fabrication process of nanobot are sulfur, hydrogen, nitrogen, oxygen, silicon, fluorine, etc., for achieving the specific purposes in nanoscale components and gears [92–94].

2. **Micro Camera:** A miniature camera is fitted inside the nanobot which help to watch the process going on inside the body. Surgeons can navigate the medical nanobot accordingly by watching the live

video on a television channel manually. Fitting a micro-camera inside nanobot is quite a complex process, so a reliable system should be made, which could ease this process [94].

3. **Payload:** Nanobot could be mounted with proper electrolytic system which will be helping for the generation of electric fields and serving as an option for fuel source. Apart from this, it could also be done that a nanobot will hold a small amount of chemicals which when comes in contact with blood, will get burn and serving as another best option for fuel source [95–97].

4. **Capacitor:** These are used for the generation of magnetic fields which will take conductive fluids from one end of an electromagnetic pump and push it out to the other end. A capacitor is little similar as battery which gives the energy or power source to start the process. For the capacitor's function, mostly non-conductive materials are used which are best suitable such as ceramic, mica, porcelain, Mylar, cellulose, Teflon, and air too [98, 99].

5. **Swimming Tail:** The aim of attaching a swimming tail is propulsion of nanobot to reach at target site and to provide direction for the same [100]. Mostly, it travels against the blood flow, so there is requirement of proper propulsion system which is created by the manipulation of arms of nanobot via creating magnetic field, but outside of the body. That magnetic field causes the vibration of the robotic arms and pushes it ahead through the blood capillaries [101].

1.5 RECOGNITION OF TARGET SITE BY NANOBOT

The various chemical moieties which are having therapeutic efficacy have ability to bind with the specific receptor molecules or a chemotactic sensor by which they show their respective therapeutic response [102]. The same is with a nanobot, after the administration of nanobot in the human body, it will recognize its actual binding site and will perform assigned task. The controlling system of medical nanobot should be well established to perform the task as soon as possible [103]. Inside the human body, the target site releases the specific type of chemicals from its surface which is firstly detected and recognized by the nanobot. Nanobot equipped with a better sensing and actuating system will have the higher affinity to recognize the source, from where the chemical is getting released [103–105]. Nanobot control design

(NCD) simulator is software, which was developed to control the movability of nanobot in viscous body fluid or environment by Brownian motion or inertial forces. By the aid of Brownian motion, a medical nanobot travels all across the body and finds the target site. After the identification of the target site, the nanobot monitors for the change in the concentration of the chemical released by the damaged body cells, which act as a signal for the confirmation of the site of action [8, 106–108]. As the nanobot detect the signal, it moves in the direction of higher concentration of the chemical and finally reaches to the exact site of action. At last, for the confirmation that the nanobot reached at target site, it releases another chemical which act as a guiding signal to reach to action site. By determining various changes in the concentrations of the chemical released by the cells, the medical nanobot detect the signal by passing through the blood capillaries, sizes in microns [109]. For the termination of the collection of nanobots at the target site, "attractant" signal by a nanorobot is terminated and then nanobots stop attracting each other and thus, responses well. Nanobot covers the affected area densely and works accurately and precisely at the action site [110].

1.6   COMMUNICATION OF NANOBOT INSIDE THE BODY

There are many ways to communicate with the nanobot administered into the body and all the methods are used to communicate with the medical nanobot [111]. One of the methods to communicate with the nanobot is by sending the acoustic signal in the form of messages. Nanobot is equipped with a specialized device which converts or decode the acoustic signals into message, similar to the ultrasound probe and works at the frequencies of 1–10 MHz [110]. This helps physicians to send a new command to nanobot, present inside the human body [87]. Each medical nanobot is already equipped with its own electric supply, computer system, sensing system and thus it can easily receive the acoustic signals and convert it into message form and work according to the command or signal send by the physician. Medical nanobot when receive the acoustic signal, it will compute the message and implement the exact response [79]. After the completion of the work, the medical nanobot will send a reply back to the physician to give the information about completion of the work. For the generation, computation, and implementation of the acoustic signal, nanobot requires very less transmission range within microns and thus it is very easily to fabricate individual signal system inside a nanobot [73]. Therefore, it becomes most convenient method to send the signal or to communicate with the medical nanobot present inside the body which easily collects the local messages forwarded by the physicians and physicians can easily monitor the medical nanobot by using an ultrasound detector [69, 109, 112].

1.7   ELIMINATION OF NANOBOTS FROM THE HUMAN BODY

Once the nanobot complete its therapeutic task, it is necessary for nanobot to get out of the body from the systemic circulation [85]. However, it is an artificial device, used for the detection and treatment of the disease inside the human body; it may be desirable to extract those devices out of the body from the circulation [75]. Some of the medical nanodevices get excreted out of the body via human excretory system; it could be excreted through feces only [113]. Few of the nanodevices get excreted by the exfusion process using aphaeresis-like process (known as nanopheresis) or maybe called body's active scavenger system. The excretion system of the nanodevices is very simple; the body's fluid system helps in the exfusion process of the nanodevices from the blood [114, 115]. Blood to be cleared may be passed from the patient to a specialized centrifugation apparatus where acoustic transmitters command nanobots to establish neutral buoyancy [116, 117]. It's very difficult to maintain the exact natural buoyancy by any other solid blood component, therefore those remaining blood components participate in the centrifugation process and helps in drawn off process. As centrifugation process gets over, all the nanobots get deposited at the bottom along with the few solid blood components, which is later on filtered through a 1 micron filter for the separation of nanobot from those suspended solid components [118, 119]. That filtered solid blood components then returned to the person's body undamaged. The separation rate of the nanobots from solid blood components could be increased by emptying all the tanks of nanobot which lowers the net density by 66% of blood plasma density or the same could be achieved by giving a command to the nanobot to blow oxygen bubble of 5 micron and allowing it to increase at the rate of 45 mm/hr under normal gravitational conditions which may adhere by surface tension to the surface [110, 120, 121].

1.8   MEDICAL NANOROBOTIC APPLICATIONS

Numerous of advancements have seen in the field of nanotechnology, it has bought a revolutionary change in the field of medicine and pharmaceuticals. They are used in the targeted delivery of drug, thus known as "pharmacytes"

[122]. They also provide necessary support for the injured organs and also help in the processing of certain chemical reactions in human body. Some of the most important uses of nanobots are as under (Figure 1.3):

- Targeted drug delivery [123];
- Cancer therapy;
- Diagnosis and treatment of diabetes;
- As surgical tools;
- Diagnosis ant testing [124];
- Respirocyte: artificial oxygen carrier nanobot;
- As artificial phagocytes;
- In cell repair and lysis;
- In treatment of atherosclerosis;
- •Neuron replacement.

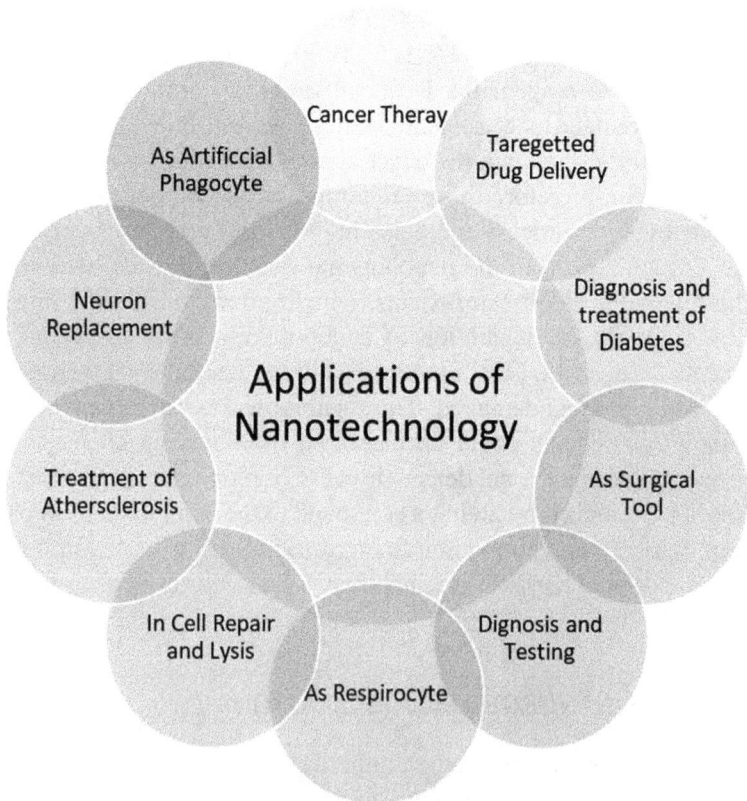

FIGURE 1.3   Various medical applications of nanotechnology.

1.8.1   NANOBOTS IN TARGETED DRUG DELIVERY

The use of nanoparticles (NPs) has made the treatment economic along with less side effects as compared to conventional dosage form [125]. They help in designing of the drug delivery system by improving its pharmacodynamic and pharmacokinetic properties. Encapsulation of drug can be done with the help of micelles; cancer can be treated with the help of Iron or Gold shells nanoparticle [39, 126]. Drug release profile is modified depending on the nature of the drug for the sake of enhancing efficiency, safety, and patient compliance. Site targeting of drugs within body can be achieved for the highly toxic drugs which decompose during their delivery to the specific site [56]. The challenging situation arises when drug is to be delivered and it passes via natural defense mechanism or barriers of the human body such as blood brain barrier (BBB) in the brain, Small intestine, colon, etc. Release rate of the drug is determined during such instances as rapid/slow release of drug may hamper the absorption of drug and can possibly give other side effects [127]. The effect on pharmacokinetic parameters of drug must be altered in such a way that it does not give any side effects. The designing of drug delivery must be done in such a way that it binds to its receptor effectively and must not influence the action of other drugs [128]. Raw material of the system must function in such a way that it does not cause any toxicity in the body and is easily eliminated from body by converting itself into fragments which can be easily excreted from the human body, i.e., it should be bioresorbable [70, 129]. In an excretory system study, it was found out that when encapsulation of mice dendrimer was done to deliver gold nanoparticle which was positively charged was found in the kidney whereas the negatively charged particle remained stuck in spleen and liver due to its small size. Therefore, NPs can be used in drug delivery and targeting but research on its toxicity parameters can increase its uses in medical and pharmaceutical industry [130].

1.8.2   NANOBOTS IN CANCER THERAPY

For the effective treatment of cancer, its early diagnosis is necessary. In addition, targeted drug delivery has to be achieved to overcome the side effects of chemotherapy [46]. Detection of tumor cells can be done with the help of biosensors as they have a property of navigation through blood which lays the building block for the therapy of treatment of cancer [56, 123]. As NPs have large surface area and volume ratio due to which functional groups can easily bound to their surfaces and to the tumor cells as well [131]. Abraxane is a nanoparticle which is albumin bound is used in the treatment of breast and lung cancer [132]. In a study carried out at Rice University and at the University of Texas in mice it was confirmed that nanoparticle can effectively deliver drug for the treatment of head and lung cancer [133]. Cremophor was used which allows intravenous delivery of hydrophobic part of paclitaxel but when nano particles were used instead of cremophor, which decreased the frequency of the dose and improved the targeting of drug along with less side effects [134].

1.8.3   NANOBOTS IN THE DIAGNOSIS AND TREATMENT OF DIABETES

Being a chronic metabolic disorder, diabetes is the most-common disease affecting 415 million population of the world out of which 46% are undiagnosed. It is also estimated that the cases of diabetes will rise to 642 million people around the globe till 2040, which creates a huge burden on the health care system in terms of health, economy, and social consequences [54]. Diabetes is a condition in which there is increased blood-glucose level. The normal uptake or utilization of glucose by the cells in prohibited which rises the level of glucose in the blood [135]. The common signs and symptoms for diabetes are painful urination, frequent urination, increase in appetite and thirst. The level of glucose in the blood decides the body's metabolic system [136]. The hSGLT3 protein has a major role in the regulation of excessive concentration of glucose by maintaining the proper functioning of skeletal muscles and gastrointestinal cholinergic nerves [57]. It can serve as the better treatment option for the diabetic patients. It also serves as a biosensor and identifies the glucose which makes it an interesting treatment aspect. The specific nanobots which are used for the detection and treatment of diabetes are having complementary metal oxide semiconductor (CMOS) nano-bioelectronics [137]. The size of nanobot is ~2 micrometer, which makes it suitable to move freely and to operate inside the body [138]. The body's immune system will not interfere with the functioning of nanobot in the detection process of glucose in blood. For the detection and the monitoring of glucose level in the body, nanobots are equipped with the special chemo-sensor which modulates the activity of hSGLT3 protein for glucose sensing activity [139]. With the help of chemo-sensor, it can easily determine glucose level and will inject the insulin as per the patient's need and will take further action as per the medical prescription, if required [140].

Nanobots flow in the blood capillaries with red blood cells and detect the level of glucose. Nanobots will maintain the glucose range around 130 mg/dl and variation of 30 mg/dl is acceptable, which depend upon the treatment process. An intelligence system is also installed in the medical nanobots, which can transfer the patient's data on his/her mobile automatically through the signal of radio frequencies. If any time, the glucose level rises beyond critical level, it will emit a signal on patient's mobile through making an alarm, and the patient will get caution regarding the same [141–143].

1.8.4   NANOBOTS IN SURGERY

Surgical nanobots are induced inside the human body through body cavities or vascular system and also by attaching it to the ends of catheters [144]. Surgical nanobots are programed in such a way that they could act as a semi-autonomous device at action site for surgery inside the body [43, 56]. Surgical nanobot performs multiple tasks inside the human body such as searching for pathology, diagnosis, and repair of the cells or tissues by either nanomanipulation or by establishing the co-ordination between itself and computer system which is under the supervision of surgeon and sends the commands in the form of ultrasound signals [145]. The researches in this field are in progress and few are already implemented in health care system. For example, a micropipette vibrating at the frequency of 100 Hz is already in use to cut dendrites from a single neuron with harming other neurons. For the regeneration of neurons, femtosecond laser surgery is used, which acts as if pair of "nano-scissors" by causing the vaporization of the local tissues and leaving adjacent tissues unharmed [61, 146, 147].

1.8.5   NANOBOTS IN DIAGNOSIS AND TESTING

Nanobots are used in the diagnosis; various blood testing's, monitoring of various disorders in the human beings as they can detect, record, and report various activities happening in the human body such as change in temperature, pressure, and any chemical modification [40].

1.8.6   RESPIROCYTE: AN ARTIFICIAL OXYGEN CARRIER NANOROBOT

These are the artificial nanomedical red cells, known as "Respirocyte" which are imaginary nanobot's floats along with the blood cells in the circulation [148]. These respirocytes fabricated by using carbon atoms arranged as diamond in the lattice system inside the spherical shell [142]. These types of nanobots are designed in a very unique way and having an empty tank inside which carries O_2 and CO_2 molecules. When it enters into the body, it releases both the gas molecules in controlled manner. These gases are stored in the empty tanks with the pressure of 1000 atmosphere. The tank of respirocytes are made up a unique material, i.e., sapphire, which is a non-flammable material having mechanical and chemical properties similar to diamond. It also consists of sensor system which defines the concentration of the gases filled inside the respirocytes. As the nanobot passes through the respiratory system, it detects the levels of both oxygen and carbon dioxide in the surrounding and as per the requirement it pumps out carbon dioxide and pumps in oxygen, and vice versa depending upon the signals decoded by sensing system of nanobot [8, 149, 150]. When deficiency of oxygen will occur in the respiratory system, the respirocytes will release the oxygen to fulfill the demand [151]. Depending upon the scenarios, the computer system sends the commands to the respirocytes present inside the human body and then respirocytes releases oxygen and absorbs carbon dioxide according to the command given [152, 153]. These respirocytes act as normal hemoglobins which delivers the oxygen all around the body parts. A respirocyte can hold 236 times more oxygen per unit volume than a natural red blood cell. 5 cm^3 dose of artificial respirocytes given in the form of aqueous suspension into the systemic circulation can exactly replace the entire O_2 and CO_2 carrying capacity of the patient's entire 5,400 cm^3 of blood. Respirocytes are having pressure sensors to identify the acoustic signals send by the physicians and by which an artificial respirocyte will perform the task inside the human body [154–156].

1.8.7 NANOBOTS AS ARTIFICIAL PHAGOCYTE (MICROBIVORE)

Microbivore is nanobot which acts by the mechanism of digest or discharge to kill unwanted pathogens which are found in the human blood [157]. The pathogens can be microbiological bacteria, virus, or fungi. When administered intravenously in the blood stream we can get rid of infections within hour when compared to any antibiotic treatment which gives results in months and increases the risk of sepsis as pathogens are not completely digested or killed by the antibiotic [158, 159]. Whereas nanobots turn the pathogens into its monomeric units like sugars, amino acids, glycerol, etc.

1.8.8   NANOBOTS AS ARTIFICIAL NEURONS

Neurons in the brain can be replaced with nanorobots known as nanotech neurons. They can function in an equivalent way like natural neurons. They function similarly like natural neurons by connecting to synapses [160]. As per literature, if the human brain is completely replaced with nanotech equivalent neuron the person has same information and the brain functions in a same way, i.e., no change in the system was observed [161, 162].

1.8.9   NANOBOTS IN TREATMENT OF ATHEROSCLEROSIS

Atherosclerosis refers to the condition in which fats, cholesterol, and other substances (plaques) deposit on the walls of the arteries which can lead to myocardial infarction, stroke, and many other health-related disorders [163, 164]. The condition can be maintained by reducing the amount of bad, i.e., LDL cholesterol in the blood. Nanorobots have capability to locate the lesions in the blood which can be treated either by modifying it pharmacologically, mechanically or chemically [135, 165].

1.8.10   CELL REPAIR AND LYSIS

A very interesting application of nanobot is the cell or tissue repairing. By getting attached with inflammatory cells or WBSc, it reaches to the action site and assists the cells or tissue in healing process [154]. These medical nanobots could transform itself into the body's inflammatory cells and white blood cells, and reaches to the inflamed tissues and help in the process of healing [166]. These movable nanobots are capable to travel all around the body parts through blood capillaries and helps in the process of tissue or organ healing, followed by extravasation, cytopenetration, and total chromatin replacement in the DNA. When nanobot will enter into the body, it will assign the situation of harmed tissue which will size up the condition by observing the cells contents and the activity, and then take action [14, 143, 167]. These medical nanobots will work along cells-by-cells and molecules-by-molecules and tissue-by-tissue which later on resulting in the whole organ repairing and thus restoring the human health by building the normal and healthy molecules [23, 35, 168].

1.9   FUTURE PERSPECTIVES

Nanotechnology has variety of applications in the different field of research and health care system. Various nanostructures have been employed for the improved diagnosis and treatment of the various life-threatening diseases such as cancer. Nanostructures such as nanobots, carbon nanotubes (CNTs), dendrimers, liposomes, micelles, and many more are used as drug carrier molecules to entrap the drug molecules inside and for the controlled release of the drug in the human body. Nanobots are the tiny nano-sized robots which are administered inside the human body and given instructions or command via ultrasound radiations and that command later on will be encoded to a message and perform the task as per the instructions given. These nanobots having ability to perform various tasks inside the human body such as cell repair, surgery, diagnosis, as an artificial respirocyte, etc. Other nanostructures such as nanorods, nanocrystals, nanowires, quantum dots, micelles, nanorobots, nanosomes, liposomes, etc., are used for the controlled release of the drug inside the body for prolongs therapeutic action. Nanobots in health care system can do their best in terms of diagnosis and the treatment of the diseases. We should move ahead with this technology for the improvement of the society. The implement of the nanotechnology in all the sectors will results in the new era and it has the higher degree of accuracy and precision. So, the current chapter details about the nanotechnology, nanobots, and various applications of the same [169, 170].

KEYWORDS

- **blood-brain barrier**
- **cancer**
- **microprocessor**
- **nanobots**
- **nanoelectromechanical**
- **nanotechnology**

REFERENCES

1. Cavalcanti, A., Shirinzadeh, B., Freitas, Jr. R. A., & Hogg, T., (2007). Nanorobot architecture for medical target identification. *Nanotechnology, 19*(1), 015103.
2. Roco, M. C., & Bainbridge, W. S., (2005). Societal implications of nanoscience and nanotechnology: Maximizing human benefit. *Journal of Nanoparticle Research, 7*(1), 1–13.

3. Roco, M. C., (2011). *The Long View of Nanotechnology Development: The National Nanotechnology Initiative at 10 Years.* Springer.

4. Roco, M. C., (2007). National nanotechnology initiative-past, present, future. *Handbook on Nanoscience, Engineering and Technology,* 2.

5. Dang, Y., Zhang, Y., Fan, L., Chen, H., & Roco, M. C., (2010). Trends in worldwide nanotechnology patent applications: 1991 to 2008. *Journal of Nanoparticle Research, 12*(3), 687–706.

6. Lok, C., (2010). Nanotechnology: Small wonders. *Nature, 467*(7311), 18.

7. Elsabahy, M., & Wooley, K. L., (2012). Design of polymeric nanoparticles for biomedical delivery applications. *Chemical Society Reviews, 41*(7), 2545–2561.

8. Freitas, R. A., (2005). Current status of nanomedicine and medical nanorobotics. *Journal of Computational and Theoretical Nanoscience, 2*(1), 1–25.

9. Kroeker, K. L., (2009). Medical nanobots. *Communications of the ACM, 52*(9), 18, 19.

10. Arnon, S., Dahan, N., Koren, A., Radiano, O., Ronen, M., Yannay, T., Giron, J., et al., (2016). Thought-controlled nanoscale robots in a living host. *PloS One, 11*(8), e0161227.

11. Barenholz, Y. C., (2012). Doxil®-the first FDA-approved nano-drug: Lessons learned. *Journal of Controlled Release, 160*(2), 117–134.

12. Douglas, S. M., Bachelet, I., & Church, G. M., (2012). A logic-gated nanorobot for targeted transport of molecular payloads. *Science, 335*(6070), 831–834.

13. McNeil, S. E., (2009). Nanoparticle therapeutics: A personal perspective. *Wiley Interdisciplinary Reviews: Nanomedicine and Nanobiotechnology, 1*(3), 264–271.

14. Kamaly, N., Xiao, Z., Valencia, P. M., Radovic-Moreno, A. F., & Farokhzad, O. C., (2012). Targeted polymeric therapeutic nanoparticles: Design, development and clinical translation. *Chemical Society Reviews, 41*(7), 2971–3010.

15. Bawarski, W. E., Chidlowsky, E., Bharali, D. J., & Mousa, S. A., (2008). Emerging nanopharmaceuticals. *Nanomedicine: Nanotechnology, Biology and Medicine, 4*(4), 273–282.

16. Choi, H. S., & Frangioni, J. V., (2010). Nanoparticles for biomedical imaging: Fundamentals of clinical translation. *Molecular Imaging, 9*(6), 291–310.

17. Tinkle, S., McNeil, S. E., Mühlebach, S., Bawa, R., Borchard, G., Barenholz, Y., Tamarkin, L., & Desai, N., (2014). Nanomedicines: Addressing the scientific and regulatory gap. *Annals of the New York Academy of Sciences, 1313*(1), 35, 56.

18. DeCharms, R. C., (2008). Applications of real-time fMRI. *Nature Reviews Neuroscience, 9*(9), 720–729.

19. Chang, E. H., (2007). Nanomedicines: Improving current cancer therapies and diagnosis. *Nanomedicine: Nanotechnology, Biology, and Medicine, 4*(3), 339.

20. Wang, R., Billone, P. S., & Mullett, W. M., (2013). Nanomedicine in action: An overview of cancer nanomedicine on the market and in clinical trials. *Journal of Nanomaterials.*

21. Tenzer, S., Docter, D., Rosfa, S., Wlodarski, A., Kuharev, J. R., Rekik, A., Knauer, S. K., et al., (2011). Nanoparticle size is a critical physicochemical determinant of the human blood plasma corona: A comprehensive quantitative proteomic analysis. *ACS Nano, 5*(9), 7155–7167.

22. Bobo, D., Robinson, K. J., Islam, J., Thurecht, K. J., & Corrie, S. R., (2016). Nanoparticle-based medicines: A review of FDA-approved materials and clinical trials to date. *Pharmaceutical Research, 33*(10), 2373–2387.

23. Peterson, C. L., (2004). Nanotechnology: From Feynman to the grand challenge of molecular manufacturing. *IEEE Technology and Society Magazine, 23*(4), 9–15.

24. Bonsor, K., (2002). *How Nanotechnology Will Work*. How Stuffs Work.
25. Drexler, K. E., (2001). Machine-phase nanotechnology. *Scientific American, 285*(3), 66, 67.
26. Roco, M. C., Williams, R. S., & Alivisatos, P., (1999). *Nanotechnology Research Directions: IWGN Workshop Report*. Vision for Nanotechnology R&D in the Next Decade; National Science and Technology Council Arlington VA.
27. Bohr, M. T., (2002). Nanotechnology goals and challenges for electronic applications. *IEEE Transactions on Nanotechnology, 1*(1), 56–62.
28. Bennett-Woods, D., (2007). Anticipating the impact of nanosience and nanotechnology in healthcare. *Nanoscale: Issues and Perspectives for the Nano Century*, 295–314.
29. Roco, M. C., (2003). Nanotechnology: Convergence with modern biology and medicine. *Current Opinion in Biotechnology, 14*(3), 337–346.
30. Maheshwari, P., & Gupta, N. V., (2012). Advances of nanotechnology in healthcare. *International Journal of Pharm. Tech. Research, 4*(3), 1221–1227.
31. Varshney, M., & Mohan, S., (2012). Nanotechnology current status in pharmaceutical science. *A, IJTA, 6*, 14–24.
32. Siddiqui, M. H., Al-Whaibi, M. H., & Mohammad, F., (2015). *Nanotechnology and Plant Sciences* (p. 303). Springer International Publishing, Cham. https://doi.org/10.1007/978-3-319-14502-0.
33. Tolles, W., & Rath, B., (2003). Nanotechnology, a stimulus for innovation. *Current Science*, 1746–1759.
34. Cozzens, S., Cortes, R., Soumonni, O., & Woodson, T., (2013). Nanotechnology and the millennium development goals: Water, energy, and agri-food. *Journal of Nanoparticle Research, 15*(11), 2001.
35. Grove, J., Vanikar, S., & Crawford, G., (2010). Nanotechnology: New tools to address old problems. *Transportation Research Record, 2141*(1), 47–51.
36. Prasad, R., Bhattacharyya, A., & Nguyen, Q. D., (2017). Nanotechnology in sustainable agriculture: Recent developments, challenges, and perspectives. *Frontiers in Microbiology, 8*, 1014.
37. McLean, C. R., Mitchell, M., & Barkmeyer, Jr. E. J., (1983). *A Computer Architecture for Small Batch Manufacturing*, https://www.nist.gov/publications/computer-architecture-small-batch-manufacturing-0?pub_id=821367.
38. Toumey, C., (2013). Nanobots today. *Nature Nanotechnology, 8*(7), 475.
39. Bhat, A., (2014). Nanobots: The future of medicine. *International Journal of Management and Engineering Sciences, 5*(1), 44–49.
40. Cerofolini, G., Amato, P., Masserini, M., & Mauri, G., (2010). A surveillance system for early-stage diagnosis of endogenous diseases by swarms of nanobots. *Advanced Science Letters, 3*(4), 345–352.
41. Lösch, A. (2008). *Yearbook of Nanotechnology in Society, Vol. 1* (eds. Fischer, E. et al.), 123–142.
42. Anselmo, A. C., & Mitragotri, S., (2015). A review of clinical translation of inorganic nanoparticles. *The AAPS Journal, 17*(5), 1041–1054.
43. Jacob, T., Hemavathy, K., Jacob, J., Hingorani, A., Marks, N., & Ascher, E., (2011). A nanotechnology-based delivery system: Nanobots novel vehicles for molecular medicine. *The Journal of Cardiovascular Surgery, 52*(2), 159–167.

44. Gutierrez, B., Villalobos, B. C., Corrales, U. Y. R., Vargas, C. S., & Vega, B. J., (2017). Nanobots: development and future, *Int J Biosen Bioelectron, 2*(5), 00037, DOI: 10.15406/ijbsbe.2017.02.00037.

45. Reddy, N. G., (2014). Nanotechnology use in medicine. *Journal of Evolution of Medical and Dental Sciences, 3*(68), 14683–14694.

46. Tucker, P., (2012). Nanobots to fight cancer. *The Futurist, 46*(3), 15.

47. Ahmad, U., & Md, F., (2016). Smart nanobots: The future in nanomedicine and biotherapeutics. *J. Nanomedine Biotherapeutic Discov., 6*, e140.

48. Jones, R., (2005). Biology, Drexler, and nanotechnology. *Materials Today, 8*(8), 56.

49. Unciti-Broceta, A., (2015). Rise of the nanobots. *Nature Chemistry, 7*(7), 538, 539.

50. Mitra, M., (2017). Medical nanobot for cell and tissue repair. *Int. Rob. Auto J., 2*(6), 00038.

51. Franci, G., Falanga, A., Galdiero, S., Palomba, L., Rai, M., Morelli, G., & Galdiero, M., (2015). Silver nanoparticles as potential antibacterial agents. *Molecules, 20*(5), 8856–8874.

52. Morones, J. R., Elechiguerra, J. L., Camacho, A., Holt, K., Kouri, J. B., Ramírez, J. T., & Yacaman, M. J., (2005). The bactericidal effect of silver nanoparticles. *Nanotechnology, 16*(10), 2346.

53. Kim, J. S., Kuk, E., Yu, K. N., Kim, J. H., Park, S. J., Lee, H. J., Kim, S. H., et al., (2007). Antimicrobial effects of silver nanoparticles. *Nanomedicine: Nanotechnology, Biology, and Medicine, 3*(1), 95–101.

54. Somanna, M., (2015). Nanobots: The future of medical treatments. *Int. J. Sci. Tech. Res., 4*, 276–278.

55. Vadlamani, L. I., (2010). *Nanobots as Therapeutic Devices*. Available at: SSRN: 1693838.

56. Devasena, U., Brindha, P., & Thiruchelvi, R., (2018). A review on DNA nanobots: A new techniques for cancer treatment. *Asian J. Pharm. Clin. Res., 11*(6), 61–64.

57. Muthukumaran, G., Ramachandraiah, U., & Samuel, D., (2015). In Role of nanorobots and their medical applications. *Advanced Materials Research* (pp. 61–67). Trans Tech Publ.

58. Ashley, S., (2001). Nanobot construction crews. *Scientific American, 285*(3), 76, 77.

59. Smalley, R. E., (2001). Nanotechnology, education, and the fear of nanobots. *Societal Implications of Nanoscience and Nanotechnology, 44*, 145.

60. Selin, C., (2007). Expectations and the emergence of nanotechnology. *Science, Technology, and Human Values, 32*(2), 196–220.

61. Mali, S., (2013). Nanotechnology for surgeons. *Indian Journal of Surgery, 75*(6), 485–492.

62. Bensaude-Vincent, B., (2007). *Nanobots and Nanotubes: Two Alternative Biomimetic Paradigms of Nanotechnology.*

63. Saini, R., & Saini, S., (2010). Nanotechnology and surgical neurology. *Surgical Neurology International, 1*.

64. Priest, S., (2008). Biotechnology, nanotechnology, media, and public opinion. In: *What Can Nanotechnology Learn from Biotechnology?* (pp. 221–234). Elsevier.

65. Rao, V., Saini, H., & Prasad, P. B., (2014). Nanorobots in medicine: A new dimension in bionanotechnology. *Trans. Netw. Commun., 2*(2), 46–57.

66. Winder, R., (2003). Nanobots: Will they rule? *Chemistry and Industry, 10*, 18, 19.

67. Roman, H. T., (2017). The nanotechnology challenge. *Technology and Engineering Teacher, 77*(4), 42, 43.

68. Bowman, D. M., Hodge, G. A., & Binks, P., (2007). Are we really the prey? Nanotechnology as science and science fiction. *Bulletin of Science, Technology, and Society, 27*(6), 435–445.

69. Plotnikova, M., (2014). In communication in nanotechnology. *Proceedings of the 16th International Conference on Mechatronics-Mechatronika* (pp. 721–724). IEEE.

70. Haberzettl, C., (2002). Nanomedicine: Destination or journey? *Nanotechnology, 13*(4), R9.

71. Divya, R., Ashok, V., & Rajajeyakumar, M., (2018). Nanobots for neurodegenerative disorders. *Journal of Neuroscience and Neuropsychology, 1*(1), 1.

72. Sakhnini, S., & Blonder, R., (2016). Nanotechnology applications as a context for teaching the essential concepts of NST. *International Journal of Science Education, 38*(3), 521–538.

73. Hauert, S., Lo, J. H., Nachum, O., Warren, A. D., & Bhatia, S. N. *Crowdsourcing Swarm Control of Nanobots for Cancer Applications.* http://citeseerx.ist.psu.edu/viewdoc/download?doi=10.1.1.708.1990&rep=rep1&type=pdf.

74. Wood, W. W., (2005). Nanobots: A new paradigm for hydrogeologic characterization. *Ground Water, 43*(4), 463, 464.

75. Loscrí, V., Mannara, V., Natalizio, E., & Aloi, G., (2012). In efficient acoustic communication techniques for nanobots. *Proceedings of the 7th International Conference on Body Area Networks* (pp. 36–39).

76. Gao, W., Sattayasamitsathit, S., & Wang, J., (2012). Catalytically propelled micro-/nanomotors: How fast can they move? *The Chemical Record, 12*(1), 224–231.

77. Khanna, V. K., (2011). *Nanosensors: Physical, Chemical, and Biological.* CRC Press.

78. Salem, A. K., Searson, P. C., & Leong, K. W., (2003). Multifunctional nanorods for gene delivery. *Nature Materials, 2*(10), 668–671.

79. Kelly, M., (2003). Startups seek perfect particles to search and destroy cancer. *Small Times,* 18.

80. Freitas, Jr. R. A., (2005). Progress in nanomedicine and medical nanorobotics. *Handbook of Theoretical and Computational Nanotechnology, 6,* 619–672.

81. Freitas, Jr. R. A., (2009). Nanomedicine and medical nanorobotics. *Biotechnology-Volume XII: Fundamentals in Biotechnology, 11,* 59.

82. Sershen, S., Westcott, S., Halas, N., & West, J., (2000). Temperature-sensitive polymer-nanoshell composites for photothermally modulated drug delivery. *Journal of Biomedical Materials Research: An Official Journal of The Society for Biomaterials, The Japanese Society for Biomaterials, and The Australian Society for Biomaterials and the Korean Society for Biomaterials, 51*(3), 293–298.

83. Quintana, A., Raczka, E., Piehler, L., Lee, I., Myc, A., Majoros, I., Patri, A. K., et al., (2002). Design and function of a dendrimer-based therapeutic nanodevice targeted to tumor cells through the folate receptor. *Pharmaceutical Research, 19*(9), 1310–1316.

84. Ozad, U., Cinpolat, A., Bektas, G., & Rizvanovic, Z., (2017). Improvement steps of plastic surgery to tissue engineering by nanotechnology. In: *Nanostructures for Novel Therapy* (pp. 409–427). Elsevier.

85. Chen, R. J., Bangsaruntip, S., Drouvalakis, K. A., Kam, N. W. S., Shim, M., Li, Y., Kim, W., et al., (2003). Noncovalent functionalization of carbon nanotubes for highly

specific electronic biosensors. *Proceedings of the National Academy of Sciences, 100*(9), 4984–4989.

86. Cai, H., Cao, X., Jiang, Y., He, P., & Fang, Y., (2003). Carbon nanotube-enhanced electrochemical DNA biosensor for DNA hybridization detection. *Analytical and Bioanalytical Chemistry, 375*(2), 287–293.

87. Kadish, K. M., & Ruoff, R. S., (2000). *Fullerenes: Chemistry, Physics, and Technology.* John Wiley & Sons.

88. Reuter, J. D., Myc, A., Hayes, M. M., Gan, Z., Roy, R., Qin, D., Yin, R., et al., (1999). Inhibition of viral adhesion and infection by sialic-acid-conjugated dendritic polymers. *Bioconjugate Chemistry, 10*(2), 271–278.

89. Emerich, D. F., (2005). *Nanomedicine-Prospective Therapeutic and Diagnostic Applications.* Taylor & Francis.

90. Emerich, D. F., & Thanos, C. G., (2005). Nanomedicine. *Current Nanoscience, 1*(3), 177–188.

91. Zekos, G. I., (2006). Patenting abstract ideas in nanotechnology. *The Journal of World Intellectual Property, 9*(1), 113–136.

92. Lancet, T., (2003). *Nanomedicine: Grounds for Optimism, and a Call for Papers.* Elsevier.

93. Steinle, E. D., Mitchell, D. T., Wirtz, M., Lee, S. B., Young, V. Y., & Martin, C. R., (2002). Ion channel mimetic micropore and nanotube membrane sensors. *Analytical Chemistry, 74*(10), 2416–2422.

94. Lee, S. B., Mitchell, D. T., Trofin, L., Nevanen, T. K., Söderlund, H., & Martin, C. R., (2002). Antibody-based bio-nanotube membranes for enantiomeric drug separations. *Science, 296*(5576), 2198–2200.

95. Mavroidis, C., & Ferreira, A., (2013). *Nanorobotics: Current Approaches and Techniques.* Springer Science and Business Media.

96. Mavroidis, C., & Ferreira, A., (2013). Nanorobotics: Past, present, and future. In: *Nanorobotics* (pp. 3–27). Springer.

97. Varadan, V. K., Chen, L., & Xie, J., (2008). *Nanomedicine: Design and Applications of Magnetic Nanomaterials, Nanosensors and Nanosystems.* John Wiley & Sons.

98. Kvennefors, A., & Persson, F., (2004). *The Technology and Applications of Bionanosensors.* Lund University, Sweden.

99. Zhu, A. Y., Yi, F., Reed, J. C., Zhu, H., & Cubukcu, E., (2014). Optoelectromechanical multimodal biosensor with graphene active region. *Nano Letters, 14*(10), 5641–5649.

100. Sattler, K. D., (2010). *Handbook of Nanophysics: Nanotubes and Nanowires.* CRC Press.

101. El-Sayed, A., & Kamel, M., (2019). Advances in nanomedical applications: Diagnostic, therapeutic, immunization, and vaccine production. *Environmental Science and Pollution Research,* 1–14.

102. Johnson, B. N., & Mutharasan, R., (2012). Biosensing using dynamic-mode cantilever sensors: A review. *Biosensors and Bioelectronics, 32*(1), 1–18.

103. Grygar, T., Marken, F., Schröder, U., & Scholz, F., (2002). Electrochemical analysis of solids: A review. *Collection of Czechoslovak Chemical Communications, 67*(2), 163–208.

104. Kang, K., Nilsen-Hamilton, M., & Shrotriya, P., (2008). Differential surface stress sensor for detection of chemical and biological species. *Applied Physics Letters, 93*(14), 143107.

105. Datar, R., Kim, S., Jeon, S., Hesketh, P., Manalis, S., Boisen, A., & Thundat, T., (2009). Cantilever sensors: Nanomechanical tools for diagnostics. *MRS Bulletin, 34*(6), 449–454.
106. Chałupniak, A., Morales-Narváez, E., & Merkoçi, A., (2015). Micro and nanomotors in diagnostics. *Advanced Drug Delivery Reviews, 95,* 104–116.
107. Pozhar, L., Kontar, E., & Hua, M. Z. C., (2002). Transport properties of nanosystems: Viscosity of nanofluids confined in slit nanopores. *Journal of Nanoscience and Nanotechnology, 2*(2), 209–227.
108. Bhatia, S. K., & Nicholson, D., (2003). Hydrodynamic origin of diffusion in nanopores. *Physical Review Letters, 90*(1), 016105.
109. Storm, A., Chen, J., Ling, X., Zandbergen, H., & Dekker, C., (2003). Fabrication of solid-state nanopores with single-nanometre precision. *Nature Materials, 2*(8), 537–540.
110. Hortelão, A. C., Patiño, T., Perez-Jiménez, A., Blanco, À., & Sánchez, S., (2018). Enzyme-powered nanobots enhance anticancer drug delivery. *Advanced Functional Materials, 28*(25), 1705086.
111. Han, M., Gao, X., Su, J. Z., & Nie, S., (2001). Quantum-dot-tagged microbeads for multiplexed optical coding of biomolecules. *Nature Biotechnology, 19*(7), 631–635.
112. McGuiness, D. T., Selis, V., & Marshall, A., (2019). Molecular-based nano-communication network: A ring topology nanobots for *in vivo* drug delivery systems. *IEEE Access, 7,* 12901–12913.
113. Jha, M. K., & Sharma, R. K. Environmental sustainability via emerging molecular nanotechnology. *I Control Pollution 24*(2), 101–110. https://www.icontrolpollution.com/articles/environmental-sustainibility-via-emerging-molecular-nanotechnology-101-110.pdf.
114. Jones, R. A. L., (2008). Rupturing the nanotech rapture. *IEEE Spectrum, 45*(6), 64–67.
115. Ahmad, U., Faiyazuddin, Md. (2016). Smart Nanobots: The Future in Nanomedicine and Biotherapeutics. *J Nanomedine Biotherapeutic Discov 6*: e140. doi:10.4172/2155-983X.1000e140.
116. Israr, M., Tiwari, A., & Gangele, A., (2014). Implementation and application of nanotechnology in industrial sector. *Int. J. of Multidisciplinary and Scientific Emerging Research, 3*(2).
117. Loboduk, M., (2012). *Molecular nanotechnology* (Doctoral dissertation, Sumy State University).
118. Preston, C. J., (2006). The promise and threat of nanotechnology: Can environmental ethics guide us? In: *Nanotechnology Challenges: Implications for Philosophy, Ethics and Society, World Scientific* (pp. 217–248).
119. Dutta, S., & Lawson, R., (2006). The coming nanotech revolution-accounting challenges. *Journal of Cost Management, 20*(3), 39.
120. Singh, M., & Naveen, B., (2014). Molecular nanotechnology: A new avenue for environment treatment. *Journal of Environmental Science, Toxicology and Food Technology, 8*(1), 93–99.
121. Fu, J., & Yan, H., (2012). Controlled drug release by a nanorobot. *Nature Biotechnology, 30*(5), 407.
122. Chude-Okonkwo, U. A., (2014). In Diffusion-controlled enzyme-catalyzed molecular communication system for targeted drug delivery. In: *2014 IEEE Global Communications Conference* (pp. 2826–2831). IEEE.

123. Krishnaswamy, D., Ramanathan, R., & Qamar, A., (2012). In Collaborative wireless nanobots for tumor discovery and drug delivery. In: *2012 IEEE International Conference on Communications (ICC)* (pp. 6203–6208). IEEE.

124. Wortmann, T., Dahmen, C., Geldmann, C., & Fatikow, S., (2010). In Recognition and tracking of magnetic nanobots using MRI. In: *2010 International Symposium on Optomechatronic Technologies* (pp. 1–6). IEEE.

125. Sajja, H. K., East, M. P., Mao, H., Wang, Y. A., Nie, S., & Yang, L., (2009). Development of multifunctional nanoparticles for targeted drug delivery and noninvasive imaging of therapeutic effect. *Current Drug Discovery Technologies, 6*(1), 43–51.

126. Gao, W., Kagan, D., Pak, O. S., Clawson, C., Campuzano, S., Chuluun-Erdene, E., Shipton, E., et al., (2012). Cargo-towing fuel-free magnetic nanoswimmers for targeted drug delivery. *Small, 8*(3), 460–467.

127. Gu, Z., Rolfe, B. E., Xu, Z. P., Campbell, J. H., Lu, G., & Thomas, A. C., (2012). Antibody-targeted drug delivery to injured arteries using layered double hydroxide nanoparticles. *Advanced Healthcare Materials, 1*(5), 669–673.

128. Mostaghaci, B., Yasa, O., Zhuang, J., & Sitti, M., (2017). Bio adhesive bacterial micro swimmers for targeted drug delivery in the urinary and gastrointestinal tracts. *Advanced Science, 4*(6), 1700058.

129. Prabha, M. R., & Juliana, M. J. M., (2019). Thought controlled nanobots for targeted drug delivery: A brief study. *Journal of the Gujarat Research Society, 21*(6), 21–24.

130. Hortelão, A. C. (2017). Urease Powered Nanobots for Drug Delivery Applications, *Nano Bio & Med*, November 22–24, 2017, Barcelona (Spain).

131. Tominaga, N., Yoshioka, Y., & Ochiya, T., (2015). A novel platform for cancer therapy using extracellular vesicles. *Advanced Drug Delivery Reviews, 95*, 50–55.

132. Sharma, A., Zhu, Y., Reddy, M., Hubel, A., Cobian, R., Tan, L., & Stadler, B., (2014). Steerable nanobots for diagnosis and therapy. In: *Universe of Scales: From Nanotechnology to Cosmology* (pp. 179–189). Springer.

133. Lenaghan, S. C., Wang, Y., Xi, N., Fukuda, T., Tarn, T., Hamel, W. R., & Zhang, M., (2013). Grand challenges in bioengineered nanorobotics for cancer therapy. *IEEE Transactions on Biomedical Engineering, 60*(3), 667–673.

134. Da Silva, L. G. V., Barros, K. V. G., De Araújo, F. V. C., Da Silva, G. B., Da Silva, P. A. F., Condori, R. C. I., & Mattos, L., (2016). Nanorobotics in drug delivery systems for treatment of cancer: A review. *J. Mat. Sci. Eng. A., 6*, 167–180.

135. Rifat, T., Hossain, M. S., Alam, M. M., & Rouf, A. S. S., (2019). A review on applications of nanobots in combating complex diseases. *Bangladesh Pharmaceutical Journal, 22*(1), 99–108.

136. Krishna, G., Mary, L. R., & Jerome, K., (2019). In nanobots for biomedical applications. *Proceedings of the 2019 9th International Conference on Biomedical Engineering and Technology* (pp. 270–279).

137. Rajesh, J., Pavithra, G., & Manjunath, T., (2018). Design & Development of Nanobots for Cancer Cure Applications in Bio-Medical Engineering, *International Journal of Engineering Research & Technology (IJERT) Ncesc, 6*(13), 1–7.

138. Thangavel, K., Balamurugan, A., Elango, M., Subiramaniyam, P., & Senrayan, M., (2014). A survey on nano-robotics in nano-medicine. *Nanotechnology, 8*, 9.

139. Sankar, K. M., (2014). *In Powering Nanorobotic Devices: Challenges and Future Strategies* (p. 90601L). Nanosensors, biosensors, and info-tech sensors and systems, International Society for Optics and Photonics.

140. Kumar, R., Baghel, O., Sidar, S. K., Sen, P. K., & Bohidar, S. K., (2014). Applications of nanorobotics. *Int. J. Sci. Res. Eng. Technol., 3*(8), 1131–1137.

141. Abeer, S., (2012). Future medicine: Nanomedicine. *JIMSA, 25*(3), 187–192.

142. Manjunath, A., & Kishore, V., (2014). The promising future in medicine: Nanorobots. *Biomedical Science and Engineering, 2*(2), 42–47.

143. Salunkhe, S. S., Bhatia, N. M., Mali, S. S., Thorat, J. D., Ahir, A. A., & Hajare, A. A., (2014). Nanorobots: Novel emerging technology in the development of pharmaceuticals for drug delivery applications. *J. Pharm. Pharm. Sci., 6*, 4728–4744.

144. McCloy, R., & Stone, R., (2001). Virtual reality in surgery. *Bmj, 323*(7318), 912–915.

145. Fullick, A., (2006). *Frontiers of Surgery*. Heinemann-Raintree Library.

146. Vasudevan, A., (2019). *They Did Surgery on a Grape*. Nature Publishing Group.

147. Jain, K. K., (2008). Nanodevices for medicine and surgery. In: *The Handbook of Nanomedicine* (pp. 183–193). Springer.

148. Hariharan, R., & Manohar, J., (2010). In: *Nanorobotics as Medicament:(Perfect Solution for Cancer), INTERACT-2010*, (pp. 4–7). IEEE.

149. Freitas, R. A., (1998). Exploratory design in medical nanotechnology: A mechanical artificial red cell. *Artificial Cells, Blood Substitutes, and Biotechnology, 26*(4), 411–430.

150. Ghai, I., & Chaudhary, H., (2014). Inventive Appliance of Nano Medicine via Artificial Red Blood Cell Respirocytes. *NSNTAIJ, 8*(4), 125–129.

151. Narasimha, L., Selvan, A., Jairam, D., & Suthakaran, R., (2014). Artificial red blood cells using nanotechnology. *Research Journal of Pharmacy and Technology, 7*(11), 1323–1329.

152. Aswini, K. (2018). Nanorobotics: A Step into New Medical Era. *International Journal of Innovative Science and Research Technology, 3*(10), 611–613.

153. Freitas, Jr. R. A., (2009). Welcome to the future of medicine. *Stud. Health Technol. Inform., 149*, 251–256.

154. Patil, L. B., Patil, S. S., Nitalikar, M. M., Magdum, C. S., & Mohite, S. K., (2016). A review on-novel approaches in nanorobotics. *Asian Journal of Pharmaceutical Research, 6*(4), 217–224.

155. Boonrong, P., & Kaewkamnerdpong, B., (2011). Canonical PSO based nanorobot control for blood vessel repair. *World Academy of Science, Engineering and Technology, 58*, 511–516.

156. Freitas, Jr. R. A. *Nanomedicine and 21st Century Health Care*. http://www.rfreitas.com/Nano/DesEvoForeword.htm.

157. Freitas, Jr. R. A., (2005). Microbivores: Artificial mechanical phagocytes using digest and discharge protocol. *J. Evol. Technol., 14*, 1–52.

158. Freitas, R. A., (2006). Pharmacytes: An ideal vehicle for targeted drug delivery. *Journal of Nanoscience and Nanotechnology, 6*(9/10), 2769–2775.

159. Shanthi, V., & Musunuri, S., (2007). Prospects for medical robots. *Azo. Journal*.

160. Sharma, S., Payal, N., Kaushik, A., & Goel, N., (2014). In blue brain technology: A subway to artificial intelligence. In: *2014 Fourth International Conference on Communication Systems and Network Technologies* (pp. 1106–1109). IEEE.

161. Saniotis, A., Henneberg, M., & Sawalma, A. R., (2018). Integration of nanobots into neural circuits as a future therapy for treating neurodegenerative disorders. *Frontiers in Neuroscience, 12*, 153.

162. Jabbari, A., & Balasingham, I., (2012). In on the modeling of a nanocommunication network using spiking neural architecture. In: *2012 IEEE International Conference on Communications (ICC)* (pp. 6193–6197). IEEE.

163. Beaudin, C., (2018). Use of biocompatible microdroplets for the treatment of atherosclerosis, heart disease and stroke. *U.S. Patent Application No. 15/691*,760.

164. Kethanvee, C., & Lin, L. (2019). Use of Robot Devices in Percutaneous Coronary Interventions and Nanobots Which are Future Devices in Treatment of Heart Diseases. *North American Academic Research, 2*(3), 171–187.

165. Murugesan, P., Awang, A. B., & Prabakar, S. (2019). Propagation Model of Molecular Communication Based Targeted Drug Delivery for Atherosclerosis Disease Therapy. SASSP 2019, Universiti Teknologi PETRONAS, Perak D Ridzuan, Malaysia.

166. Naoghare, P. K., & Song, J. M., (2012). Nanomedicine technologies for cell-based drug screening. *Nanomedicine in Diagnostics,* 151.

167. Kwon, E. Y., Kim, Y. T., & Kim, D. E., (2009). Investigation of penetration force of living cell using an atomic force microscope. *Journal of Mechanical Science and Technology, 23*(7), 1932–1938.

168. Martin-Loeches, I., Forster, R., & Prina-Mello, A., (2018). Intensive care medicine in 2050, nanotechnology. Emerging technologies and approaches and their impact on critical care. *Intensive Care Medicine, 44*(8), 1299–1301.

169. Bhushan, B., (2017). *Springer Handbook of Nanotechnology*. Springer.

170. Maynard, A. D., Aitken, R. J., Butz, T., Colvin, V., Donaldson, K., Oberdörster, G., Philbert, M. A., et al., (2006). Safe handling of nanotechnology. *Nature, 444*(7117), 267–269.

CHAPTER 2

Impact of Nanotechnology in Pharmaceutical Sciences

MOUSMEE SHARMA[1] and PARTEEK PRASHER[1,2]

[1]*Department of Chemistry, Uttaranchal University, Arcadia Grant, Dehradun – 248007, Uttarakhand, India,*
E-mail: mousmee.sharma90@gmail.com (M. Sharma)

[2]*Department of Chemistry, University of Petroleum and Energy Studies, Energy Acres, Dehradun – 248007, Uttarakhand, India*

ABSTRACT

Efficient drug loading capacity, targeted delivery, improved bioavailability, and minimal toxicity cater to contemporary pharmaceutical development's basic principles. However, the ensuing drug development efforts suffered severe setbacks such as multidrug resistance, drug destabilization in the systemic circulation, and crossing blood-brain barriers, which necessitated identifying novel pharmacophores and innovative strategies for effective pharmaceutical designing. In addition, it also requires a precise understanding of the drug pharmacokinetics, pharmacodynamics, and drug metabolism for ensuring a high success rate to improve over the existing therapeutics. Nanotechnology revolutionized the pharma sector by delivering highly desirable results, considering the salient features of any drug development paradigm. From target-specific drug delivery, highly efficient gene transfection in molecular medicine to biomolecular imaging and biosensing in molecular diagnostics, the nanoformulations present a robust candidature as next-generation pharmaceuticals. In this chapter, we present a succinct discussion on the impact of nanotechnology in pharmaceutical sciences, emerging trends, and further challenges.

2.1   INTRODUCTION

Targeted drug delivery represents the benchmark of blockbuster drugs and the most desirable parameter while engineering the drug delivery prototypes for the representative pharmaceuticals, which customarily effect the physiological surroundings of the actual site of drug deliberation [1]. Crossing the physiological barriers such as biological membranes, blood-brain barrier [2], and physicochemical parameters such as lipophilicity pose significant challenge for an effective drug delivery [3]. In addition, the toxicity profile of the administered drug affecting the patient compliance [4], and its controlled release present another significant hurdle in contemporary drug delivery paradigm [5]. Nevertheless, the scarcity of state-of-the-art *in vivo* bioimaging and biosensing probes further hinder the mechanistic appraisal of local drug metabolism. The expression of multidrug-resistance in lethal microbes further necessitated the development of innovative strategies for targeting multiple pathways associated with deleterious expression of a target biological event [6]. The identification of 'hits' and appraisal of 'lead molecules' further requires a comprehensive delivery to the target site for rationalizing the therapeutic prototype of novel pharmaceutics. Nanotechnology affords profitable solutions to counter the existing exigencies associated with pharmaceuticals design, development, and delivery [7, 8]. The vital paradigm shift from conventional medication to nanomedicine, nano-bioprobing, and nano-bioimaging, and nano-theranostics played a decisive role in pharma revolution [9]. Nanosized, biodegradable drug delivery vectors ensured prompt delivery of cargo pharmaceuticals at the diseased tissues and cells, with improved patience compliance [10]. Interestingly, the nanoparticles (NPs) improve the uptake and clearance of therapeutic molecules and active pharmaceutical ingredients from the cerebrospinal fluid, hence validating an effective crossing of the blood-brain barrier (BBB) [11]. Metal NPs endowed with intrinsic oligodynamic and biocidal potency reportedly act in synergism with the representative antibiotics and demonstrate a robust candidature for the development of next-generation antibiotics, which potentially overcome multidrug-resistance [12]. Hence, it justifies the resilience of nanotechnology for laying a sturdy foundation of modern-age pharmaceutical sciences.

2.2   NANOTECHNOLOGY IN DRUG DELIVERY

A successful drug delivery system must ensure the protection of a labile therapeutic molecule from physiological degradation, while maintaining its optimal pharmacokinetic profile and prompting an affluent epithelial diffusion. Nanoscale drug-delivery vehicles such as liposomes, nanoemulsions, micelles, and polymeric NPs serve as contemporary vectors systems for effectively delivering the extremely hydrophilic or highly lipophilic cargo drug molecules across the anatomical membranes.

2.2.1 LIPOSOMES

The liposomes bear a minimum diameter of 30 nm and constitute a biodegradable, physiologically inert, and non-immunogenic lipid bilayer enclosing an aqueous core, capable of accommodating the hydrophilic molecules within the core, whereas the hydrophobic drug molecules place themselves inside the bilayer [13]. The liposomes possess prolonged shelf life in systemic circulation with considerable biological benevolence and efficiently release the payload drugs including anticancer agents, enzymes, vaccines, peptide hormones to the target site [14]. The lysosome interacts with the target cells mainly via adsorption followed by endocytosis into cytoplasm [15]. A direct delivery of the payload material may occur via the fusion of lysosomal lipids with the anatomical phospholipid bilayer, or via lipid exchange process [16]. Drug delivery by liposomes depends on critical physicochemical parameters such as size, composition, and loading efficiency, which decide the modes of delivery by enhanced drug solubility (minoxidil, amphotericin B), protection of labile therapeutic molecules (ribozymes, oligonucleotides, cytosine arabinose), modified physiological distribution/pharmacokinetic profile of cargo drug, and its enhanced intracellular uptake (antineoplastic agents) [17]. The liposome may instigate the innate immune response while entering the cells, which subsequently results in the inactivation of the encapsulated cargo therapeutic thereby necessitating the development of biocompatible liposomal surfaces [18]. Surface ligand modification of liposomes by anchoring with peptides, antibodies, and transferrin enhance their potency to recognize and bind to the target cells effectively [19]. The immunoliposomes that contain antibody functionalized liposomal formulations afford targeted drug delivery to tumor-morbid cells and tumor-associated structures; however, their stability within the reticulo endothelial system is still questionable [20]. Likewise, liposomes possess size limitation to transit effectively cross transcapillary passages; nevertheless, the liposomes reportedly accumulate in several tumors compared to the normal healthy tissues [21]. In some reports, the endosomal pH sensitive liposomes prepared from glutamic acid backbone based cationic amphiphile for delivery of anticancer drugs paclitaxel and curcumin attenuated the cancer growth in animal models [22].

Similarly, the coencapsulation of disulfiram and doxorubicin via liposomal bilayer downregulates the activity of P-glycoprotein, which poses significant challenges in the successful activity of the anticancer drug [23]. Jiang et al. [24] developed liposomes decorated with pH-responsive cell-penetrating peptide and hyaluronic acid (HA), which accumulates at the target tumor cells where it detaches HA eventually generating cationic liposome. The physiological pH of the cancer cells facilitates the cellular uptake of cationic liposome and the proton sponge effect promotes its endosomal/lysosomal escape thereby prompting an efficient intracellular delivery. Baek et al. [25] designed liposome carriers conjugated to RNA aptamer for efficiently delivering the anticancer drug doxorubicin for *in vivo* targeting of LNCaP prostate epithelial cells, with a superior selectivity without affecting the healthy cells. These findings suggested the mitigation of toxicity of the encapsulated chemotherapeutics within aptamosomes. Besides, several reports suggest the heightened efficacy of pH-responsive liposomes for mitochondrial-targeted anticancer drug delivery [26]. The first stage charge conversion in mildly acidic tumor microenvironment followed by proton sponge effect and subsequent hydrolysis of linker enhance the surface positive charge thereby enhancing the efficiency of the carrier liposome in targeted drug delivery [27]. Hence, liposomes afford a considerable drug delivery profile with high target specificity, while overcoming the undesirable physiological toxicity of the payload drug by enhancing the drug bioavailability, and circulation time.

2.2.2 POLYMERIC NANOPARTICLES (NPS)

Polymeric NPs present considerable biocompatibility, non-invasiveness, and biomimetic character hence serving as highly efficient smart drug delivery vehicles [28]. Notably, the hemocompatibility of polymeric NPs depends on factors ranging from surface area and charge, hydrophobicity, and hydrophilicity [29]. Similarly, the histocompatibility of polymeric NPs depends on the type of biomaterial utilized for their synthesis [30]. Such as poly(lactic-co-glycolic acid) (PLGA)-based NPs display show a trivial reaction with tissues; however, the same polymeric NPs instigate inflammatory response in the loose connective tissues surrounding the nerves [31]. The polymeric NPs act as colloidal drug delivery systems, hence improving the therapeutic index of the active pharmaceutical substance by altering the physiological distribution [32]. The polymeric NPs prompt the passive targeting of the carrier drug to the intracellular sites and inflammation affected areas by enhanced permeability and retention effect [33]. In addition, the utilization of polymers as drug delivery systems ensures biodegradability after a prolonged shelf life in systemic circulation [34]. The polyelectrolyte polymers containing free -COOH, -SO$_3$H, or -NH$_2$ functionalities serve as pH-sensitive drug delivery systems for effectively transporting the cargo drug molecules at the tumor microenvironment displaying abnormal pH levels compared to the health cells [35]. Interestingly, the temperature-sensitive polymers contain hydrophobic functionalities such as -CH$_3$, -C$_2$H$_5$, -C$_3$H$_7$, and possess ability to generate secondary H-bond forming complexes, which leads to 'zipper effect' thereby extending their applications for on/off controlled drug release [36]. The polymers containing azobenzene, triphenylmethane, cinnamonyl, and quantum dots display photosensitivity [37]. Some studies suggested the exposure of near infrared light to trigger the release of proteins from a composite material generated from gold nanoshells and poly(NIPAAM-co-AAm) [38]. Several diffusion and erosion mechanisms govern the kinetics of drug release from polymeric NPs. In case of nonerodible polymers, the diffusion mechanism decides the kinetics of pharmaceutical elution that prompts an unavoidable torrential release of drugs. Conversely, for the biodegradable polymers, both diffusion and degradation contribute to drug elution response [39]. Nevertheless, the polymer erosion presents higher complexities compared to its degradation as the former depends on multiple factors including swelling, dissolution, diffusion, and degradation of the constituent monomers and oligomers at an unspecified rate. Rejinold et al. [40]; reported curcumin-loaded biocompatible, thermoresponsive polymeric NPs generated from chitosan-g-poly (N-vinylcaprolactam) for anticancer drug delivery. The polymeric NPs displayed extraordinary stability after curcumin loading with specific toxicity towards the target cancer cells and lower critical solution temperature (LCST) based release mechanism for the cargo drug. The test polymeric NPs proved non-hemolytic in the concentration range of 0.002–10 mg/ml. Katiyar et al. [41] reported poly($_{D,L}$-lactide-co-glycolide) polymer NPs with average size 150 nm, loaded with rapamycin and piperine for breast cancer treatment. The test NPs demonstrated prolonged drug release *in vitro* for 14 days following zero order kinetics with non-Fickian transport, subsequently followed by Higuchi kinetics with Fickian diffusion. The co-loading of polymeric NPs with piperine improved the absorption profile and bioavailability of rapamycin by 4.8-folds. Zhang et al. [42] reported pH-sensitive polymeric NPs or co-delivery of doxorubicin and curcumin for cancer therapy. The combination of anticancer drugs with pro-apoptotic and anti-angiogenic activities proved highly useful for the treatment of human hepatocellular carcinoma (HCC). The co-loading further ensured improved encapsulation efficiency, and better release in the acidic microenvironment of tumor cells. Zhang et al. [43] developed ROS-activated polymeric NPs for self-reporting drug delivery. The amphiphilic block copolymers consisting of a PEG segment and an oxidation-responsive hydrophobic block served as building block for the synthesis of polymeric NPs. The degradation of polymeric NPs displayed colorimetric change of fluorophore thereby encouraging the detection of ROS via ratiometric fluorescent approach. The polymeric NPs effectively encapsulated the anticancer drug doxorubicin released under ROS generated oxidative stress in tumor microenvironment.

2.2.3   NANOEMULSIONS

Nanoemulsions represent heterogeneous oil in water dispersions, having an effective oil droplet size less than 500 nm in diameter, where the oil core incorporates poorly soluble drugs [44]. The amphiphilic surfactants such as phosphatidylcholine and co-surfactants such as deoxycholic acid, steray-lamine, and dioleoyltrimethylammoniumpropane stabilize the droplets of nanoemulsions and afford a desired surface charge [45]. In addition, the presence of surface ligands facilitate the cellular uptake of nanoemulsions droplets via receptor-mediated endocytosis, and the presence of contrast agents in oil phase provides visualization properties [46]. The encapsulation of drug molecule in nanoemulsions facilitates a direct transfer of the cargo across the membranes of gastrointestinal tract thereby preventing their dissolution in the aqueous environment [47]. Contrary to the GI tract, the central nervous system offers negligible paracellular permeability as the endothelial cells lining the brain capillaries form tight junctions thereby restraining the paracellular pathway for transport of hydrophilic drugs [48]. Nanoemulsions play critical role in successfully delivering the lipophilic drug molecules across the BBB, promote their controlled release, and safeguard the cargo drugs from enzymatic degradation [49]. The cellular transfer of the encapsulated drug molecules occurs by the direct fusion of oil droplets with membranes by lipid exchange or by endocytosis [50]. The zeta potential of the nanoemulsions oil droplets also cater to the *in vivo* activity, physiological clearance, and interactions with anatomical membranes. The negatively charged nanoemulsions readily remove from the systemic circulation and demonstrate superior uptake by the tissues compared to the positively charged nanoemulsions [51]. Ma et al. [52] developed core-matched nanoemulsions functionalized by vitamin E and tocopherol poly(ethylene glycol)succinate (TPGS) for co-delivery of hydrophilic and hydrophobic anticancer drugs paclitaxel and 5-fluorouracil respectively across multi-drug-resistant (MDR) human epidermal carcinoma cell lines. The nanoemulsions displayed very high drug entrapment efficiency (EE) and negative zeta potential. The nanoemulsions promoted an upregulation of tumor suppressor p53 and β-tubulin whereby displaying a significant inhibition of the progression of cell cycle of the cancer cells with minimal cytotoxicity towards the healthy cells. Notably, the co-delivery of two chemotherapeutic drugs having different mechanisms prompts the desirable simultaneous interruption of composite anticancer pathways, thereby resulting in heightened therapeutic response. Shakeel et al. [53] reported the transdermal delivery of hydrophilic anticancer drug caffeine via water-in-oil nanoemulsions with an improved safety profile. The animal models after exposure to nanoemulsions displayed morphological changes in stratum corneum however; the infiltration of inflammatory cells or edema did not appear, thereby indicating the absence of skin irritation. Najlah et al. [54] reported the nanoemulsions encapsulation of hydrophobic anticancer drug paclitaxel for beneficial applications against glioma cell lines. The nanoemulsions loaded with paclitaxel showed concentration-dependent cytotoxicity against U87-MG cell and SVG-P12 cell lines. Tripathi et al. [55] reported α-linolenic acid potentiated, and folate functionalized nano-emulsion for targeted delivery of doxorubicin to the breast cancer cells. The drug-loaded nanoemulsions attenuated the proliferative and metastatic effects of 7,12-dimethylbenz[a]anthracene (DMBA) in cancer cells mainly by the activation of mitochondria-mediated apoptotic pathways and by successful downregulation of anti-apoptotic and pro-migratory proteins. Pereira et al. [56] reported oral delivery of anticancer omega-3 fatty acid derivatives via nanoemulsions. The nanoemulsions displayed a considerable bioavailability for displaying anticancer effect in animal models. Hence, the drug-loaded nanoemulsions display considerable biological tolerance and demonstrate anti-inflammatory, antioxidant, and anticancer properties. Hence, the nanoemulsions constitute a vital tool in nanotechnology driven drug delivery paradigm for effecting efficient, targeted drug delivery of hydrophobic anticancer therapeutics with considerably high biocompat-ibility, biodegradability, high compliance, thermodynamic stability, optical clarity, sustained, and controlled drug release, ease of preparation, and non-immunogenicity.

2.3 NANOTECHNOLOGY IN MOLECULAR MEDICINE

2.3.1 *METAL NANOPARTICLES (NPS)*

The metallic NPs possessing exceptional physicochemical, magneto-optical, and biocidal properties and unique ability to bioconjugate to the biomacromolecules, present a robust candidature in the development of molecular medicine in pharmaceutical sciences [57]. The surface engineered metal NPs possessing functional head groups serve as the point of conjugation to the biomolecule of interest [58]. Sulfo-NHS coupling between the surface-functionalized nanoparticle and biomolecule generates amine-reactive esters of carboxylate groups, which find applications in chemical labeling, solid phase immobilization, peptide modifications, and enzyme remodeling [59]. Other covalent crosslinking/bioconjugation approaches such as EDC-NHS coupling between the surface-decorated metal nanoparticle and biomolecules afford several immunodiagnostic applications [60]. The bioconjugated noble metal NPs afford characteristic surface Plasmon resonance that extends their applications in biological probing and for the development of diagnostic sensors for successfully detecting biomarkers by immunoassays [61]. Such as, antibody decorated noble metal NPs pose several applications in the development of highly sensitive fluorescence detection systems for biomarker detection [62]. Similarly, the magnetic properties of superparamagnetic metal oxide nanaoparticles serve as the image contrast enhancing agents in computed tomography and magnetic resonance imaging (MRI) techniques [63]. These NPs also assist in the *in vivo* detection of drug pharmacokinetics, drug activity, and metabolism inside the cells. Besides, the metal oxide NPs such as ZnO, TiO_2, and CuO NPs present beneficial wound healing applications by upregulating tissue engineering, and modulating the cell signaling pathways [64].

2.3.2 *NUCLEIC ACID NANOPARTICLES (NPS)*

The spherical nucleic acid (SNA) NPs demonstrate nanostructures comprising appropriately oriented dense oligonucleotide layers arranged in three-dimensional geometry [65]. The SNA NPs reformed the contemporary gene regulation and therapy, drug delivery, and molecular diagnostics, and RNA interference (RNAi) therapy paradigm [66]. The arrangement of nucleic acids over the nanoparticle surface and their properties depends on the type of the latter, such as quantum dots, SiO_2 NPs, Fe_3O_4 NPs, proteins, and liposomes that are widely used as SNA core [67]. Moreover, the shape and size of the NPs also determine the density of oligonucleotides present on the surface of NPs. Smaller sized NPs afford improved density of the surface oligonucleotide layer due to their higher radius of curvature [68]. It provides a natural deflection angle between the neighboring oligonucleotides hence generating supplementary space around the individual strands. A higher oligonucleotide density around NPs promotes their superior loading over the core hence prompting an affluent entry across the target cells [69]. Notably, the utilization of liposome, PLGA (polylactic-co-glycolic acid), and proteins support control release of the cargo drug for a prolonged therapeutic effect. Further, the introduction of chemically modified nucleotides or nucleic acids into the composition of SNA NPs enhance their thermo-chemical stability, which also promote the desirable activation of the innate immune system for the development of vaccines and tumor immunotherapy [70]. Importantly, the clinical application of SNA NPs obligates the appraisal of their immunologic response as the foreign nucleic acids reportedly trigger the innate immune response in cells [71]. Nevertheless, the negative charge on the nucleic acids restrain the cellular entry of SNA NPs, hence necessitating carrier molecules, which may further amplify the immunostimulatory properties of SNA NPs.

2.3.3   GENE THERAPY

Gene therapy incorporates induction of engineered genes, which code for functional proteins whose expression plays regulatory role in the advancement of diseases. An effective gene therapy counters challenges including physicochemical stability of cargo nucleic acid, and the carrier vector in the extracellular space [72]. In addition, the anatomical membranes, uptake by endocytosis, endosomal escape, and nuclear localization further complicate the representative transfection procedures [73]. NPs validate high transduction efficiency of targeted specific delivery of therapeutic genes, with considerable biocompatibility [74]. Superparamagnetic NPs coated with transfection agents, complex with nucleic acids via polymeric linkers in the presence of magnetic field, which further raises their transfection efficiency by 40-fold as compared to the standard conditions. Polymer-based NPs relying on cationic polymers present robust gene delivery features [75]. On complexing with DNA, they undergo self-assembly, and condensation hence neutralizing the DNA molecules to form nanosized polyplexes, which provide protection to the cargo nucleic acids against nuclease activity [76]. Dendrimers present attractive nanomaterials for highly effective transfection of desired nucleic acids. The candidature of dendrimers appears due to their complex, branched, and globular architecture with inner cavities available to encapsulate the guest molecules [77]. Unique characteristics of dendrimers such as monodispersity, ability to conjugate with target molecules via their terminal head groups, and their extraordinary uniformity further support their utility as transfection agents [78]. Nonetheless, the polysaccharide-based nanomaterials also present utility as transfection agents but minimal solubility, low buffering capacity at physiological pH, and poor endosomal escape followed by an inefficient cytoplasmic decoupling of the complexed nucleic acid limit their applications [79].

2.4 NANOTECHNOLOGY IN THERANOSTICS

2.4.1 *PHOTOTHERMAL THERAPY*

The doctrine of theranostics nanomedicine inculcates ferrying of bioimaging and diagnostic pharmaceutics by nanoplatforms to monitor their therapeutic response *in vivo* [80]. The nanoplatforms include quantum dots, metallic, and magnetic NPs, and carbon nanotubes (CNTs), that offer an enhanced-permeability-and-retention (EPR) effect in the imaging of malignant cancer cells [81]. However, the undue instigation of innate immune response and insignificant half live in systemic circulation limit the clinical efficacy of these nanoplatforms. Photothermal therapy presents an exciting approach in theranostics nanotechnology for annihilation of cancer cells by employing heat [82]. The optical absorption of near-infra light by nanostructures comprising graphene oxide sheets, metal NPs, and CNTs plays a critical role on advancement of photothermal therapy in anticancer theranostics [83]. The conventional oncologic approaches challenge the patience compliance and produce insignificant effect on advanced stages of malignant cancers. Photothermal therapy and photodynamic therapy present the most advanced strategies for selectively obliterating the cancer tissues by instigating thermal or oxidative stress [84]. In photothermal therapy, the near-infrared lasers topically illuminate the morbid cells by converting light energy to heat via optical absorption, hence causing ablation of the target unhealthy tissue. The photothermal therapy aims at uniformly raising the temperature of tumor cells and tissues, while preventing unnecessary damage to the surrounding healthy tissues [85]. The therapeutic temperature maintained at the center of tumor mainly at $\geq 50°C$, hence producing a thermal gradient across the tumor tissue, which eventually reaches this therapeutic temperature throughout the cancer microenvironment [86]. The temperature gradient provides for beneficial therapeutic effect over broader range of cancer expression.

2.4.2 SUPERPARAMAGNETIC IRON OXIDE NANOPARTICLES (SPIONS)

SPIONs possess intrinsic ability to guide biomacromolecules including anti-bodies, nucleotides, peptides, and drugs molecules towards morbid cells under the influence of an external magnetic field [87]. The drug loaded SPIONs enable effective endothelial penetration of cargo molecules in cancer tissues, and for obliterating tumor stroma fibroblasts. Further applications of SPIONs extend in diagnosis, multimodal imaging, chemotherapy, and gene delivery. Notably, the SPIONs afford prolonged retention in systemic circulation, high drug-loading ability, appreciable biodegradability, and minimal physiological toxicity [88]. The SPIONs reportedly produce magnetic hyperthermia by locally transforming the high-frequency external field energy to thermal energy for a selective annihilation of tumor cells. Magnetic hyperthermia presents highly desirable results compared to the laser-triggered photothermal therapy due to the indefinite tissue penetration ability and lower damage to the skin [89]. The most attractive application of SPIONs is their utility as contrast enhancing agents in non-invasive techniques such as MRI, and computed tomography, as they possess high magnetization in the presence of an external magnetic field, which nullifies after the removal of applied field [90]. The SPIONs improve the negative contrast in phantom MRI images by T_2 relaxation of protons present in cellular water content. Similarly, the utility of SPIONs in tomographic methods such as magnetic particle imaging (MPI) affords the detection of their three-dimensional distribution in target tissues [91]. However, the toxicity of SPIONs presents a limiting factor in their theranostics application as they instigate oxidative stress, post tissue penetration by producing ferric/ferrous ions that react with cellular levels of hydrogen peroxide. Further, the factors such as size, magnetic content, and surface coating of SPIONs decide their biocompatibility [92]. Such as small-sized SPIONs with surface positive charge induce enhanced toxicity and interfere with physiological iron metabolism.

2.4.3 QUANTUM DOTS

Quantum dots represent nanosized colloidal semiconductor nanocrystals with exceptional optical properties. They afford a prolonged fluorescence and extraordinary photostability hence prompting the monitoring of vital cellular processes such as probing cellular migration, differentiation, and progressive metastasis [93]. The quantum dots encapsulated in carrier organic and inorganic moieties act as optical imaging probes for critical biological events [94]. The conjugates of quantum dots internalized in the target cells via endocytosis, electroporation, and microinjection customarily display variations in resulting cellular labeling patterns [95]. Hence, the appraisal of internalization mechanisms serves as critical feature for effective cellular delivery of quantum dots. In addition, the surface oxidation, and leaching out of heavy metals from the core or quantum dots promote a considerable cytotoxicity, which also depends on several other physicochemical parameters such as size, surface coating material, surface charge, level of dosage, and exposure time [96]. Moreover, the quantum dots may alter the diffusivity of carrier biomolecule, and potentially disturb the native protein conformation. The appreciable surface-to-volume ration of quantum dots favors the development of smart nanoplatforms that inculcate both imaging, and therapeutic/ diagnostic properties [97]. The smaller sized quantum dots, with thinner coating prove beneficial to QD-mediated Forster resonance energy transfer (QD-FRET) constructs amplify the energy transfer efficiency [98]. Nevertheless, the quantum dots possess superior optical properties compared to the organic fluorophores, which proves advantageous for FRET configurations, provides broad range absorption, and strong resistance to photobleaching [99]. The quantum dots also possess fairly symmetric and tunable spectra due to the quantum confinement effect. Besides, the recent development in the doctrine of quantum dots focuses on biological and medicinal applications with enhanced sensitivity of detection.

2.5 CONCLUSION AND FUTURE PERSPECTIVES

The nanotechnology has boosted the pharmaceutical sector by providing state-of-the-art approaches towards the intricacies that malign the progress of contemporary doctrines of drug design, development, and delivery. The nanotechnologies based on surface functionalized metal NPs, SPIONs, and quantum dots enable the *in vivo* imaging and molecular diagnosis with high efficiency. The bioconjugation of nanomaterials to the biomacromolecules of interest opens up new ventures in molecular medicine and targeted drug delivery. The development of liposomes, nucleic acid NPs, core shell nanosystems, promote an effective drug delivery at the target site, in addition to an efficient transfection where the encapsulating nanosystem protects the cargo molecules, nucleic acids, and therapeutics from physiological degradation by enzymes. Strategies such as photothermal therapy and magnetic hyperthermia ensure effective annihilation of malignant tumors without affecting the surrounding healthy cells. Similarly, the gene therapy by nanosystems validates reversal of faulty gene expression by pinpointed engineering of deleterious nucleic acids. Besides several advantages, the clinical implications of nanomaterials face persistent challenges related to the associated physiological toxicity, and triggering of innate immune response, which further manifests unwanted biological effects. The metal-based nanomaterials instigate redox stress by supporting the production of a cascade of reactive oxygen species (ROS) that further demonstrate undesirable malfunctioning of essential peptides, enzymes, and organelles. Lastly, the biological acceptability of nanomaterials for efficient translocation of cargo molecules across the anatomical membranes, and BBB is highly questionable. Hence, it is highly anticipated of a profitable nanotechnology doctrine in pharmaceutical sciences to appraise the representative anomalies that decelerate the clinical progression of nanomaterials.

KEYWORDS

- **enhanced-permeability-and-retention**
- **hepatocellular carcinoma**
- **hyaluronic acid**
- **lower critical solution temperature**
- **magnetic particle imaging**
- **multi-drug-resistant**

REFERENCES

1. Wang, B., Hu, L., & Siahaan, T. J., (2016). *Drug Delivery: Principal and Applications.* John Wiley and Sons.
2. Pandit, R., Chen, L., & Gotz, J., (2019). The blood-brain barrier: Physiology and strategies for drug delivery. *Adv. Drug Deliv. Rev.* doi: 10.1016/j.addr.2019.11.009.
3. Lipinski, C. A., (2016). Rule of five in 2015 and beyond: Target and ligand structural limitations, ligand chemistry structure and drug-discovery project decisions. *Adv. Drug Deliv. Rev., 101*, 34–41.

4. Malaviya, P., Shukla, D., & Vasavada, A. R., (2019). Nanotechnology-based drug delivery, metabolism and toxicity. *Curr. Drug. Metabol., 20*, 1167–1190.

5. Park, K., (2014). Controlled drug delivery systems: Past forward and future back. *J. Control. Release, 190*, 3–8.

6. Kunjachan, S., Rychlik, B., Storm, G., Kiessling, F., & Lammers, T., (2013). Multidrug resistance: Physiological principles and nanomedical solutions. *Adv. Drug Deliv. Rev., 65*, 1852–1865.

7. Wacker, M. G., (2019). Frontiers in pharmaceutical nanotechnology. *Beilstein J. Nanotechnol., 10*, 2538–2540.

8. Prasher, P., Sharma, M., Mudila, H., Gupta, G., Sharma, A. K., Kumar, D., Bakshi, H. A., et al., (2020). Emerging trends in clinical implications of bio-conjugated silver nanoparticles in drug delivery. *Colloid Interface Sci. Commun., 35*, Article: 100244.

9. Saxena, S. K., & Khurana, S. M. P., (2020). *NanoBioMedicine*. Springer, Singapore.

10. Jain, K. K., (2020). *Drug Delivery Systems*. Humana, New York, NY.

11. Tosi, G., Duskey, J. T., & Kreuter, J., (2020). Nanoparticles as carriers for drug delivery of macromolecules across the blood-brain barrier. *Expert Opin. Drug Deliv., 17*. doi: 10.1080/17425247.2020.1698544.

12. Prasher, P., Singh, M., & Mudila, H., (2018). Green synthesis of silver nanoparticles and their antifungal properties. *BioNanoScience, 8*, 254–263.

13. Daraee, H., Etemadi, A., Kouhi, M., Alimirzalu, S., & Akbarzadeh, A., (2016). Application of liposomes in medicine and drug delivery. *Artif. Cells Nanomed. Biotechnol., 44*, 381–391.

14. Rip, J., (2016). Liposome technologies and drug delivery to the CNS. *Drug Disc. Today, 20*, 53–58.

15. Vahed, S. Z., Salehi, R., Davaran, S., & Sharifi, S., (2017). Liposome-based drug co-delivery system in cancer cells. *Mater. Sci. Engg. C., 71*, 1327–1341.

16. Zylberberg, C., & Matosevic, S., (2016). Pharmaceutical liposomal drug delivery: A review of new delivery systems and a look at the regulatory landscape. *Drug Deliv., 23*, 3319–3329.

17. Lee, Y., & Thompson, D. H., (2017). Stimuli-responsive liposomes for drug delivery. *WIREs Nanomed. Nanobiotechnol., 9*, Article: e1450.

18. Szebeni, J., & Moghimi, S. M., (2009). Liposome triggering of innate immune responses: A perspective on benefits and adverse reactions. *J. Liposome Res., 19*, 85–90.

19. La-Beck, N. M., Liu, X., & Wood, L. M., (2019). Harnessing liposome interactions with the immune system for the next breakthrough in cancer drug delivery. *Front. Pharmacol., 10*, Article: 220.

20. Sun, X., Yan, X., Jacobson, O., Sun, W., Wang, Z., Tong, X., Xia, Y., et al., (2017). Improved tumor uptake by optimizing liposome-based RES blockade strategy. *Theranostics, 7*, 319–328.

21. Deshpande, P. P., Biswas, S., & Torchilin, V. P., (2013). Current trends in the use of liposomes for tumor targeting. *Nanomed. (London Eng.), 8*, 1509–1528.

22. Moku, G., Gulla, S. K., Nimmu, N. V., Khalid, S., & Chaudhuri, A., (2016). Delivering anticancer drugs with endosomal pH-sensitive anticancer liposomes. *Biomater. Sci., 4*, 627–638.

23. Rolle, F., Bincoletto, V., Gazzano, E., Lollo, G., Stella, B., Riganti, C., & Arpicco, S., (2020). Coencapsulation of disulfiram and doxorubicin in liposomes strongly reverses multidrug resistance in breast cancer cells. *Int. J. Pharmaceutics, 580*, Article: 119191.

24. Jiang, T., Zhang, Z., Zhang, Y., Lv, H., Zhou, J., Li, C., Hou, L., & Zhang, Q., (2012). Dual-functional liposomes based on pH-responsive cell-penetrating peptide and hyaluronic acid for tumor-targeted anticancer drug delivery. *Biomater., 33*, 9246–9258.

25. Baek, S. E., Lee, K. H., Park, Y. S., Oh, D. K., Oh, S., Kim, K. S., & Kim, D. E., (2014). RNA aptamer-conjugated liposome as an efficient anticancer drug delivery vehicle targeting cancer cells *in vivo. J. Control. Release, 196*, 234–242.

26. Mo, R., Sun, Q., Xue, J., Li, N., Li, W., Zhang, C., & Ping, Q., (2012). Multistage pH-responsive liposomes for mitochondrial-targeted anticancer drug delivery. *Adv. Mater., 24*, 3659–3665.

27. Nisini, R., Poerio, N., Mariotti, S., Santis, F. D., & Fraziano, M., (2018). The multirole of liposomes in therapy and prevention of infectious diseases. *Front. Immunol., 9*, Article: 155.

28. Kumari, A., Yadav, S. K., & Yadav, S. C., (2010). Biodegradable polymeric nanoparticles based drug delivery systems. *Colloid Surf. B., 75*, 1–18.

29. Thasneem, Y. M., Sajeesh, S., & Sharma, C. P., (2011). Effect of thiol functionalization on the hemo-compatibility of PLGA nanoparticles. *J. Biomed. Mater. Res., 99A*, 607–617.

30. Colson, Y. L., & Grinstaff, M. W., (2012). Biologically responsive polymeric nanoparticles for drug delivery. *Adv. Mater., 24*, 3878–3886.

31. Mao, C., Jiang, L. C., Luo, W. P., Liu, H. K., Bao, J. C., Huang, X. H., & Shen, J., (2009). Novel-blood compatibility polyurethane ionomer nanoparticles. *Macromol. 42*, 9366–9368.

32. El-Say, K. M., & El-Sawy, H. S., (2017). Polymeric nanoparticles: Promising platform for drug delivery. *Int. J. Pharm., 528*, 675–691.

33. Kumari, P., Ghosh, B., & Biswas, S., (2016). Nanocarriers for cancer-targeted drug delivery. *J. Drug Target., 24*, 179–191.

34. Karlsson, J., Vaughan, H. J., & Green, J. J., (2018). Biodegradable polymeric nanoparticles for therapeutic cancer treatments. *Ann. Rev. Chem. Biomol. Eng., 9*, 105–127.

35. Feng, Q., Wilhelm, J., & Gao, J., (2019). Transistor-like ultra-pH-sensitive polymeric nanoparticles. *Acc. Chem. Res., 52*, 1485–1495.

36. Zhang, J., Li, J., Shi, Z., Yang, Y., Xie, X., Lee, S. M. Y., Wang, Y., et al., (2017). pH-sensitive polymeric nanoparticles for co-delivery of doxorubicin and curcumin to treat cancer via enhanced pro-apoptotic and anti-angiogenic activities. *Acta Biomater., 58*, 349–364.

37. Amiri, M., & Alizaddeh, N., (2020). Highly photosensitive near infrared photodetector based on polypyrrole nanoparticle incorporated with CdS quantum dots. *Mater. Sci. Semicond. Process, 111*, Article: 104964.

38. Sershen, S. R., Westcott, S. L., Halas, N. J., & West, J. L., (2000). Temperature-sensitive polymer-nanoshell composites for photothermally modulated delivery. *J. Biomed. Mater. Res., 51*, 293–298.

39. Zhao, K., Li, D., Shi, C., Ma, X., Rong, G., Kang, H., Wang, X., & Sun, B., (2016). Biodegradable polymeric nanoparticles as the delivery carrier for drug. *Curr. Drug Deliv., 13*, 494–499.

40. Rejinold, N. S., Muthunarayanan, M., Divyarani, V. V., Sreerekha, P. R., Chennazhi, K. P., Nair, S. V., Tamura, H., & Jayakumar, H., (2011). Curcumin-loaded biocompatible thermoresponsive polymeric nanoparticles for cancer drug delivery. *J. Colloid Interf. Sci., 360*, 39–51.

41. Katiyar, S. S., Muntimadugu, E., Rafeeqi, T. A., Domb, A. J., & Khan, W., (2016). Co-delivery of rapamycin- and piperine-loaded polymeric nanoparticles for breast cancer treatment. *Drug Deliv., 23*, 2608–2616.

42. Zhang, P., Wang, Y., Lian, J., Shen, Q., Wang, C., Ma, B., Zhang, Y., et al., (2017). Engineering the surface of smart nanocarriers using a pH-/thermal-/GSH-responsive polymer zipper for precise tumor targeting therapy *in vivo*. *Adv. Mater., 29*, Article: e1702311.

43. Zhang, M., Song, C. C., Su, S., Du, F. S., & Li, Z. C., (2018). ROS-activated ratiometric fluorescent polymeric nanoparticles for self-reporting drug delivery. *ACS Appl. Mater. Interface, 10*, 7798–7810.

44. Jaiswal, M., Dudhe, R., & Sharma, P. K., (2015). Nanoemulsion: An advanced mode of drug delivery system. *3 Biotech., 5*, 123–127.

45. Singh, Y., Meher, J. G., Raval, K., Khan, F. A., Chaurasia, M., Jain, N. K., & Chourasia, M. K., (2017). Nanoemulsion: Concepts, development and applications in drug delivery. *J. Control. Rel., 252*, 28–49.

46. Hormann, K., & Zimmer, A., (2016). Drug delivery and drug targeting with parenteral lipid nanoemulsions: A review. *J. Control. Rel., 223*, 85–98.

47. Ge, L., He, X., Zhang, Y., Chai, F., Jiang, L., Webster, T. J., & Zheng, C., (2018). A dabigatran etexilate phospholipid complex nanoemulsion system for further oral bioavailability by reducing drug-leakage in the gastrointestinal tract. *Nanomed: Nanotechnol. Biol. Med., 14*, 1455–1464.

48. Ali, A., Ansari, V. A., Ahmad, U., Akhtar, J., & Jahan, A., (2017). Nanoemulsion: An advanced vehicle for efficient drug delivery. *Drug Res., 67*, 617–631.

49. Karami, Z., Zanjani, M. R. S., & Hamidi, M., (2019). Nanoemulsions in CNS drug delivery: Recent developments, impacts and challenges. *Drug Disc. Today., 24*, 1104–1115.

50. Megumi, N. Y., Edna, T. M. K., Raimar, L., & Nadia, A. B. C., (2017). Challenges and future prospects of nanoemulsion as a drug delivery system. *Curr. Pharm. Des., 23*, 495–508.

51. Ganta, S., Talekar, M., Singh, A., Coleman, T. P., & Amiji, M. M., (2014). Nanoemulsions in translational research-opportunities and challenges in targeted cancer therapy. *AAPS PharmSciTech., 15*, 694–708.

52. Ma, Y., Liu, D., Wang, D., Wang, Y., Fu, Q., Fallon, J. K., Yang, X., et al., (2014). Combinational delivery of hydrophobic and hydrophilic anticancer drugs in single nanoemulsions to treat MDR in *Cancer. Mol. Pharm., 11*, 2623–2630.

53. Shakeel, F., & Ramadan, W., (2010). Transdermal delivery of anticancer drug caffeine from water-in-oil nanoemulsions. *Colloid. Surf. B, 75*, 356–362.

54. Najlah, M., Kadam, A., Wan, K. W., Ahmed, W., Taylor, K. M. G., & Elhissi, A. M. A., (2016). Novel paclitaxel formulations solubilized by parenteral nutrition nanoemulsions for application against glioma cell lines. *Int. J. Pharm., 506*, 102–109.

55. Tripathi, C. B., Parashar, P., Arya, M., Singh, M., Kanoujiya, J., Kaithwas, G., & Saraf, S. A., (2018). QbD-based development of α-linolenic acid potentiated nanoemulsion for targeted delivery of doxorubicin in DMBA-induced mammary gland carcinoma: *In vitro* and *in vivo* evaluation. *Drug. Deliv. Transl. Res., 8*, 1313–1334.

56. Pereira, G. G., Rawling, T., Pozzoli, M., Pazderka, C., Chen, Y., Dunstan, C. R., Murray, M., & Sonvico, F., (2018). Nanoemulsion-enabled oral delivery of novel anticancer ω-3 fatty acid derivatives. *Nanomater., 8*, 825.

57. Prasher, P., Singh, M., & Mudila, H., (2018). Oligodynamic effect of silver nanoparticles: A review. *BioNanoScience, 8*, 951–962.
58. Singh, M., & Prasher, P., (2018). Ultrafine silver nanoparticles: Synthesis and biocidal studies. *BioNanoScience, 8*, 735–741.
59. Bartczak, D., & Kanaras, A. G., (2011). *Preparation of Peptide-Functionalized Gold Nanoparticles Using One Pot EDC/Sulfo-NHS Coupling* (Vol. 27, pp. 10119–10123). Langmuir.
60. Yuce, M., & Kurt, H., (2017). How to make nanobiosensors: Surface modification and characterization of nanomaterials for biosensing applications. *RSC Adv., 7*, 49386–49403.
61. Oliveira, J. P., Prado, A. R., Keijok, W. J., Antunes, P. W. P., Yapuchura, E. R., & Guimaraes, M. C. C., (2019). Impact of conjugation strategies for targeting of antibodies in gold nanoparticles for ultrasensitive detection of 17β-estradiol. *Sci. Rep., 9*, Article: 13859.
62. Cardoso, M. M., Peca, I. N., & Roque, A. C., (2012). Antibody-conjugated nanoparticles for therapeutic applications. *Curr. Med. Chem., 19*, 3103–3127.
63. Wahajuddin, & Arora, S., (2012). Superparamagnetic iron oxide nanoparticles: Magnetic nanoplatforms as drug carriers. *Int. J. Nanomed., 7*, 3445–3471.
64. Erathodiyil, M., & Ying, J. Y., (2011). Functionalization of inorganic nanoparticles for bioimaging applications. *Acc. Chem. Res., 44*, 925–935.
65. Li, H., Zhang, B., Lu, X., Tan, X., Jia, F., Xiao, Y., Cheng, Z., et al., (2018). Molecular spherical nucleic acids. *Proc. Natl. Acad. Sci., 115*, 4340–4344.
66. Williams, S. C. P., (2013). Spherical nucleic acids: A whole new ball game. *Proc. Natl. Acad. Sci., 110*, 13231–13233.
67. Mokhtarzadeh, A., Vahidnezhad, H., Youssefian, L., Baradaran, B., & Uitto, J., (2019). Applications of spherical nucleic acid nanoparticles as delivery systems. *Trends Mol. Med., 25*, 1066–1079.
68. Kapadia, C. H., Melamed, J. R., & Day, E. S., (2018). Spherical nucleic acid nanoparticles: Therapeutic potential. *Bio. Drugs, 32*, 297–309.
69. Wu, X. A., Choi, C. H. J., Zhang, C., Hao, L., & Mirkin, C. A., (2014). Intracellular fate of spherical nucleic acid nanoparticle conjugates. *J. Am. Chem. Soc., 136*, 7726–7733.
70. Bousmail, D., Amrein, L., Fakhoury, J. J., Fakih, H. H., Hsu, J. C. C., Panasci, L., & Sleiman, H. F., (2017). Precision spherical nucleic acids for delivery of anticancer drugs. *Chem. Sci., 8*, 6218–6229.
71. Choi, C. H. J., Hao, L., Narayan, S. P., Auyeng, E., & Mirkin, C. A., (2013). Mechanism for the endocytosis of spherical nucleic acid nanoparticle conjugates. *Proc. Natl. Acad. Sci., 110*, 7625–7630.
72. Davis, S. S., (1997). Biomedical applications of nanotechnology: Implications for drug targeting and gene therapy. *Trends Biotechnol., 15*, 217–224.
73. Labhasetwar, V., (2005). Nanotechnology for drug and gene therapy: The importance of understanding molecular mechanisms of delivery. *Curr. Opin. Biotechnol., 16*, 674–680.
74. Angell, C., Xie, S., Zhang, L., & Chen, Y., (2016). DNA nanotechnology for precise control over drug delivery and gene therapy. *Small, 12*, 1117–1132.
75. Balcells, L., Fornaguera, C., Brugada-Vila, P., Guerra-Rebollo, M., Meca-Cortes, O., Martinez, G., Rubio, N., et al., (2019). SPIONs' enhancer effect on cell transfection: An unexpected advantage for an improved gene delivery system. *ACS Omega, 4*, 2728–2740.

76. Laurent, S., Saei, A. A., Behzadi, S., Panahifar, A., & Mahmoudi, M., (2014). Superparamagnetic iron oxide nanoparticles for delivery of therapeutic agents: Opportunities and challenges. *Expert Opin. Drug Deliv., 11*, 1449–1470.

77. Gorzkiewicz, M., Konopka, M., Janaszewska, A., Tarasenko, I. I., Sheveleva, N. N., Gajek, A., Neelov, I. M., & Maculevicz, B. K., (2020). Application of new lysine-based peptide dendrimers D3K2 and D3G2 for gene delivery: Specific cytotoxicity to cancer cells and transfection *in vitro. Bioorg. Chem., 95*, Article: 103504.

78. Somani, S., Laskar, P., Altwaijry, N., Kewcharoenvong, P., Irving, C., Robb, G., Pickard, B. S., & Dufes, C., (2018). PEGylation of polypropylenimine dendrimers: Effects on cytotoxicity, DNA condensation, gene delivery and expression in cancer cells. *Sci. Rep., 8*, Article: 9410.

79. Khan, W., Hosseinkhani, H., Ickowicz, D., Hong, P. D., Yu, D. S., & Domb, A. J., (2012). Polysaccharide gene transfection agents. *Acta Biomater., 8*, 4224–4232.

80. Beik, J., Abed, Z., Ghoreishi, F. S., Nami, S. H., Mehrzadi, S., Zadeh, A. S., & Kamrava, S. K., (2016). Nanotechnology in hyperthermia cancer therapy: From fundamental principles to advanced applications. *J. Control. Rel., 235*, 205–221.

81. Tian, Z., Yao, X., Ma, K., Niu, X., Grothe, J., Xu, Q., Liu, L., et al., (2017). Metal-organic framework/graphene quantum dot nanoparticles used for synergistic chemo- and photothermal therapy. *ACS Omega, 2*, 1249–1258.

82. Liu, Y., Bhattarai, P., Dai, Z., & Chen, X., (2019). Photothermal therapy and photoacoustic imaging via nanotheranostics in fighting cancer. *Chem. Soc. Rev., 48*, 2053–2108.

83. Bao, Z., Liu, X., Liu, Y., Liu, H., & Zhao, K., (2016). Near-infrared light-responsive inorganic nanomaterials for photothermal therapy. *Asian J. Pharm. Sci., 11*, 349–364.

84. Wei, W., Zhang, X., Zhang, S., Wei, G., & Su, Z., (2019). Biomedical and bioactive engineered nanomaterials for targeted tumor photothermal therapy: A review. *Mater. Sci. Eng. C, 104*, Article: 109891.

85. Khot, M. I., Andrew, H., Svavarsdottir, H. S., Armstrong, G., Quyn, A. J., & Jayne, D. G., (2019). A Review on the scope of photothermal therapy-based nanomedicines in preclinical models of colorectal cancer. *Clin. Colorect. Cancer, 18*, 200–209.

86. Wu, X., Mu, L., Chen, M., Liang, S., Wang, Y., She, G., & Shi, W., (2019). Bifunctional gold nanobipyramids for photothermal therapy and temperature monitoring. *ACS Appl. Bio Mater., 2*, 2668–2675.

87. Kralj, S., Potrc, T., Kocbek, P., Marchesan, S., & Macovec, D., (2017). Design and fabrication of magnetically responsive nanocarriers for drug delivery. *Curr. Med. Chem., 24*, 454–469.

88. Jarockyte, G., Daugelaite, E., Stasys, M., et al., (2016). Accumulation and toxicity of superparamagnetic iron oxide nanoparticles in cells and experimental animals. *Int. J. Mol. Sci., 17*, Article: 1193.

89. Ghartavol, R. V., Borojeni, A. A. M., Ghartavol, Z. V., et al., (2020). Toxicity assessment of superparamagnetic iron oxide nanoparticles in different tissues. *Artif. Cells Nanomed. Biotechnol., 48*, 443–451.

90. Mahmoudi, M., Sant, S., Wang, B., Laurent, S., & Sen, T., (2011). Superparamagnetic iron oxide nanoparticles (SPIONs): Development, surface modification and applications in chemotherapy. *Adv. Drug Deliv. Rev., 63*, 24–46.

91. Nguyen, T. D. T., Pitchaimani, A., Ferrel, C., Thakkar, R., & Aryal, S., (2017). Nano-confinement-driven enhanced magnetic relaxivity of SPIONs for targeted tumor bioimaging. *Nanoscale, 10*, 284–294.

92. Pongrac, I. M., Pavicic, I., Milic, M., Brkic, A. L., Horak, D., Vinkovic, V. I., & Gajovic, S., (2016). Oxidative stress response in neural stem cells exposed to different superparamagnetic iron oxide nanoparticles. *Int. J. Nanomed., 11*, 1701–1715.

93. Jamieson, T., Bakhshi, R., Petrova, D., Pocock, R., Imano, M., & Seifalian, A. M., (2007). Biological applications of quantum dots. *Biomater., 28*, 4717–4732.

94. Abbasi, E., Kafshdooz, T., Bakhtiary, M., Nikzamir, N., Nikzamir, M., Mohammadian, M., & Akbarzadeh, A., (2016). Biomedical and biological applications of quantum dots. *Artif. Cells, Nanomed. Biotechnol., 44*, 885–891.

95. Zhao, M. X., & Zeng, E. Z., (2015). Application of functional quantum dot nanoparticles as fluorescence probes in cell labeling and tumor diagnostic imaging. *Nanoscale Res. Lett., 10*, Article: 171.

96. Winnik, F. M., & Maysinger, D., (2013). Quantum dot cytotoxicity and ways to reduce it. *Acc. Chem. Res., 46*, 672–680.

97. Jha, S., Mathur, P., Ramteke, S., & Jain, N. K., (2018). Pharmaceutical potential of quantum dots. *Artif. Cells Nanomed. Biotechnol., 46*, 57–65.

98. Bajwa, N., Mehra, N. K., Jain, K., & Jain, N. K., (2016). Pharmaceutical and biomedical applications of quantum dots. *Artif. Cells Nanomed. Biotechnol., 44*, 758–768.

99. Santos, M. C. D., Algar, W. R., Medintz, I. L., & Heidelbrandt, N., (2020). Quantum dots for Förster resonance energy transfer (FRET). *TrAC Trends Anal. Chem., 125*, Article: 115819.

100. Prasher, P., Singh, M., & Mudila, H., (2018). Silver nanoparticles as antimicrobial therapeutics: Current perspectives and future challenges. *3 Biotech., 8*, Article: 411.

CHAPTER 3

Impact of Nanotechnology on Human Life

PRIYANKA, SHIPRA, and VIRINDER KUMAR SINGLA

COEM, Punjabi University Neighborhood Campus, Rampura Phul, Punjab, India, E-mail: shipracoem@pbi.ac.in (Shipra)

ABSTRACT

Nanotechnology has massive capability in several areas and is developed as a technology to direct the way on the way to sustainable progress in the approaching years. The indispensable subject of nanotechnology is to use particles of nanometer dimensions. In this chapter, we throw light on the basic concept of nanotechnology. This chapter answers the questions like: Who is the father of Nanotechnology? How it is developed? Gigantic spotlight on nanotechnology in past decades have led to its open expansion and thus vast use of nanoparticles (NPs). Along with this we discussed in this chapter about how many types of nanomaterials are there? This chapter also deals with how nanotechnology affects human life in both ways: positive and negative in various fields like health, environmental, social, and economic fields. Some measures are also discussed by which this technology can be used more effectively with minimum negative effect.

3.1 INTRODUCTION

What counts today as "nanotechnology" is important, groundbreaking work that is taking place at research centers all over the world. Nanotechnology, monitoring, and processing products and procedure on the level of iotas, or tiny particle arrays. Nanotechnology is generally considered for particle handling within the size range < 100 nm frequently they make an inspection with a human hair, which is about 80,000 nm wide. The innovative work of nanotechnology is dynamic universally, and nanotechnologies are as of now utilized in several items, including sunscreens, beauty care products, materials, and athletic gear. Nanotechnology is also urbanized for use in conveyance tranquilization and other medical and biological areas. Further, nanotechnology is furthermore fashioned for apply in conservation fields (www.rroij.com; academicjournals.org; [1]).

3.2   HISTORY

"There's plenty of room at the bottom," was addressed by American physicist Richard Feynman at the American Physical Society convention, held at Caltech on December 29, 1959, to give inspiration to the ground of nanotechnology. The "nano-innovation" was first utilized by the Japanese researcher Norio Taniguchi of Tokyo University of Science to depict semiconductor actions, for example, flimsy film affidavit and particle bar processing showing trademark organize on the appeal for a nanometer. His report was, "nano-innovation, for the most part, comprises of the preparing of partition, solidification, and disfigurement of materials by one particle or one atom." However, the word was not utilized again up to 1981 when Eric Drexler, who was uneducated about Taniguchi's former use of the term, circulated his first manuscript on nanotechnology in 1981 [1]. For the period of 1980s, the prospects of nanotechnology were settled, as a substitute of debatable. In 1980, Drexler practiced Feynman's stimulating 1959 talk "There's plenty of room at the bottom" while setting up his underlying logical paper regarding the matter, "Sub-atomic engineering: A way to deal with the improvement of general abilities for sub-atomic control," spread in the communication of the National Academy of Sciences held in 1981. The expression "nanotechnology" (which matched Taniguchi's "nano-innovation") was in parallel applied by Drexler in his 1986 book "Engines of Creation: The Coming Era of Nanotechnology," which projected the likelihood of a nanoscale "constructing agent" which would have the option to fabricate a duplicate of itself and of different things of self-assertive multifaceted nature. He likewise first distributed the expression "dim goo" to depict what may occur if a theoretical self-repeating machine, fit for free activity, were built, and discharged. Drexler's visualization of nanotechnology is commonly called "Sub-atomic Nanotechnology" in addition "sub-atomic assembling."

Administration financing was outperformed by company expenditure on nanotechnology Research and Development, with the majority of the sponsoring opening from undertakings arranged in the United States of America, Germany, and Japan. The standard 5 affiliations that recorded the most cultured licenses on nanotechnology Research and Development in the span from 1970 to 2011 were Nippon Steel (1,490 1st licenses), Samsung Electronics (2,578 1st licenses), Toshiba (1,298 1st licenses), IBM (1,360 1st licenses) and Canon (1,162 1st licenses). The elementary five affiliations that scattered the mainly consistent papers on nanotechnology explore someplace in the scope of 1970 and 2012 were the Chinese Academy of Sciences, The French National Centre for Scientific Research, Russian Academy of Sciences, Osaka University, and University of Tokyo.

3.3   TYPES OF NANO-MATERIALS

As stated by Siegel, Nanostructured Materials are assigned as 0-dimensional, 1-dimensional, 2-dimensional, 3-dimensional nanostructures [2]. Nanomaterials can be generated with dissimilar dimensionalities of control, as defined by Richard W. Siegel: 0 (nuclear groups, fibers, and bunch collections), 1 (multilayers)), 2 (covered layers), and 3 (nanophase assets constituted of equiaxed nanometer estimated grains). Most recent nanomaterials can be organized into four types, which are discussed in subsections:

3.3.1   CARBON-BASED MATERIALS

For the most part, these nanomaterials are made from carbon, appearing mostly as hollow spheres, cylinders, or ellipsoids. Ellipsoidal and Circular nanomaterials of carbon are referred to as fullerenes. Such particles have many possible applications, including better coatings and films, more beached and lighter fabrics, and utility in gadgets [3].

3.3.2   METAL-BASED MATERIALS

These types of nanomaterials include, for example, titanium dioxide, quantum specks, nanogold, nanosilver, and metal oxides [4]. A quantum speck is a deeply packed semiconductor gem that contains 100s or 1000s of particles, and the thickness of which is up to a few 100 nanometers on demand. Altering the dimension of quantum dabs alters their optical characteristics.

3.3.3   DENDRIMERS

Such nanomaterials are polymers working from extensive units that are nanosized. There are several chain closures on the outside of a dendrimer which can be adapted to carry out specific synthetic capacities. However, since 3-dimensional dendrimers hold cavities in which unlike particles can be situated [5].

3.3.4 COMPOSITES

Composites unite nanoparticles (NPs) with dissimilar NPs or with larger materials of accumulation type. For example, NPs are now applied to products running from vehicle parts to bundling equipment to enhance mechanical, fire-resistant, and dry properties [6]. Nanomaterials are of fascination, as they evolve fascinating optical, attractive, electrical, and various properties at this size. Such through properties in gadgets, medicines, and various fields have the potential for extraordinary effects. Common forms of nanomaterials include nanotubes, dendrimers, quantic dabs, and completeness [7].

3.4   IMPACT OF NANOTECHNOLOGY ON HUMAN LIFE

3.4.1 HEALTH IMPACT

The use of nanotechnological materials and devices may greatly impact the human health. Considering that nanotechnology is an up-and-coming field, so there is substantial question about the extent to which nanotechnology can advantage or pose risks to human health [8]. The health impacts of nanotechnology can be divided into two sides of a coin: the possible for medical applications to heal disease with nanotechnological advances, and the possible health dangers posed by nanomaterial touch [9].

3.4.1.1 MEDICAL APPLICATION

NPs could be successfully used to transmit qualities to cells, to treat malignant growth, just as in immunization. In many areas of therapeutic applications, especially in disease treatment, nanotechnology is expected to be promising [10]. Careful consultation, radiotherapy, and chemotherapy are the approaches for treating malignant growth that are now available.

Although they are usually persuasive, they frequently cause serious fundamental harm and pulverize several tissues that surround the tumor. Despite the fact that complete recovery is not constantly feasible, recent cited symptoms are considered worthy due to the lack of better alternatives. Among the potential outcomes of treatment that have recently evolved, there are not many that have been explicitly structured to make them less harmful to the sound tissues of the patient. One is the nano shells that are currently experiencing preliminaries of clinical practice [11].

These very thin, gold-secured NPs can directly target malignant growth cells by penetrating into the tissues anywhere, and can also be tuned to ingest in the near infrared locale (NIR). Once immersed, nanoshells will accumulate in the neoplastic tissue and cause tumor removal either through the skin once illuminated with a NIR laser. Expanding peculiarity of the neoplasm may be done through the use of exceptional tumor focusing on moieties. Because of the way NPs scatter light, such as dull field microscopy and optical intelligibility tomography can also be used in imaging strategies.

3.4.1.2   NANOMEDICINE

Nanomedicine is the restorative use of nanotechnology Nanomedicine is one of the up-and-coming types of treatment that focuses on elective strategies for the conveyance of drugs and their effectiveness, while also minimizing reactions to sound tissues. Because of the entangled idea of disease sedate obstruction and the need to think about an assortment of components, growing new treatment alternatives is amazingly troublesome. In the near future, nanomedicine aims to express a substantial arrangement of testing instruments and clinically helpful gadgets [12]. The National Nanotechnology proposal hopes fresh technologies in the medical trade that could integrate propelled conveyance systems, new treatments; the theoretical field of atomic nanotechnology accepts that phone fix machines could upset drug and the restorative field. As the nanomedicine business keeps on developing, it is required to significantly affect the economy. Several nanotechnology-based medications that is monetarily accessible or in creature medical preliminaries include:

- Abraxane, that is used to treat bosom cancer, non-little cell lung disease (NSCLC) and pancreatic cancer, is authorized by the U.S. Nourishment and Drug Administration (FDA), is made up of nano-sized particle.
- Doxil was at first authorized by FDA for utilization on Kaposi's sarcoma related to HIV. Now ovarian disease and different myeloma are being treated with Doxil. This medicine is sheathed in liposomes. Liposomes are of self-gathering, spherical, shut colloidal structures that are made up of lipid bilayers which cover a watery space.

- Onivyde, which was endorsed by FDA in October 2015 is used to treat pancreatic malevolent growth.
- Rapamune is endorsed by the FDA in 2000 to prevent organ dismissal later than operation of changing organs. These nano sized crystal segments take into account expanded medication solvency and disintegration rate, prompting improved ingestion and high bioavailability.

Immunizations are frequently viewed as among the best accomplishments of current prescription. Introduction of antibodies caused an extensive reduction in the rate and death pace of numerous irresistible infections and now and again even prepared for the annihilation of certain risky ailments. Ongoing revelations demonstrate that the nanomaterials may end up being productive antigen conveyance frameworks, essentially on the grounds that they can give supported and controlled discharge profiles.

3.4.1.3   FOOD SYSTEM

Diet nanotechnology has invaded into various parts of consumer items, for illustration, sustenance bundling, added substances, and sustenance safeguarding. Nanotechnology permits improving the viability and the sustenance estimation of nourishment fixings by expanding the security and bioavailability of bioactive nourishment fixings. By utilizing nanoencapsulation advances corruption of labile mixes, for example, nutrients, and cancer prevention agents, can be lessened. Abundant tale nano sized materials have been produced for advancing nourishment superiority and security, yield development, in addition to checking natural circumstances.

Nano sized materials are additionally being utilized to create environ-intellectually amicable composts to bring down ozone harming substance emissions during crop generation. Broadly utilized carbon-based manures like ammonium bicarbonate and urea in various soil types in spite of their effectiveness are powerless to decomposition and hydrolysis prompting raised degrees of results, for example, nitrogen, ammonium carbonate and ammunition in soils.

3.4.1.4   HEALTH HAZARDS

Built nanomaterials (ENMs) can possibly profit creature and human wellbeing. In any case, there are vulnerabilities encompassing the long haul impacts of applying ENMs to nourishment and pharmaceutical items devoured by people and creatures. Poisonous quality of NPs relies upon nanoparticle dose, their surface properties, covering, structure, size, and capacity to total. The NPs can enter the body through numerous courses; dermally, by ingestion, inward breath, infusion or by implantation. Some NPs break down effectively and their impacts on living life forms are equivalent with the impacts of the substance they are made of. Notwithstanding, different NPs don't corrupt or break down promptly [12]. On the off chance that NPs have poor solvency they can cause malignancy, they may cumulative in natural conditions and undergoes of specific thought. NPs can be stored all through the human respiratory tract, and a significant piece of breathed in nano sized particles stock up in lungs. NPs move from the lungs to various organs, for example, the spleen, the liver and most probably to the baby in expecting ladies. Carbon nanotubes (CNTs), both single-and multi-walled, are genuine instances of ENMs known to have noteworthy potential for poisonous quality. Carbon nanotubes (CNT), both single-and multi-walled, are genuine instances of ENMs known to have noteworthy potential for poisonous quality, contingent upon introduction time [13].

Another potential course of breathed in NPs inside the body is the olfactory nerve; NPs may cross the mucous film inside the nose and afterward arrive at the cerebrum through the olfactory nerve. Materials which were lacking any other person's input are not critical could be fatal in case that these particles are breathed in as nano particles. The effect of inhaled NPs in the body may cause lung exasperation and heart related problems. Concentrates in citizens illustrate that consuming diesel deposit causes a common fiery retort and changes the outline that instructs the mechanical capacities in the cardiovascular structure. The pneumonic damage and irritation coming about because of the inward breath of nanosize urban particulate gives off an impression of being because of the oxidative pressure that these particles cause in the cells [14].

3.4.2   NATURAL IMPACT

The ecological effect of nanotechnology can be divided into two points of view: probability for nanotechnological progress, and potentially new category of contamination that nanotechnological materials cause as these poisonous particles release to the earth.

3.4.2.1   ENVIRONMENTAL APPLICATION

Nanotechnology provides possible benefits in terms of physical, cultural, and environment. Furthermore, nanotechnology can help to lessen the human impression on nature by providing answers for the use of vitality, pollution, and emanations of green gas. Nanotechnology provides notable natural advantages, including:

- Enhanced facility to recognize in addition to dispense with infectivity by improving water, air, along with earth superiority;
- Elevated correctness fabricating via diminishing quantity of fritter;
- Amputation of ozone depleting substances plus dissimilar toxins commencing the air;
- Depleted necessity for large motorized flora.
- The nanoscale items that use graphene in a mechanical use or research can profit nature in a few different ways:
- Graphene-based nanocomposites minimize aircraft heaviness and the effect of decreasing weight consequences drop of large quantity of gas.
- Graphene meager sheets and Bucky papers are used to avoid the immediate and roundabout impacts of lightning strikes instead of metal networks around the fuselage of the aircraft [9].
- Graphene's influential properties create the efficiency of cutting-edge renewable forms of power supply, for example, decreasing the heaviness of sharp edges of a breeze turbine, and increasing the efficacy of speaking vitality.

3.4.2.2   ENVIRONMENTAL POLLUTION

The knowledge of the natural nanotechnology-related impacts and hazards is somewhat restricted and contradictory. The possible natural harm can be summarized as follows by nanotechnology [11]:

- Lofty strength prerequisites orchestrating nano sized particles constructing far above the ground vitality appeal.
- Dissemination of contaminated, diligent nano sized substances beginning environmental damage.
- Environmental deterioration of other life cycle arranges additionally not understandable.
- Lack of all set architects and laborers causing additional worries.

Graphene dependent compounds may likewise hurt earth. The hazardous character of graphene is obscure, and hard to expel graphene from squander [3, 8].

3.4.3   SOCIAL AND ECONOMIC IMPACT

Nanotechnology has tested cultural and realistic encounters with unusual results, and increased low open awareness about their dangers. Whereas nanotechnology enables huge-scale firms to evolve and converge in the industry, it could weaken littler firms' financial practicability.

3.4.3.1   INTELLECTUAL PROPERTY ISSUE

Nanotech technologies expand research and drug control points, and broaden the boundaries of protected innovation law. To the degree that the USPTO issues a multiplication of large and conceivably covering nanotechnology licenses, enhancing a nanotechnology patent brush could impede the authorization procedure needed for further growth. In the event that the temporary worker declines the bureaucratic office's solicitation, the organization can give a permit to the candidate itself if "the contractual worker or chosen one has not taken, or isn't relied upon to take inside a sensible time, compelling strides to accomplish useful use of the subject development in such field of utilization" or if "activity is important to meet necessities for open utilize indicated by Federal guidelines and such prerequisites are not sensibly fulfilled by the contract based worker, appointee, or licensees." Theoretically, the walk in right shows the intensity of the legislature to forestall the nonuse of licenses with regards to patent accumulating or blocking licenses used to smother rivalry. Such college-based nanotechnology inquiries into focuses are in a prime position to validate offers for noteworthy portions of the latest Nanotechnology Act subsidy; they will potentially have an expanded government permit resistance in order to successfully perform incremental and innovative work without being unduly hindered by a patent shrubbery of nanotechnology.

3.4.3.2   EFFECT ON UNEMPLOYMENT

The nanomaterials and nano-gadget parts are small, but scaling down is not the main goal. Nanotechnology will open up material efficiency and open up large-scale market opportunities. When you are unlikely to be aware of the possible uses of your exploration, this can lead you to become a visionary business. Nanotechnology can help us escape jobless production trends. Yet, only if you set up the right motivators and bolster instruments [8].

3.4.3.3 ECONOMIC GROWTH

In many nations, nanotechnology's financial estimate is calculated with respect to research, preparation, inquiry about travel, and the selling of products and procedures. Such measurements will usually differ. Nanotechnology measurements are difficult to characterize. We can list licenses but not all licenses are put on the market. We may determine inquiries by reviewing published articles and documents, but variations in their importance and value are usually investigated [14]. Nanotechnology degrees received by college understudies provide a valuable metric to test the estimation of scholastic movement alongside the analysis of undertakings and yields. Be that as it may, monetary measurements are progressively intricate. It is considerably harder for government organizations and policymakers to determine the impact on the venture from nanotechnology speculations. Regularly, in nanotechnology speculations, an administration office is approached to legitimize the value obtained from millions or billions of dollars. These ventures can be assessed in terms of creating work, decreasing assembly costs, developing new organizations, engaging with business enterprises, and making new items or administrations-but once in a while there is a straight-line metric that decipheres legitimately into business esteem the subsidizing of nanotechnology activities. Considering the above contemplations, we will attempt to evaluate the business capability of nanotechnology applications, in light of nanomaterials, nanolithography devices and nanoelectronics and nano sensors and that, as per an investigation gave by BCC Research in 2016, should reach 90.5 billion dollar by 2021 from 39.2 billion dollar out of 2016 at a compound yearly development rate (CAGR) from 2016 to 2021 is 18.2%

3.5 CONCLUSION

In the earlier period, nanotechnology has made tremendous strides. Nanotechnology is expected to modify health care, dentistry, and personal living extremely than previous innovations. Nanotechnology, as with all innovations, holds a major prospective for mishandling and exploitation on a size. These have the ability to fetch important benefits, such as superior health, improved utilization of normal possessions, and compact emissions. Micro- and nanoplastics are emerging with being there in almost all ecological matrices as environmental pollutants of great scale.

There is a need to identify a common set of parameters to determine the deadly effects of a variety of NPs on diverse flora and under different ecological situations. Future research needs to focus on leveraging the advantages of technology that takes into account cost constraints and requirements.

KEYWORDS

- **carbon nanotubes**
- **engineered nanomaterials**
- **Food and Drug Administration**
- **nanoparticles**
- **nanotechnology**
- **near-infrared locale**

REFERENCES

1. Charles, P., Poole, Jr., & Owens, F. J., (2003). *Introduction to Nanotechnology.* John Wiley & Sons, Inc.
2. Dagani, R., (2003). *Nanomaterials: Safe or Unsafe?* (pp. 30–33). Chem. & Eng. News.
3. Marcano, D. C., Kosynkin, D. V., Berlin, J. M., et al., (2010). Improved synthesis of graphene oxide. *ACS Nano, 4,* 4806–4814.
4. Emerich, D. F., & Thanos, C. G., (2003). Nanotechnology and medicine. *Expert Opinion on Biological Therapy, 3,* 655–663.
5. Kotchey, G. P., (2011). The enzymatic oxidation of graphene oxide. *ACS Nano, 5,* 2098–2108.
6. Bhatt, I., & Tripathi, B. N., (2011). Interaction of engineered nanoparticles with various components of the environment and possible strategies for their risk assessment. *Chemosphere, 82,* 308–317.
7. Fenoglio, I., (2009). Non-UV-induced radical reactions at the surface of TiO_2 nanoparticles that may trigger toxic responses. *Chemistry: A European Journal, 15,* 4614–4621.
8. Dunphy, K. A., (2006). Environmental risks of nanotechnology: National nanotechnology initiative funding, 2000−2004. *Environmental Science and Technology, 40,* 1401–1407.
9. Thompson, M., (2018). Social capital, innovation and economic growth. *Journal of Behavioral and Experimental Economics, 73,* 46–52.
10. Singh, N. A., (2017). Nanotechnology innovations, industrial applications and patents. *Environmental Chemistry Letters, 15*(2), 185–191.
11. Farokhzad, O. C., & Langer, R., (2009). Impact of nanotechnology on drug delivery. *ACS Nano, 3,* 16–20.

12. Finglas, P. M., Yada, R. Y., & Toldra, F., (2014). Nanotechnology in foods: Science behind and Future perspectives. *Trends. Food Sci. Technol., 40,* 125, 126.
13. Purohit, R., (2017). Social, environmental and ethical impacts of nanotechnology. *Materials Today: Proceedings, 4*(4), 5461–5467.
14. Verdejo, R., Bernail, M. M., Romasanta, L. J., & Lopez-Manchado, M. A., (2011). Graphene filled polymer nanocomposites. *Journal of Materials Chemistry, 21,* 3301–3310.

CHAPTER 4

Potential Applications of Nanomedicines in the Treatment of Diabetic Neuropathy

ANKITA SOOD, PANKAJ PRASHAR, ANAMIKA GAUTAM,
BIMLESH KUMAR, INDU MELKANI, SAKSHI PANCHAL,
SACHIN KUMAR SINGH, MONICA GULATI,
NARENDRA KUMAR PANDEY, PARDEEP KUMAR SHARMA, LINU DASH,
ANUPRIYA, and VARIMADUGU BHANUKIRANKUMAR REDDY

*School of Pharmaceutical Sciences, Lovely Professional University,
Phagwara – 144411, Punjab, India, Tel.: +919888720835,
Fax: +91 1824501900; E-mails: Bimlesh.12474@pu.co.in;
bimlesh1pham@gmail.com (B. Kumar)*

ABSTRACT

Nanotechnology-mediated developed medicine is now an emerging trend for the treatment of neuropathic pain (NeP). It creates rapid advancement to resolve the pharmaceutical challenges associated with various drugs. Among all kinds of NeP, the most common and challenging kind of pain in NeP is associated with diabetes. About 230 million people have been reported as patients of diabetes, and this has become a global challenge due to partial/unsatisfactory relief from pain. Nanomedicine has provided a lot of attention to the medical science, pharmaceutical as well as biosciences in the treatment of chronic disease state due to uniqueness in physiochemical and biological advancement. Here, we are going to highlight the emerging role of nanomedicine in diabetic neuropathy (DPN), road backs of nanomedicine with its toxicity along with this we discussed about the future aspects of nanomedicine in the diabetic NeP.

4.1 INTRODUCTION

Diabetes mellitus exist globally and reported more severe in developing countries [1]. The incidence of diabetes is increasing quickly, and it is anticipated that by 2030, this will almost double. Nevertheless, the largest rise in the incidence of diabetes is anticipated in the near future in Asia and Africa, expected to increase 50% by the year 2030 [2]. In the 21st century, diabetic neuropathy (DPN) has become one of the world's largest health care problems. DPN is the main cause of disability due to foot amputation and ulceration, fall-related injury, and gait disturbance. It decreases the quality of life and, at the same time, increases the health cost related to diabetes [3–5]. The annual cost for the treatment of diabetes is $6.632 per patient that becomes ($12.492) two-fold increase in patients having diabetic peripheral neuropathy, and those having severe kinds of DPN experienced a fourfold increase in cost ($30.755). It is the third most common neurological disorder, ranging from 100,000 individuals per year to around 54 million [6, 7]. Past population-based studies of type 1 and type 2 diabetic patients recorded prevalence rates of neuropathy ranges between 8.54% and 13.46%as given in Figure 4.1 [8, 9].

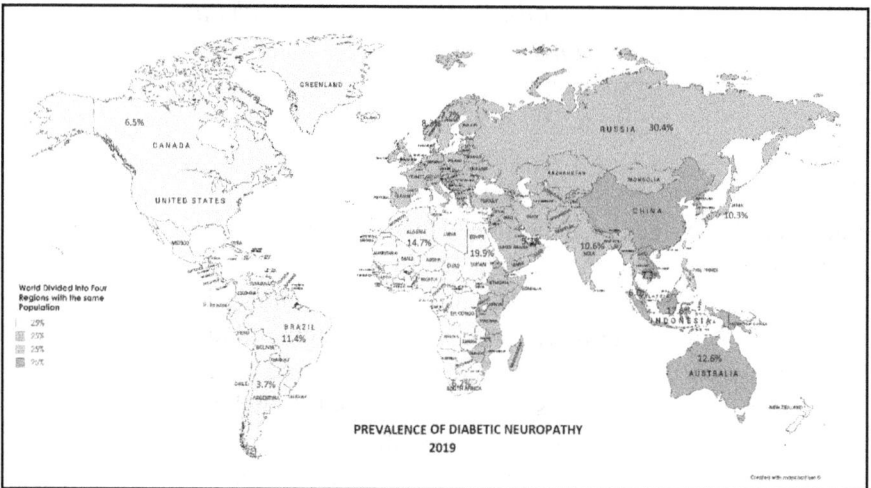

FIGURE 4.1   Prevalence of diabetes.

India has one of the world's highest type-2 diabetes mellitus prevalence. It is estimated that nearly 2 million Indians with type-2 diabetes mellitus will be present in the country by the year 2030. There is an increase in the proportion of people of 60 years of age or older in tandem with a decline in the proportion of young people, so that the proportion of elderly people is projected to be increased 15–25% by 2050 [10]. In India, peripheral neuropathy is exacerbated by poor foot hygiene, unsuitable shoes, and excessive bare foot walking and drug non-compliance. The prevalence (60.4%) and frequency (8.76%) of sensory peripheral neuropathy in diabetic patients observed is higher and may continue to increase as age progresses [3, 11–13]. Hence, proper identification of the problem and guidance may help to reduce development. To prevent progression by long-term glycemic control, DPN remains a challenge for a doctor as well as the clinical pharmacist. Counseling, diagnosis of substance-related problems and initiation of appropriate drug therapy play a key role in reducing prevalence.

4.2 DIABETIC NEUROPATHY (DPN)

Diabetes is a class of metabolic diseases which are identified by hyperglycemia induced by deficiency in the secretion of insulin, function of insulin or both [14]. Chronic diabetic hyperglycemia is associated with long-term damage, deficiency, and abnormality in various organs, particularly in kidneys, eyes, heart, nerves, and blood vessels. Diabetes progression includes many pathogenic pathways which vary from inflammatory degradation of the pancreatic β-cells with subsequent insulin deficiency to insulin resistance (Figure 4.2) [15].

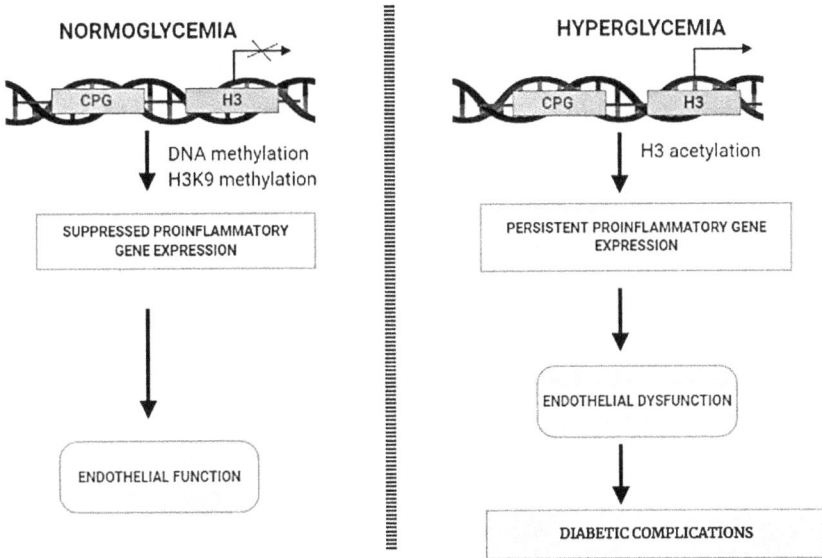

FIGURE 4.2   Difference between normoglycemic and hyperglycemic state.

In diabetes, the inadequate effect of insulin on target tissues leads to defects in carbohydrate, fat, and protein metabolism. Significant hyperglycemia causes polyuria, reduction of body weight, polydipsia, blurred vision, and sometimes polyphagia [16, 17]. Chronic hyperglycemia may also be linked with reduced development and vulnerability to certain infections. Long term diabetic complications involve retinopathy with a possible vision loss, nephropathy leads to kidney damage, peripheral neuropathy with a possibility of foot ulcers, amputations, and autonomous neuropathy leads to GIT and CVS disorders [18–20]. The frequency of cardiovascular, peripheral arterial, cerebrovascular, and atherosclerotic disorders is intensified in patients with diabetes [21]. There are two primary types of diabetes, type 1 and type 2, but diabetes, like gestational diabetes, medication-induced diabetes, genetic disorders, endocrinopathies, and pancreatic exocrine disorders caused by diabetes can also be observed. A number of problems are associated with diabetes. A diabetic ketoacidosis from extremely high blood glucose (hyperglycemia) and coma due to low blood glucose (hypoglycemia) are severe metabolic disorders associated with death. Complications of diabetes are divided into two groups based on the type of blood vessel damaged due to increased level of glucose which includes microvascular and macrovascular complications. Microvascular disorders include eye disease or retinopathy, kidney disease referred to as nephropathy, and nerve injury or neuropathy. The most common macrovascular complications involve acute cardiovascular disease contributing to myocardial infarction and stroke [2, 22].

Among these complications of diabetes, the DPN is the most serious one. More than half of diabetes sufferers experience neuropathy, with a life-long threat of one or more, lower limb amputations reported up to 15% of population. DPN is a condition that includes the somatic and autonomic divisions of the peripheral system. Nevertheless, it is becoming increasingly evident that injury may also happen to the spinal cord and to the upper central nervous system [23, 24]. Neuropathy is a key factor in injured wound recovery, erectile dysfunction, and cardiovascular disease in diabetes. The occurrence of vascular disorders, including thickening of capillary basement membrane, endothelial hyperplasia, and hypoxia, have historically been characterized by neuropathy progressions. Advanced neuropathy attributable to the degradation of the nerve fiber in diabetes is distinguished by altered vibration and heat threshold sensitivities that enhance the sensory perception deficit. In 40–50% of patients with DPN having hyperalgesia, paraesthesia, and allodynia [25]. Many diabetic persons are also suffering from these symptoms without clinical evidence of neuropathy (Elder 10–20%) which can affect the quality of life. There is a greater reduction in the rate of neuronal conduction with impairment of the nerve endpoints in diabetes.

That's why loss of feeling and sensation is often seen first in the feet and then ascend in other places, particularly in hands. This is generally called a "glove and stocking," which involves numbness, dysesthesia (pins and needles), sensory deprivation, and nighttime discomfort. In DPN progressive motor impairment may trigger the digits of the hands and toes [26]. The autonomous nervous system is also affected by diabetes apart from motor neural dysfunction. Orthostatic hypotension is one serious abnormality of autonomic control in people with diabetes because it makes person unable to alter the heart rhythm and vascular tone to regulate the blood flow to the brain. The period of diabetes and the absence of glycemic control, along with other symptoms, are major risk factors for neuropathy in both essential diabetic types [27]. Various pathways are involved in the pathogenesis of DPN like Polyol flux pathway, AGEs formation, mitochondrial dysfunction, Protein Kinase C activity and increased diacylglycerol level [28–33]. All these pathways directly or indirectly increase the production of superoxides that are responsible for nerve damage (Figure 4.3(A)). Out of these pathways Polyol flux pathway play a most important role in pathogenesis of DPN. In this pathway when the level of glucose is increased inside the body then this excess of glucose is converted into sorbitol. Aldose reductase is the most important enzyme of this pathway that is used during the conversion of glucose into sorbitol. As sorbitol is not able to cross the cell membrane it starts accumulating in and around the nerve and ultimately cause neuro-inflammation leads to nerve damage. In addition, when glucose is converted into sorbitol NADPH is used as a cofactor.

This NADPH is also required for the conversion of oxidized form of glutathione into reduced form. However, all the NADPH is utilized in conversion of glucose into sorbitol so the NADPH level is decreased. This cause increase level of oxidized form of glutathione leads to the generation of free radicals that further cause oxidative stress and nerve damage. Moreover, sorbitol also is converted into fructose with the help of enzyme sorbitol dehydrogenase. In this step, NAD+ is reduced to NADPH following the generation of free radicals by the action of enzyme NADPH oxidase. In this way, all these steps lead to oxidative stress, and this oxidative stress is responsible for neuropathy as given in Figure 4.3(B) [29].

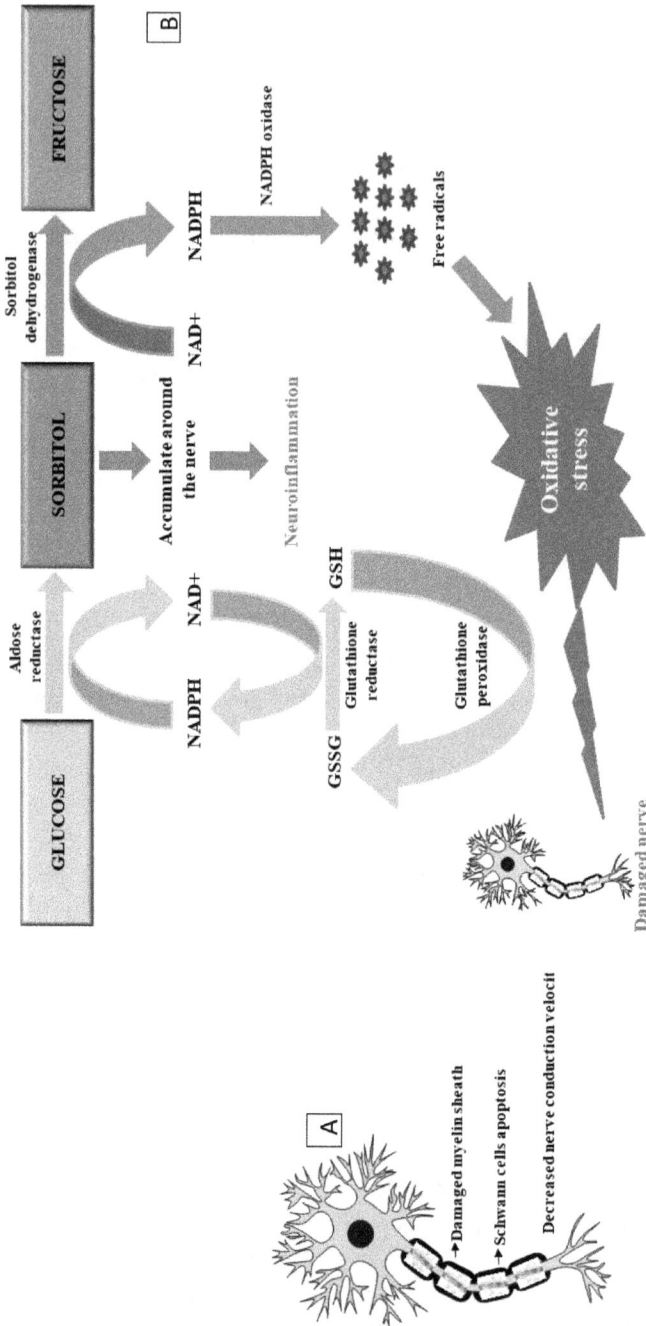

FIGURE 4.3   (A) Represent structure of damaged nerve; (B) Pathogenesis of diabetic neuropathy (Polyol flux pathway).

4.3 CONVENTIONAL TREATMENT OF DIABETIC NEUROPATHY (DPN)

4.3.1 *HYPERGLYCAEMIC CONTROL*

Glucose management tends to be the only disease-modifying therapy for DPN since uncontrolled diabetes contributes to a high oxidative stress that can be overcome once patients gain glycemic control [34]. On the basis of assumption that oxidative stress is responsible for the complications of diabetes, antioxidant therapy is the most effective in this condition. Pain management is another basis for DPN treatment that can greatly improve the quality of life of ill patient. In the past two decades, immense effort was made to find appropriate therapies of DPN by randomized placebo-controlled trials. Based on these findings, a variety of drug classes are considered viable in the management of DPN [35, 36].

4.3.2 *PAIN MANAGEMENT THERAPY*

The TCAs are used for several years as a first-line treatment to alleviate DPN discomfort, although they are not explicitly approved. TCAs act by blocking sodium and calcium channels, in order to suppress neuronal hyperexcitability that is responsible for pain in nerve damage. However, due to high risk of side effects and numerous contraindications of TCAs, SSRIs are preferred over TCAs (Figure 4.4) [37].

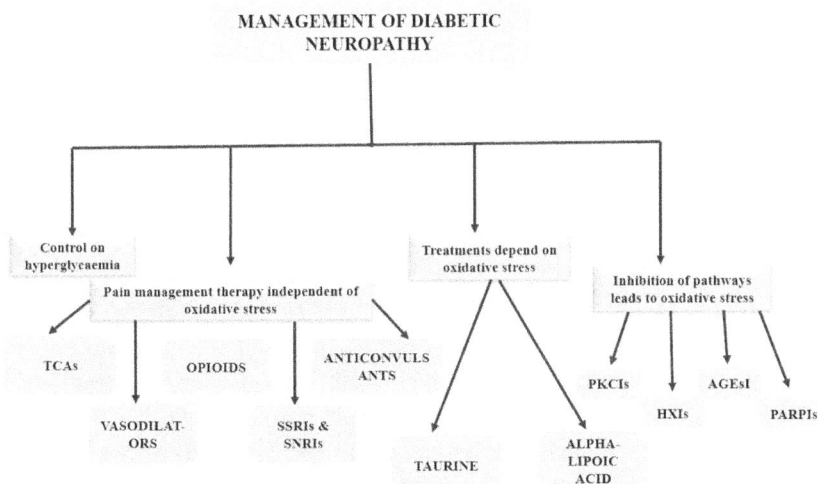

FIGURE 4.4   Treatments of diabetic neuropathy.

The main mechanism of SSRIs is to suppress serotonin reuptake and unlike TCAs, they do not have postsynaptic receptor blocking effects and no membrane stabilization like quinidine. There is weak evidence that SSRIs have a positive impact because they are not approved to relieve DPN pain [38]. Antidepressants like duloxetine and venlafaxine have dual specific suppression of serotonin and norepinephrine reuptake (SNRIs) are considered even stronger pain reliever that SSRIs. SNRIs obstruct noradrenaline and 5-HT carriers and prevent the reuptake of monoamine into presynaptic neuron. In the case of an ineffective reaction or counter indications to TCAs, gabapentine, and pregabalin will be used as first-line treatment for DPN discomfort. Pregabalin and gabapentin bind to the alpha 2 and delta subunit on the presynaptic neuron and inhibit activated neurotransmitters release. Pregabalin is prescribed for the diagnosis of severe DPN on the basis of new guidelines [39]. Sodium valproate is also helpful in reducing pain and should be taken into account for the treatment of severe DPN. However, it may escalate glycemic regulation and weight gain. In the management of severe DPN, oxcarbazepine, lacosamide, and lamotrigine will generally not used. Topiramate still lacks sufficient protection for use in DPN [40]. Opioids by binding with receptors on neuronal cell membranes inhibit neurotransmitter production and reduce pain. Oxycodone, tramadol, and morphine are some examples of opioids used in severe pain. Some atypical antipsychotics are also used for intense DPN pain. Several animal studies compliant with clinical studies support fewer dopamine-selective antipsychotics. However, these compounds can cause negative metabolic effects, including weight gain and resistance to insulin [41].

4.3.3   REDUCTION OF OXIDATIVE STRESS

Studies that show beneficial effects of antioxidants in both animal models and patients are the best indication for the function of oxidative stress in DPN. Though all antioxidants that can inhibit or postpone the development of DPN cannot be investigated, there are many others that could be identified, such as amino acids, acetyl-L-carnitine (ALC), curcumin, taurine, vitamin E, ascorbic acid, β-carotene, and lipoic acid [42]. Taurine is one of the most effective antioxidants. It inhibits the neuronal calcium channels and improves blood supply to neurons that is responsible for its analgesic properties in DPN patients. ALC is significantly decrease pain in patients of DPN. D-L-α-lipoic acid (ALA) has been extensively assessed in DPN subjects and it has shown good effects as a potent antioxidant. However, there is insufficient evidence on the basis of new guidelines to suggest it for DPN treatment [43].

4.3.4   INHIBITION OF PATHWAYS LEADS TO OXIDATIVE STRESS

Since, the continuous tight control of glucose in most situations is challenging and still a problem; experts have taken care of additional therapy targeting the symptoms of hyperglycemia. To this end it is considered effective to deal with certain known pathways triggered as a result of an elevated oxidative stress and glucose flux to regulate DPN [43]. Activation of PKC caused by hyperglycemia induced oxidative stress is thought to lead to the generation of free radicals that cause diabetic microvascular complications. Several PKC inhibitors have shown antioxidant activity, like ruboxistaurine. Ruboxistaurin seems to have been remarkably effective in decreasing the progression of DPN, and but its official approval is still pending. As noted earlier, activation of the polyol pathway by hyperglycemia will create oxidative stress underlying DPN [44]. A main enzyme in DPN pathogenesis is aldose reductase. ARIs decrease the glucose flux via polyol or sorbitol channels, which reduces sorbitol/fructose intracellular accumulation. Fidarestat (new ARIs) have been shown to regulate DPN with subsequent reduction in oxidative stress. Similarly, epalrestat long-term therapy, an ARI, can effectively delay progression of DPN and reduce symptoms, particularly in individuals with minimal microangiopathy and good glycemic control. AGE aggregation and RAGE activation contribute to oxidative stress and damage to peripheral nerve. Benfotiamine, a thiamine derivative (vitamin B1), decreases the formation and accumulation of AGEs thus reduce the oxidative stress in people with DPN depend upon dose and the duration of treatment. It also inhibits the hexosamine pathway that implicates in the pathogenesis of DPN thus this drug also effective in pain associated with DPN [45]. Aspirin can also counteract this pathway and decrease oxidative stress in DPN due to its antioxidant potential and a free radical scavenger activity. PARP activation is also implies in the pathogenesis of DPN, and its suppression illuminates multiple physiological situations related to the oxidative stress in DPN. Nicotinamide has been demonstrated to function as a PARP inhibitor and antioxidant in animals, which enhances early DPN complications [46].

4.4   NANOMEDICINE: AN EFFECTIVE APPROACH TO TREAT DIABETIC NEUROPATHY (DPN)

Nanotechnology can be characterized as a science and engineering engaged in development, production, and implementation of materials and devices of a nanometer (1 billionth of a meter), the smallest functional entity in one direction [47]. When this science is specifically applied to medical problems, it is called "nanomedicine." The classification of nanomedicine can be (a) (or 'nanometrology') either involving the measurement of extremely small quantities of analytes (e.g., microphysiography) and of very small measurements devices (e.g., quantum dots). (b) Therapy as all Nano-level material manipulations and construction eventually involve concern treatments (e.g., artificial nanopancreas) [48, 49]. Neuropathic pain (NeP) is a complicated and persistent pain disorder related to nerve damage. Around 10% of the global population suffers from this disorder, but the current treatment methods are insufficient to provide pain relief and are correlated with serious adverse effects. In order to overcome these drawbacks, several researchers are working on developing new drugs of high potency and fewer side effects, alternative cell and gene-based treatments, and the improved formulation of previously approved medicines. Due to its unique physiochemical and biological properties, nanomedicine received considerable interest in the treatment of different pathological conditions [50–52].

The National Health Institutes (NIH) of the United States reported that nanomedicine is a science that utilizes the unique properties of designed nanomaterials to detect, treat, and prevent specific medical conditions. Nanomedicine allow the development of new physical, chemical, and biological nanomaterials which enabled them for monitoring, diagnostic, and therapeutic applications (controlled releases of drugs, targeted supply and tissue engineering) [53]. It is well established that modification of the molecular and cellular biological process is the cause of many medical conditions. The specific physiological, electrical, and optical properties of nanomaterials allow cellular and molecular diagnosis and aid in early detection of the condition. For example, the production of ultrasensitive nanodevices for the identification of different NeP-implied biomarkers would serve as a significant move towards effective treatment, thereby enhancing the relevant outcomes. In nanotechnology developments such as nanoparticles (NPs) (gold and silver), quantum dots, nanochips, nano-arrays, nanomachine, nanosensors, etc., have markedly increase diagnostic ability (Figure 4.5).

Nanomedicines have the ability to improve the efficiency and precision of the treatment of body fluid samples. Compared with traditional delivery systems, therapeutic use of nanomedicine seems to be very encouraging. Due to their unique character, they can raise the therapeutic index by increasing the pharmacokinetics and pharmacodynamic properties of drugs [54, 55].

The capacity of NPs to distribute through oral routes is one of the new benefits of Nanomedicine, because they can improve the bioavailability, solubility, and permeability of potent drugs that are hard to deliver orally

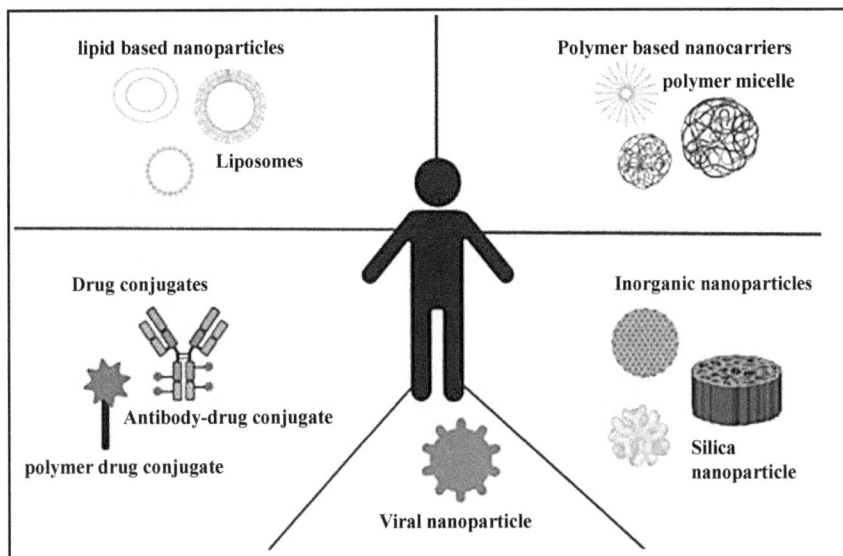

FIGURE 4.5 Different types of nanomedicine used in diabetic neuropathy.

and essential for parental administration. The lipid or polymer matrix encapsulating these medicines enables them to circumvent the gastrointestinal barrier and to achieve systemic circulation [55]. As the nanomaterials have similar dimensions to those of proteins and macromolecules, they can infiltrate the biological membranes and can therefore easily be manipulated using a particular ligand (antibodies, aptamers, etc.), in drug targeting. Moreover, the surface properties of these nanomaterials are also easily handled [56]. Nanotechnology can therefore be applied strategically in the development of new drug delivery technologies that can broaden the drug market.

4.5 TYPES OF NANOMEDICINES USED IN DIABETIC NEUROPATHY (DPN)

4.5.1 NIOSOMES

Niosomes are nanometric-size synthetic vesicles, which consist of non-ionic surfactants and are arranged in concentric bilayers that are stabilized with cholesterol. Niosomes may serve as drug reservoirs for sustained release. The structure of noisome comprise of hydrophilic, lipophilic, and amphiphilic moieties that allow them to tolerate medicines with varying solubility. Niosomes are biocompatible and have very low toxicity due to their non-ionic nature. Niosomes and liposomes have similar ability of drug delivery and both increase the effectiveness of medications associated with free drugs. Niosomes are stronger than liposomes because the former demonstrates strong chemical stability. Niosomal dispersion in a hydrated system is emulsified to control its distribution in a nonaqueous system. Insulin-loaded niosomes were vaginally administered *in vivo*. Recent studies show that when span 40 and 60 niosomes loaded with insulin (developed by evaporation of liquid phase and sonification) were administered to alloxan induced diabetic rats then this leads to blood glucose reduction with increased insulin bioavailability. Metformin is generally used as a fist line medication for type II diabetes, but 2–3 times of daily use leads to serious side effects such as lactic acidosis, gastric irritation, allergic reactions, and chest pain. So metformin release can be controlled in order to reduce its negative consequences and also reducing its dosing frequency by making niosomes. Thin-film hydration approach is used for loading metformin to develop niosomes that are composed of cholesterol (prevent drug leakage), dicetyl phosphate, and non-ionic surfactant (Figure 4.6(A)) [57].

4.5.2   LIPOSOMES

Liposomes have also received considerable interest as drug mechanisms because of their biocompatibility, low toxicity, biodegradability, and specific availability at target site. Liposomes are tiny spherical vesicles formed of phospholipids (having both hydrophilic and hydrophobic character). They differ significantly according to their lipid content, size, methods of preparation and charge. Both hydrophilic and hydrophobic medicines can be encapsulated by liposomes. In addition, they inhibit the quick degradation of the encapsulated medication and released the drug at desired target. Moreover, by the use of bilayer materials, the rigidity and surface charge of liposomes is calculated. On the basis of size of vesicles and number of bilayers they are classified into two group's unilamellar vesicles and multilamellar vesicles. Furthermore, unilamellar vesicles can be categorized as small and large unilamellar vesicles. These vesicles are comprised of only one phospholipid bilayer that encloses the aqueous solution. On the other hand, multilamellar vesicles constitute condensed phospholipid spheres isolated by aqueous layers. Liposomes are therefore prepared by aqueous dispersion of lipid. In addition, dispersion is done by using many methods, such as mechanical methods (micro emulsification and sonication) and solvent dispersion methods (Figure 4.6(B)) [58]. Now a day's Stealth liposomes are generally favored to traditional liposomes because of their longevity in the bloodstream and they can resist the mononuclear phagocytic processes that trap and clear liposomes. Studies also show that the treatment of liposomes with PEG decreases the action of macrophages on liposomes and eventually facilitates their long-term presence in blood. Antidiabetic drug, gliclazide reduces blood sugar levels by insulin induction but the frequent gliclazide therapy lowers patient compliance. So, continuing release formulation of gliclazide was prepared by researchers in the form of liposomes to improve patient's acceptance for the drug. Liposomes were produced by utilizing various cholesterol concentration in ethanol injection without sonication. The results showed that all official criteria were met in the formulations. The release of medications was sluggish and lasted >12 hours. These formulations were accompanied by first-order kinetics. Liposomes can therefore be a great option in the delivery of peptides (insulin, GLP-1) and other medications to decrease the blood sugar level, thereby making it possible to improve hyperglycemia and its complication [59, 60].

4.5.3 DENDRIMERS

The modern polymer architectures are dendrimers which are considered for their well-established structures, flexibility in drug delivery and resemblance with biomolecules. Dendrimers have highly symmetrical structure with branches or arms identical to the trees. The dendrimer structure is comprised of a central core made up of an atom or groups of atoms from which divisions of other atoms expand by means of a sequence of chemical reactions. Dendrimers are generally synthesized either by convergent or divergent processes [61, 62]. In divergent, the synthesis process starts at the center and progresses step by step into the periphery. Generally, 1st generation dendrimer is synthesized by reaction of a multifunctional core molecule with monomer molecules which contain 1 or >1 dormant groups. In the convergent process, the synthesis begins from the periphery and at the end of the dendrimer the origin is the outermost layer (Figure 4.6(C)). These hyperbranched macromolecules with a large number of functional groups are suitable for a wide range of therapeutic applications.

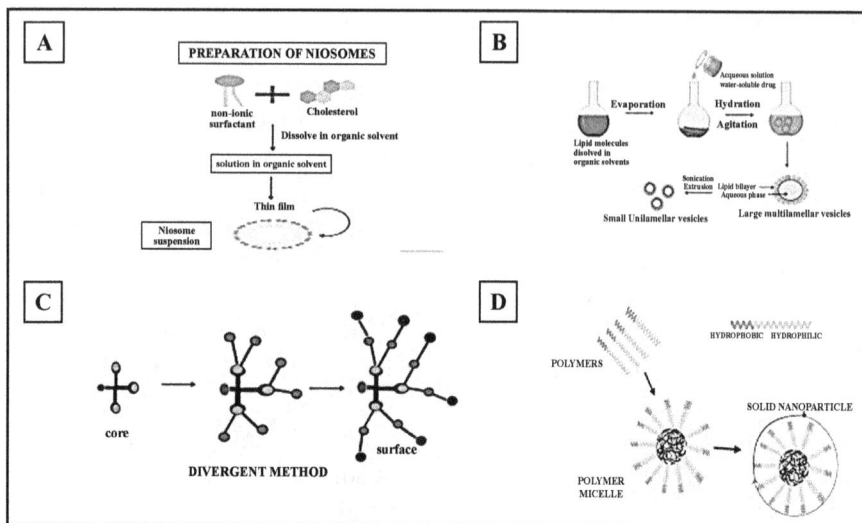

FIGURE 4.6 (A) Preparation of niosomes; (B) formation of liposomes; (C) formation of dendrimers; (D) formation of polymer micelles.

Such nanostructured macromolecules are used to encapsulate the large molecular weight hydrophilic/hydrophobic compounds through covalent bonding and host-guest interactions [63, 64]. Dendrimer's most important use resides in their ability to deliver managed and target-oriented drugs. Drugs associated with delivery vehicles are distinguished by greater stability, increased bioavailability, and half-life. In fact, constant medication release through the conjugate of drug-dendrimer decreases systemic toxicity and prevents tumor tissue aggregation. Poly(amidoamine) (PAMAM) dendrimers are the most widely used for medicinal and biochemical purposes. PAMAM dendrimers impaired the production of protein fibril without interacting with the measured concentration of peptides and are good hormonal and gene delivery systems to regulate hyperglycemia. PAMAM G4 has same action as of hypoglycemic drugs in decreasing blood glucose level [65]. Various parameters (HbA1c, AGEs, and aminotransferases) were also regulated to physiological values during treatment with PAMAM G4. In spite of their vast application, their use in biological systems is restricted due to toxicity problems related to them [66].

4.5.4 POLYMERIC MICELLES

Micelles are produced by dispersion of the hydrophobic and hydrophilic molecules in solution. The parameters regulating micelle formation are

amphiphiles concentration, hydrophobic-hydrophilic domain size, temperature, and solvent used. Micelles are produced by self-assembly and phase
begins only when a minimum concentration, generally recognized as the
critical micellar concentration, is achieved [51, 67]. Moreover, the temperature at which amphiphylic molecules exist is called the critical micellar
temperature, and under this temperature, micelles can collapse. Because of
the high durability, low cytotoxicity, and controlled and continuous drug
delivery, polymeric micelles are of great interest. By adjusting the ratio of
monomers in block copolymers, appropriate micelles can be obtained. Most
hydrophobic medications can be conveniently incorporated micelle cores.
Nanoscopic core element and hydrophilic layer limit the removal of micelles
from the body and eventually increase the drug's bioavailability [68].
Polymeric micelles prefer targeted treatment and sustainable drug delivery
because of the inner core's high drug load capacity. The hydrophobic
micellar center is ideal for the encapsulation of hydrophobic medicines.
Polymeric micelles can be produced through several processes including
electro-static interaction, hydrophobic interactions, and metal complexing.
Polymeric micelles are mainly prepared in a selective/non-selective solvent
by dissolving the copolymer layer, whereas drug-loaded micelles are mainly
prepared through direct dissolution, solvent evaporation, and dialysis (Figure
4.6(D)). Several strategies utilize micelles for the targeted treatment, such as
improved permeability, retention effect and ligand-based micelles [69].

4.5.4.1 NANOEMULSION

Nanoemulsions are submicron-sized colloidal system also called microemulsion which is thermodynamically and kinetically stable isotropic dispersion,
composed of two immiscible liquids such as water and oil, stabilized by
a single-phase interfacial layer consisting of appropriate surfactant and
co-surfactants [70]. Nanoemulsions are divided into three types: Water-in
Oil form (water is dispersed in oil phase), O/W type (oil is distributed in the
aqueous phase), and Bio-continuous (water and oil are mixed together with
in the system). Transformation of these three forms may be accomplished by
modifying the emulsion components. Multiple emulsions are often a form
of nanoemulsions, where both O/W and W/O emulsions exist in one cell
simultaneously. Various nanoemulsion preparation approaches such as high-
pressure homogenization, microfludization, random emulsification, solvent
evaporation, and creation of hydrogel have been introduced. The double
emulsion solvent evaporation method typically produces multiple emulsions
(Figure 4.7(A)).

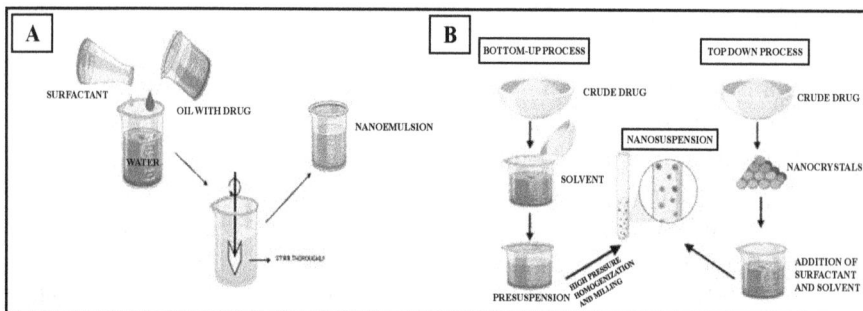

FIGURE 4.7   (A) Preparation of nanoemulsions; (B) preparation of nanosuspension.

A variety of methods were used to classify nanoemulsions [71]. The nanoemulsions are developed using two main techniques, (a) persuasion method and (b) brute force method. These are also considered to enhance the processing of antidiabetic lipophilic substances. α-eleostearic acid nanoemulsions were provided orally to a diabetic model of rat which enhance the antidiabetic abilities, increase cell uptake and antioxidant effect. Streptozotocin-induced model of diabetic rats have been treated with nanoemulsions loaded with alpha-tocopherol that have been defensive in various species [72]. Li et al. developed insulin-loaded nanoemulsions encapsulated with polyelectrolyte cross-linking chitosan/alginate. Conformational consistency in the cross-linking phase was verified although nanoemulsions maintained integrity in the synthetic gastric medium during the *in vitro* leakage test [73, 74].

4.5.4.2   NANOSUSPENSION

Nanosuspensions are another form of delivery system that can be used for oral administration with a solid state of the active ingredients. It is generally characterized as a very finely dispersed, biphasic colloid that comprises of solid drug particles smaller than 1 micron. This suspension does not include matrix content and is stabilized by polymers and surfactants. One of the biggest problems with poorly soluble drugs is very low bioavailability [51, 75]. The question is even more complicated for drugs that are insoluble in both aqueous and non-aqueous media and are listed in BCS class II. Nanosuspension approach is an enticing and effective option for solving such problems. Nanosuspension comprises of the water-insoluble pharmaceutical substance without a dispersed matrix material which is suspended in dispersion Nanosuspension preparation is easy and can be applied to all medicines that insoluble in water. A nanosuspension not only addresses the problems of low solubility and bioavailability, but also changes the pharmacokinetic profile and enhances the safety and efficacy of drug.

Two methods are used to prepare nanosuspensions: "bottom-up technology" and "top-down technology" (Figure 4.7(B)). The bottom-up technique is an assembly process for shaping NPs, including precipitation, microemulsion, melt emulsification, whereas the top-down technology, includes decomposing large particles into NPs, examples of which is milling method. Vaculikova et al. enhanced solubility of glibenclamide (antihyperglycemic drug belongs to BCS class II) by developing nanosuspensions using emulsion solvent evaporation method where dichloromethane is used as a solvent and carboxymethyl dextran sodium salt used as a stabilizer. Glimepiride is another oral hypoglycemic drug that belongs to BCS Class II, develop as nanosuspensions to enhance its solubility [76]. Diabetic rat model created by nicotinamide-streptozotocin were tested with glimepiride nanosuspensions showed improved pharmacokinetic profile and antihyperglycemic action [77, 78]. In nanosuspensions, phyto-ingredients have been developed to boost antidiabetic potency in contrast to traditional formulations. The usage of a nanosuspension process to prepare gymnemic acids indicates increased bioavailability as well. The scientists have investigated the role of nanosuspensions in humans, finding an improved antidiabetic activity and an increased influence on glucose decrease. In addition, the increased antidiabetic activity in T2DM animal models in a lower dose was shown by Berberine nanosuspensions [79].

4.5.5 THERAPEUTIC TARGETS OF NANOMEDICINES IN DIABETIC NEUROPATHY (DPN)

About all approved drugs used to treat DPN have some drawbacks such as inadequate transmission to the target site, serious adverse effects and short-term intervention. Present pharmacotherapy focuses mainly on site-specific targeting, minimizes the adverse effects, and prolongs the medication release. Convectional formulations are not very effective due to their accelerated metabolism and excretion from the body, so high doses should also be given at a daily interval of time, which is the main cause of adverse effects [80]. NPs have been widely studied in recent decades as an advanced delivery system. Nanotechnology approaches utilizing multiple nanomaterials could be an appealing and innovative solution to the care of DPN [81, 82]. NPs also proven to be the most powerful delivery method for drugs, biosimilars, genes immunologicals, etc. Several medications have been used to combat DPN, but none of them provide complete recovery from the disease, while these medications have several limitations. Different methods, like nanoformulations, are used to prevent such limitations [52, 83].

4.5.5.1   TARGET TO PAIN

Numerous studies have shown that glial cells play an important role in NeP development Normally, these glial cells are non-proliferating, but these cells become stimulated following nerve injury due to activation of certain mediators, such as ATP, glutamate, and substance P that induce mechanical allodynia and hyperalgesia (neurological effects of chronic pain) [84]. Several reports have shown that the mitogen-activated protein kinases play an important role in the development of DPN and some studies have also documented the upregulation of Rheb (i.e., involved in m-TOR pathway) in astrocytes if the spinal cord is damaged [85, 86]. Therefore, inhibitors of this pathway may have a major therapeutic effect for DPN.

Zoledronic acid, effective in bone demineralization works by inhibiting the isoprenylation of all small GTPases, including Ras family proteins, may help in treatment NeP by blocking the MAPK cascade, but its poor pharmacokinetic profile leads to accumulation of drugs in bones and decreases plasma concentration which restricts its use in the treatment of DPNP [87]. Caraglia and colleagues investigated Zoledronic acid encapsulated liposomes that can easily be introduced into the CNS, thereby resolving its constraints (Figure 4.8).

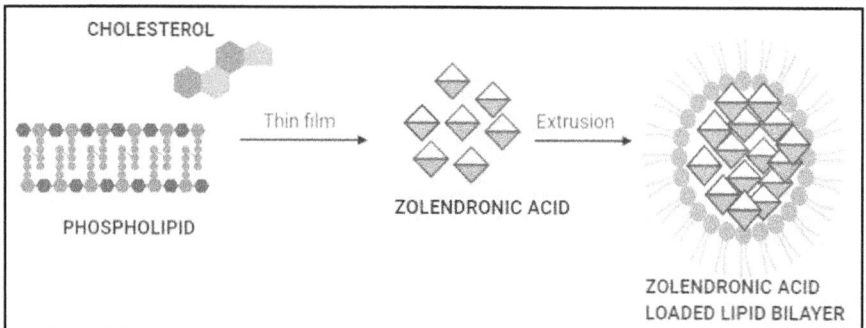

FIGURE 4.8   Encapsulation of zolendronic acid in liposomes.

The results of this study presented a mechanistic and reliable solution of nanomedicine in diabetic NeP. Another medication used in the management of seizures and NeP, lamotrigine works by inhibiting the sodium and calcium channels. However, due to poor pharmacokinetic profiles and improper distribution, a high dose is required that leads to extreme rashes and Stevens-Johnson syndrome [88]. Hence, its use is restricted in DPNP. In 2014, Lalani and colleagues devised surface-engineered NPs that were used to resolve such adverse effects and also to identify a brain target. They utilized ligands such as transferrin and lactoferrin, which could distribute a significant amount of drug via receptor-mediated endocytosis in the brain. The functional NPs of lactoferrin were found to be superior to transferrin NPs in brain targeting [89]. During the intravenous use of this solution in an animal model of DPNP, strong antinociceptive behavior with a higher brain concentration was observed for 48 hours. This study indicated that NPs strengthen the weak pharmacokinetic profile of medications by targeting the brain and expanding therapeutic effects in low dose [90].

Nefopam hydrochloride is a non-opioid analgesic which prevents the reuptake of three different neurotransmitters (serotonin, noradrenaline, and dopamine). It also works on H3 receptor of histamine, blocking the sodium channels and modulating glutamine transmissions. However, some adverse effects such as diarrhea, vomiting, dizziness, sweating, and patient non-compliance, are recorded when Nefopam hydrochloride has been given to the patients of DPNP [91]. Therefore, a reliable dosage form that assures sustained drug release and reduces metabolism will be an excellent strategy for DPNP application.

Peripherally controlled synthetic cannabinoids have a huge potential in recent years to substitute drugs in the diagnosis of DPN. Synthetic cannabinoid agonists such as CB13, 212-2, WIN55 and CP55, 940 are CB1 and CB2 receptor agonists have high antiallodynic and anti-hyperalgesic activity serves as a second-line treatment of DPN but due to their unpredictable nature these cannot be easily developed in any dosage type. CB13 belongs to class II of the Biopharmaceutical Classification System (low solubility and high permeability) which therefore limits its therapeutic use because of its hydrophobic nature thus cannot be delivered by oral route [92]. Work thus demands a new formulation that could encapsulate the hydrophobic medications and be delivered conveniently by oral route. In this context, Martin-Banderas and colleagues developed NPs comprising CB13. *In vitro* and *ex vivo* studies showed that these NPs accumulated in liver and spleen, to prevent these NPs were PEGylated with Polyethylene Glycol (PEG).

The benefits of PEGylation are to improve stability, blood circulation, NPs opsonization, inhibit protein plasma binding and reducing immunogenic reactions. One of the conflicting concerns is local anesthetics in the care of DPN [93]. S Shankarappa and D. Kohane proposed that prolonged release of local anesthetics could provide a viable solution in treating DPN conditions. In recent times, a lot of attention has been given to locally active agents, which may minimize pain severity by working locally and thus avoid systemic side effects such as constipation, sedation, urinary retention, abuse, itching, and opioid mortality. The management of chronic pain conditions initiated with local anesthetics such as procaine, lidocaine, and bupivacaine. However, they act only for short duration and cannot provide the intended results by prolonging the nerve block, their use is limited [94]. Various drug delivery systems such as injectable particles utilizing liposomes and lipospheres, micro-polymeric, and NPs, injectable liquids including cyclodextrins, hydrogels, and fluid polymers, hybrid formulations such as polymeric microparticle-hydrogel have been explored to overcome this problem [51].

4.5.5.2 TARGET TO INFLAMMATION

Recent studies suggest that the inflammatory response plays a crucial role in DPN pathogenesis. Macrophages, neutrophils, and glial cells can be stimulated by prolonged hyperglycemia. Activated immune cells could also induce TNF-α, IL-1β and IL-6 expression. Many experiments have shown that the TNF-α is a main pro-inflammatory cytokine in DPN and may trigger both nuclear factor kappa B and apoptotic pathways that cause reduction in cell viability and nerve damage. Nevertheless, anti-inflammatory pharmaceutical drugs are restricted by their evident side effects and tissue sensitivity in clinical applications [95]. Endogenous tiny inflammatory molecules can therefore be useful for the management of DPN. miRNAs are minute endogenous entities contain 18–25 nucleotides. They can attach with specific RNA by complementary 3'UTR base and can modify the transcription or translation of gene expression adversely. A possible nucleic acid vector is supposed to carry out the intracellular transmission of miRNA. The vectors transfected by miRNAs for animals actually contain adenoviruses and retroviruses [96]. The transfection efficiency of these could be high; but their immunogenicity, potential risk of tumorigenesis and lack of tissue specificity have hindered the therapeutic use of viral vectors.

Nanotechnology also plays an important role in these conditions. Since NPs are not immunogenic, several nanocarrier medications are also used in the clinical practice. NPs can be classified into organic NPs and inorganic NPs at present. The optimal nanocarrier for miRNAs will meet the required specifications. First, stable after direct contact with miRNAs and high load rate transition of miRNAs. Secondly, the nanocarrier is designed to selectively encapsulate miRNA and secure miRNAs from intracellular RNA-degradation by various degrading enzymes. The polyplex must be capable of resisting the degradation of endosomes and release the miRNAs into the target cells following phagocytization (Figure 4.9). In addition, after nano-miR-146a-5p, the inflammatory reactions and apoptosis of DPN rats were inhibited. This result indicated that nano-miR-146a-5p would safeguard the sciatic nerve by inhibiting inflammatory reaction [97].

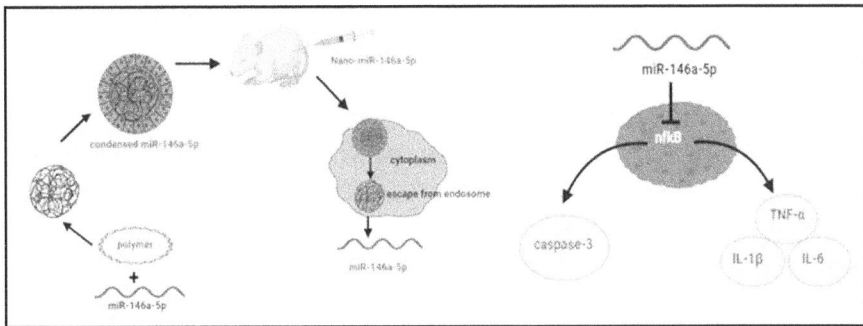

FIGURE 4.9 Nanomedicines targeting inflammatory pathway.

4.5.5.3 TARGET TO FREE RADICALS

Oxidative stress performs a prominent role in the pathogenesis of several complications of diabetes. The big issue with diabetic patients is slow recovery and still threatens its permanent cure. The healing impaired is defined by retarded cellular infiltration and granulation of tissue formation, lowered collagen, and decreased angiogenesis. The mechanism of such modification is because of high degree of ROS generation leading to early programed cell death of inflammatory cells, which then in turn adversely affects the fibroblasts, endothelial cells and metabolism of collagen [98]. Some of the NPs now prepare in a way that functions as free radical scavenger. There have recently been surprising evidence for certain nanomaterials, like metallic NPs, having enzyme-like antioxidant properties that can scavenge free radicals and reduce ROS concentrations. Such NPs are generally referred to as nanoantioxidants. These nano-antioxidants act either as an antioxidant delivery system or having underlying antioxidant properties [99]. The specific molecular mechanisms that assess the antioxidant ability of metallic NPs remain largely unknown. These NPs are most likely to have a high surface-to-volume ratio, oxygen vacancy defects, electronic configuration and catalytic and redox potential. For, e.g., CeO_2 NPs that modulate main antioxidant pathways such as Nrf2 [100]. Moreover, for these processes, interactions of NPs and cellular macromolecules especially nucleic acids, proteins, and lipids could be extremely important. A large number of interactions between NPs and proteins are expected because of protein structure diversity and the kinetics of nanoparticle protein interactions that can be significantly influenced by NP structure, availability of protein and interaction duration. Because of these effects, the NPs may primarily affect the cellular redox process by inducing or inhibiting the production of ROS under certain conditions. Therefore, the hermetic phenomenon is known as the "biphasic dose-response interaction, which is defined by a low-dose stimulus and high-dose inhibition" play a significant role in modulating the redox effect of NPs. Under oxidative stress conditions, specific NPs can reduce oxidative damage to important biomolecules and therefore provide various health benefits and disease prevention. Cerium oxide (CeO_2) is one of them that are effective due to its excellent ability of engulfing free radicals. CeO_2 NPs provide unique advantages relative to other antioxidants including: the ability to regenerate free radicals engulfing action by randomly going from reduction to oxidation, and it also has a wide number of available sites that are used to scavenge the free radicals. It easily crosses the blood—the brain barrier because of its small size that makes CeO_2 NPs a good candidate for treatment of DPN. In several *in vitro* and *in vivo* tests, safety results of CeO_2 NPs have been studied [101–103].

The findings verified that CeO_2 NPs perform an important role in the attenuation of damage caused by oxidative stress. Oxidative stress performs a leading role in the etiology of various complications of diabetes. The capacity of gold NPs to suppress lipid peroxidation leads to inhibit ROS generation that further re-establish the imbalances in the level of antioxidants and liver enzymes which cause cell malfunction and degradation. Gold nanoparticles are recognized for their enormous use in the area of therapy and diagnosis.

Recent studies verified the anti-oxidative and anti-hyperglycemic functions of gold nanoparticles in streptozotocin-induced diabetic mice by inhibiting the ROS generation and engulfing free radicals that ultimately increase the level of antioxidant enzymes (Figure 4.10). Gold nanoparticles have shown non-toxic and defensive effects on the organs without causing deadly results in the model of mouse, thus constantly regulating the disease progression. The potential application of gold nanoparticles to inhibit oxidative stress and have less adverse effects has opened the door to a new field of cost-effective alternative in DPN [104, 105].

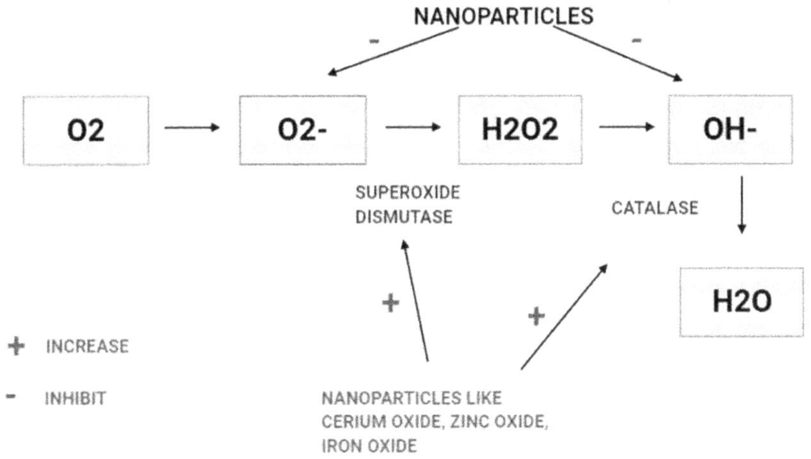

NANOPARTICLES

FIGURE 4.10   Nanoparticles targeting oxidative stress.

4.5.5.4   *Target to Glucose*

Imbalances of natural metabolism of body, because of massive amount of molecular or reactive oxygen species (ROS), contribute to a high level of glucose in the blood (hyperglycemia), which leads to oxidative stress and persistent complications of diabetes. The control of diabetic disorders by insulin therapy has several disadvantages such as insulin resistance and anorexia nervosa, and fatty liver. Therefore, many experiments of nano-size particles are going on nowadays to overcome these constraints in diabetes treatment. A total of 285 million people worldwide are suffering from severe, progressive, and often debilitating diabetes because of the pancreas failure to control blood glucose levels [106]. During these past decades, subcutaneous route is preferred for insulin ingestion, which mostly fails to resemble the glucose homeostasis as seen in normal individuals because here insulin is supplied to peripheral circulation instead of portal circulation and direct to the liver that is the physiological pathway for normal subjects. In fact, many routine doses of insulin-related to poor patient compliance related to subcutaneous route administration. Therefore, several experiments have been done to determine the safer and better route of insulin delivery and nanotechnology has been implemented in medicine to resolve this problem. Diabetic patients obtain oral insulin can be better than subcutaneous route, not only because it relieves pain and trauma-induced by SC injections, but it also imitates the physiological fate of insulin. Moreover, the administration of oral protein medications, such as insulin, has certain disadvantages like difficulty in absorption due to low pH of stomach and its digestive enzymes. Intestinal epithelium is also a major obstacle in absorption of macromolecules such as proteins and nucleic acids, before they can enter the target cell to produce its action [107].

The paracellular availability of hydrophilic macromolecules was therefore enhanced by the use of nanotechnology in diabetic study. Prodrugs (insulin-polymer conjugation), liposomes, niosomes, nanosuspensions, dendrimers, stable lipid nanoparticles, and nanoparticles of biodegradable polymers are the techniques that can be used for the oral delivery of insulin. PEGylation (i.e., medication conjugation to polyethylene glycol (PEG)) is the most common drug formulation used for improved solubility, permeability, and durability in prodrug technology. Insulin/PEG medications show great benefits in oral distribution [108]. Scott Moncrieff et al. created a system of bile-salt and fatty acid micelles containing sodium glycocholate and linoleic acid that dramatically increased enteral absorption of insulin. However, micelles did not appear to be suitable for hydrophilic drug delivery. Rather, liposomes can be much more efficient in the production of oral insulin. A glycocholate-containing liposomal delivery system has been recently developed with a view to oral insulin transmission, and it indicates that insulin ion is better protected from pepsin, trypsin, and acymotrypsin enzyme degradation [109]. US FDA approved NPs of biodegradable polymers have also been tested for oral delivery of insulin. NPs may not be suitable for delivering hydrophilic medicines; hence, how hydrophilic medications like insulin can be used for oral delivery remains a challenge. Consequently, the increase in the paracellular transport of hydrophilic drugs has been focused on. Different intestinal permeation enhancers like chitosan (CS) were used to promote hydrophilic macromolecular absorption. A carrier network is therefore required to protect protein drugs from the unpleasant environment of the stomach and small intestines, if delivered orally. CS NPs have also been more effective in enhancing intestinal absorption of protein molecules. Insulin-loaded NPs coated with mucoadhesive CS will extend their presence in the small intestine, penetrate the mucus layer and then mediate the opening of close ties between epithelial cells, thus becoming unstable and degrade because of their pH sensitivity and degradability. Insulin released from the damaged NPs could then enter from the paracellular pathway to the bloodstream and reach the destination [110]. The most effective approach for oral insulin is to use a microsphere method. Microspheres serve as inhibitors of the protease by shielding encapsulated insulin in the matrix from enzyme degradation and facilitating permeation through the crossing of the epithelial layer following oral administration [111].

4.5.6   TOXICITY OF NANOPARTICLES (NPS)

The nanotoxicology can be described as the analysis of side effects of nanomedicines on living organisms and ecosystems and prevention of these effects. The decreased nanomaterial size increases the cell surface area exponentially that contributes to complicated interactions with bio-nano-interfaces if physiologically exposed. Size is indeed a major factor in the toxicology assessment because the NPs are able to translocate across cell boundaries or neuronal pathways. The cell entry mechanism is also decided by the thickness, since the smallest nanomaterial will diffuse over the cell membranes. Studies have shown that nanomedicines can induce or inhibit the immune response by stimulating the immune system which is most commonly observed. NMPs communicate with both the natural and adaptive immune systems. Complementary pseudoallergic reactions are also the common side effects of FDA approved lipid and polymer-based nanomedicines administered *in vivo* via intravenous (i.v.) route [112]. Since then, i.v. administration indicates direct contact with blood and thus increases the chance of damage to the host's immune system. Antigenicity of NPs contributes to a chain of responses that leads to autoimmune processes or adverse drug reactions like allergic reactions or hypersensitivity reactions. *In vivo* mouse models after administration of inorganic NPs, adverse drug reactions that triggered by the adaptive immune system via activation of T cells or B cells contributing to cascades of cell-mediated and humoral responses were observed. Size and surface to volume ratios are critical factors leading to nanotoxicity [113]. Between NP size and toxicity there is a clear inverse correlation. NMP-related toxicity has impeded their therapeutic usage due to problems related to toxicity as well as the absence of established safety rules for nanomedicine. Advances in nanomedicine also led to the development of new tests for NMP toxicity evaluation. Another new test from the American Society for Testing Materials (ASTM) has measured the impact on NMPs on the development of macrophage colonies of the mouse and is a systematic method for the measurement of immunotoxicity responses of NMPs. The other possible methods for *in vivo* and *in vitro* research are complement stimulation, production of pro-inflammatory cytokines and activation of lymphocytes. However, no *in vitro* experiments that evaluate human-specific NMP impact on the immune system are presently available on the commercial scale [114].

4.6   FUTURE ASPECTS

Nanomedicine aim in the near future is to deliver a versatile collection of research tools and clinically important devices. The NTI (National Technology Initiative) expects new practical applications in the pharmaceutical industry that can include innovative medication delivery systems, novel treatments, and *in vivo* imaging. The US National Institutes of Health finance nanomedicine studies, including the sponsorship of a 5-year program for the creation of four nanomedicine centers in 2005. Although the potential future outcomes of nanomedicines look very positive, their technological, social, and health consequences need to be regulated wholeheartedly and safely [115]. The further development of nanomedicines requires a detailed scientific and medical understanding of the social and economic problems that obstructing their marketing. For this, unproductive barriers need to be replaced with effective-filters to allow nanomedicines more affordable, stable, and convenient to use in the long term [68].

4.7   CONCLUSION

Available treatment is not sufficient to provide complete relief from diabetic pain. Hence, these are scope for development of many new strategies which could be extrapolated in such a way that it must reach to clinics. Nanomedicine offers significant potential for treating critical kinds of pain as it can resolve the pharmaceutical challenges of various drugs and can provide faster and targeted treatment. This manuscript highlights the emerging role of nanomedicine in designing future pharmacologic treatments specifically towards pain associated with diabetes.

KEYWORDS

- **acetyl-L-carnitine**
- **American Society for Testing Materials**
- **cerium oxide**
- **chitosan**
- **diabetic neuropathy**
- **nanoparticles**

REFERENCES

1. Asmat, U., Abad, K., & Ismail, K., (2016). Diabetes mellitus and oxidative stress: A concise review. *Saudi Pharmaceutical Journal, 24*(5), 547–553.
2. Zheng, Y., Ley, S. H., & Hu, F. B., (2018). Global etiology and epidemiology of type 2 diabetes mellitus and its complications. *Nature Reviews Endocrinology, 14*(2), 88.
3. Feldman, E. L., Callaghan, B. C., Pop-Busui, R., Zochodne, D. W., Wright, D. E., Bennett, D. L., Bril, V., et al., (2019). Diabetic neuropathy. *Nature Reviews Disease Primers, 5*(1), 1–18.
4. Feldman, E. L., Nave, K. A., Jensen, T. S., & Bennett, D. L., (2017). New horizons in diabetic neuropathy: Mechanisms, bioenergetics, and pain. *Neuron, 93*(6), 1296–1313.
5. Sardu, C., De Lucia, C., Wallner, M., & Santulli, G., (2019). Diabetes mellitus and its cardiovascular complications: New insights into an old disease. *Journal of Diabetes Research, 2019.*
6. Bellows, B. K., Nelson, R. E., Oderda, G. M., & LaFleur, J., (2016). Long-term cost-effectiveness of initiating treatment for painful diabetic neuropathy with pregabalin, duloxetine, gabapentin, or desipramine. *Pain, 157*(1), 203–213.
7. Ruiz-Negrón, N., Menon, J., King, J. B., Ma, J., & Bellows, B. K., (2019). Cost-effectiveness of treatment options for neuropathic pain: A systematic review. *Pharmacoeconomics, 37*(5), 669–688.
8. Pradeepa, R., & Mohan, V., (2017). Prevalence of type 2 diabetes and its complications in India and economic costs to the nation. *European Journal of Clinical Nutrition, 71*(7), 816–824.
9. Kiyani, M., Yang, Z., Charalambous, L. T., Adil, S. M., Lee, H. J., Yang, S., Pagadala, P., et al., (2020). Painful diabetic peripheral neuropathy: Health care costs and complications from 2010 to 2015. *Neurology: Clinical Practice, 10*(1), 47–57.
10. Yorek, M., Malik, R. A., Calcutt, N. A., Vinik, A., & Yagihashi, S., (2018). Diabetic neuropathy: New insights to early diagnosis and treatments. *Journal of Diabetes Research, 2018.*
11. Callaghan, B. C., Viswanathan, V., Reynolds, E. L., & Feldman, E. L., (2018). *The Metabolic Drivers of Neuropathy in India.* Am Diabetes Assoc.
12. Pop-Busui, R., Boulton, A. J., Feldman, E. L., Bril, V., Freeman, R., Malik, R. A., Sosenko, J. M., & Ziegler, D., (2017). Diabetic neuropathy: A position statement by the American diabetes association. *Diabetes Care, 40*(1), 136–154.
13. Unnikrishnan, R., Anjana, R. M., & Mohan, V., (2016). Diabetes mellitus and its complications in India. *Nature Reviews Endocrinology, 12*(6), 357.
14. Chatterjee, S., Khunti, K., & Davies, M. J., (2017). Type 2 diabetes. *The Lancet, 389*(10085), 2239–2251.
15. Association, A. D., (2018). Classification and diagnosis of diabetes: Standards of medical care in diabetes. *Diabetes Care, 41*(1), S13–S27.
16. Holt, R. I., Cockram, C., Flyvbjerg, A., & Goldstein, B. J., (2017). *Textbook of Diabetes.* John Wiley & Sons.
17. Katsarou, A., Gudbjörnsdottir, S., Rawshani, A., Dabelea, D., Bonifacio, E., Anderson, B. J., Jacobsen, L. M., et al., (2017). Type 1 diabetes mellitus. *Nature Reviews Disease Primers, 3*(1), 1–17.
18. Asif, M., & Batool, S., (2020). Association between diabetic neuropathy, fall risks and balance in diabetes type 2 patients. *Rawal Medical Journal, 45*(1), 27–30.

19. Khoury, C. C., Chen, S., & Ziyadeh, F. N., (2020). Pathophysiology of diabetic nephropathy. In: *Chronic Renal Disease* (pp. 279–296). Elsevier.

20. Nare, S., Mohan, S., Satagopan, U., Natarajan, S., & Kumaramanickavel, G., (2019). Diabetic retinopathy: Clinical, genetic, and health economics (an Asian perspective). In: *Advances in Vision Research* (Vol. II, pp. 345–356). Springer.

21. Petrie, J. R., Guzik, T. J., & Touyz, R. M., (2018). Diabetes, hypertension, and cardiovascular disease: Clinical insights and vascular mechanisms. *Canadian Journal of Cardiology*, *34*(5), 575–584.

22. Alaboud, A. F., Tourkmani, A. M., Alharbi, T. J., Alobikan, A. H., Abdelhay, O., Al Batal, S. M., Alkhashan, H. I., & Mohammed, U. Y., (2016). Microvascular and macrovascular complications of type 2 diabetic mellitus in central, Kingdom of Saudi Arabia. *Saudi Medical Journal*, *37*(12), 1408.

23. Domingueti, C. P., Dusse, L. M. S. A., Das, G. C. M., De Sousa, L. P., Gomes, K. B., & Fernandes, A. P., (2016). Diabetes mellitus: The linkage between oxidative stress, inflammation, hypercoagulability and vascular complications. *Journal of Diabetes and its Complications*, *30*(4), 738–745.

24. Chawla, A., Chawla, R., & Jaggi, S., (2016). Microvascular and macrovascular complications in diabetes mellitus: Distinct or continuum? *Indian Journal of Endocrinology and Metabolism*, *20*(4), 546.

25. Lotfy, M., Adeghate, J., Kalasz, H., Singh, J., & Adeghate, E., (2017). Chronic complications of diabetes mellitus: A mini review. *Current Diabetes Reviews*, *13*(1), 3–10.

26. Papatheodorou, K., Banach, M., Bekiari, E., Rizzo, M., & Edmonds, M., (2018). Complications of diabetes 2017. *Journal of Diabetes Research*, *2018*.

27. Hébert, H. L., Veluchamy, A., Torrance, N., & Smith, B. H., (2017). Risk factors for neuropathic pain in diabetes mellitus. *Pain*, *158*(4), 560.

28. Yagihashi, S., & Mizukami, H., (2018). Diabetic neuropathy. In: *Diabetes and Aging-Related Complications* (pp. 31–43). Springer.

29. Yan, L. J., (2018). Redox imbalance stress in diabetes mellitus: Role of the polyol pathway. *Animal Models and Experimental Medicine*, *1*(1), 7–13.

30. You, L., (2017). Progress in pathogenesis of diabetic peripheral neuropathy. *Journal of Shanghai Jiao Tong University (Medical Science)*, *37*(10), 1441–1445.

31. Qi, W., Keenan, H. A., Li, Q., Ishikado, A., Kannt, A., Sadowski, T., Yorek, M. A., et al., (2017). Pyruvate kinase M2 activation may protect against the progression of diabetic glomerular pathology and mitochondrial dysfunction. *Nature Medicine*, *23*(6), 753.

32. Kawanami, D., Matoba, K., & Utsunomiya, K., (2016). Signaling pathways in diabetic nephropathy. *Histology and Histopathology*, *31*(10), 1059–1067.

33. Rezaee, R., Monemi, A., Sadeghi-Bonjar, M. A., & Hashemzaei, M., (2019). Berberine alleviates paclitaxel-induced neuropathy. *Journal of Pharmacopuncture*, *22*(2), 90.

34. Hei, C., Liu, P., Yang, X., Niu, J., & Li, P. A., (2017). Inhibition of mTOR signaling confers protection against cerebral ischemic injury in acute hyperglycemic rats. *International Journal of Biological Sciences*, *13*(7), 878.

35. Al Hroob, A. M., Abukhalil, M. H., Alghonmeen, R. D., & Mahmoud, A. M., (2018). Ginger alleviates hyperglycemia-induced oxidative stress, inflammation and apoptosis and protects rats against diabetic nephropathy. *Biomedicine and Pharmacotherapy*, *106*, 381–389.

36. Alam, U., (2020). *Diabetic Neuropathy Collection: Treatment of Diabetic Neuropathy*. Springer.

37. Kremer, M., Salvat, E., Muller, A., Yalcin, I., & Barrot, M., (2016). Antidepressants and gabapentinoids in neuropathic pain: Mechanistic insights. *Neuroscience, 338,* 183–206.

38. Obata, H., (2017). Analgesic mechanisms of antidepressants for neuropathic pain. *International Journal of Molecular Sciences, 18*(11), 2483.

39. Khdour, M. R., (2020). Treatment of diabetic peripheral neuropathy: A review. *Journal of Pharmacy and Pharmacology.*

40. Onuțu, A. H., Dîrzu, D. S., & Petrişor, C., (2018). Serotonin reuptake inhibitors and their role in chronic pain management. In: *Serotonin,* IntechOpen.

41. Snyder, M. J., Gibbs, L. M., & Lindsay, T. J., (2016). Treating painful diabetic peripheral neuropathy: An update. *American Family Physician, 94*(3), 227–234.

42. Al-Waili, N., Al-Waili, H., Al-Waili, T., & Salom, K., (2017). Natural antioxidants in the treatment and prevention of diabetic nephropathy; a potential approach that warrants clinical trials. *Redox Report, 22*(3), 99–118.

43. Thakur, P., Kumar, A., & Kumar, A., (2018). Targeting oxidative stress through antioxidants in diabetes mellitus. *Journal of Drug Targeting, 26*(9), 766–776.

44. Didangelos, T., & Veves, A., (2020). Treatment of diabetic cardiovascular autonomic, peripheral and painful neuropathy. Focus on the treatment of cardiovascular autonomic neuropathy with ACE inhibitors. *Current Vascular Pharmacology.*

45. Atef, M., El-Sayed, N. M., Mostafa, Y. M., & Ahmed, A. A. M., (2019). Recent updates in treatment of diabetic neuropathy. *Records of Pharmaceutical and Biomedical Sciences, 3*(2), 15–27.

46. Sharma, D., Bhattacharya, P., Kalia, K., & Tiwari, V., (2017). Diabetic nephropathy: New insights into established therapeutic paradigms and novel molecular targets. *Diabetes Research and Clinical Practice, 128,* 91–108.

47. Bhushan, B., (2017). Introduction to nanotechnology. In: *Springer Handbook of Nanotechnology* (pp. 1–19). Springer.

48. Pelaz, B., Alexiou, C., Alvarez-Puebla, R. A., Alves, F., Andrews, A. M., Ashraf, S., Balogh, L. P., et al., (2017). *Diverse Applications of Nanomedicine.* ACS Publications.

49. Ventola, C. L., (2017). Progress in nanomedicine: Approved and investigational nanodrugs. *Pharmacy and Therapeutics, 42*(12), 742.

50. Bannister, K., Sachau, J., Baron, R., & Dickenson, A. H., (2020). Neuropathic pain: Mechanism-based therapeutics. *Annual Review of Pharmacology and Toxicology, 60,* 257–274.

51. Bidve, P., Prajapati, N., Kalia, K., Tekade, R., & Tiwari, V., (2020). Emerging role of nanomedicine in the treatment of neuropathic pain. *Journal of Drug Targeting, 28*(1), 11–22.

52. Kuthati, Y., Navakanth, R. V., Busa, P., Tummala, S., Davuluri, V. N. G., & Wong, C. S., (2020). Scope and applications of nanomedicines for the management of neuropathic pain. *Molecular Pharmaceutics.*

53. Chan, W. C., (2017). Nanomedicine 2.0. *Accounts of Chemical Research, 50*(3), 627–632.

54. Xu, G., Zeng, S., Zhang, B., Swihart, M. T., Yong, K. T., & Prasad, P. N., (2016). New generation cadmium-free quantum dots for biophotonics and nanomedicine. *Chemical Reviews, 116*(19), 12234–12327.

55. Premkumar, K., (2018). *Nanoparticles for Nanomedicine.*

56. Peng, F., Zhang, W., & Qiu, F., (2019). Self-assembling peptides in current nanomedicine: Versatile nanomaterials for drug delivery. *Current Medicinal Chemistry.*

57. Yeo, P. L., Lim, C. L., Chye, S. M., Ling, A. P. K., & Koh, R. Y., (2018). Niosomes: A review of their structure, properties, methods of preparation, and medical applications. *Asian Biomedicine, 11*(4), 301–314.
58. Beltrán-Gracia, E., López-Camacho, A., Higuera-Ciapara, I., Velázquez-Fernández, J. B., & Vallejo-Cardona, A. A., (2019). Nanomedicine review: Clinical developments in liposomal applications. *Cancer Nanotechnology, 10*(1), 11.
59. De Paula, E., Oliveira, J. D., De Lima, F. F., De Morais, R. L. N., (2020). 16 liposome-based delivery. *Controlled Drug Delivery Systems, 299.*
60. Zang, X., Lee, J. B., Deshpande, K., Garbuzenko, O. B., Minko, T., & Kagan, L., (2019). Prevention of paclitaxel-induced neuropathy by formulation approach. *Journal of Controlled Release, 303,* 109–116.
61. Caminade, A. M., (2017). Phosphorus dendrimers for nanomedicine. *Chemical Communications, 53*(71), 9830–9838.
62. Shcharbin, D., Shcharbina, N., Dzmitruk, V., Pedziwiatr-Werbicka, E., Ionov, M., Mignani, S., De La Mata, F. J., et al., (2017). Dendrimer-protein interactions versus dendrimer-based nanomedicine. *Colloids and Surfaces B: Biointerfaces, 152,* 414–422.
63. Kretzmann, J., Evans, C., Norret, M., & Iyer, K. S., (2017). Supramolecular assemblies of dendrimers and dendritic polymers in nanomedicine. In: *Comprehensive Supramolecular Chemistry II* (pp. 237–256). Academic Press (Elsevier Inc.).
64. Shcharbin, D., Shcharbina, N., Dzmitruk, V., Pedziwiatr-Werbicka, E., Ionov, M., Mignani, S., De La Mata, F. J., et al., (2017). Dendrimer-protein interactions versus dendrimer-based nanomedicine. *B. Biointerfaces, 152,* 414–422.
65. Kim, Y., Park, E. J., & Na, D. H., (2018). Recent progress in dendrimer-based nanomedicine development. *Archives of Pharmacal Research, 41*(6), 571–582.
66. Gok, O., Kambhampati, S. P., Smith, E., Kannan, S., & Kannan, R. M., (2018). In: *Design of a Novel PAMAM-Based Nanomedicine with Sustained NAC Release for Treatment of Neuroinflammation* (pp. i–iii). 2017 21st National Biomedical Engineering Meeting (BIYOMUT), IEEE.
67. Cagel, M., Tesan, F. C., Bernabeu, E., Salgueiro, M. J., Zubillaga, M. B., Moretton, M. A., & Chiappetta, D. A., (2017). Polymeric mixed micelles as nanomedicines: Achievements and perspectives. *European Journal of Pharmaceutics and Biopharmaceutics, 113,* 211–228.
68. Islam, N., Abbas, M., & Rahman, S., (2017). Neuropathic pain and lung delivery of nanoparticulate drugs: An emerging novel therapeutic strategy. *CNS and Neurological Disorders-Drug Targets (Formerly Current Drug Targets-CNS and Neurological Disorders), 16*(3), 303–310.
69. Ganugula, R., Deng, M., Arora, M., Pan, H. L., & Kumar, M. R., (2019). Polyester nanoparticle encapsulation mitigates paclitaxel-induced peripheral neuropathy. *ACS Chemical Neuroscience, 10*(3), 1801–1812.
70. Stevens, A. M., Liu, L., Bertovich, D., Janjic, J. M., & Pollock, J. A., (2019). Differential expression of neuroinflammatory mRNAs in the rat sciatic nerve following chronic constriction injury and pain-relieving nanoemulsion NSAID delivery to infiltrating macrophages. *International Journal of Molecular Sciences, 20*(21), 5269.
71. Bonferoni, M. C., Rossi, S., Sandri, G., Ferrari, F., Gavini, E., Rassu, G., & Giunchedi, P., (2019). Nanoemulsions for "nose-to-brain" drug delivery. *Pharmaceutics, 11*(2), 84.
72. Bonferoni, M., Riva, F., Invernizzi, A., Dellera, E., Sandri, G., Rossi, S., Marrubini, G., et al., (2018). Alpha tocopherol loaded chitosan oleate nanoemulsions for wound

healing. Evaluation on cell lines and *ex vivo* human biopsies, and stabilization in spray dried Trojan microparticles. *European Journal of Pharmaceutics and Biopharmaceutics, 123*, 31–41.

73. Singh, Y., Meher, J. G., Raval, K., Khan, F. A., Chaurasia, M., Jain, N. K., & Chourasia, M. K., (2017). Nanoemulsion: Concepts, development and applications in drug delivery. *Journal of Controlled Release, 252*, 28–49.

74. Hu, X. B., Tang, T. T., Li, Y. J., Wu, J. Y., Wang, J. M., Liu, X. Y., & Xiang, D. X., (2019). Phospholipid complex based nanoemulsion system for oral insulin delivery: Preparation, *in vitro*, and *in vivo* evaluations. *International Journal of Nanomedicine, 14*, 3055.

75. Tekade, R. K., Maheshwari, R., Soni, N., Tekade, M., & Chougule, M. B., (2017). Nanotechnology for the development of nanomedicine. In: *Nanotechnology-Based Approaches for Targeting and Delivery of Drugs and Genes* (pp. 3–61). Elsevier.

76. Zhang, H., Chen, W., Zhao, Z., Dong, Q., Yin, L., Zhou, J., & Ding, Y., (2017). Lyophilized nanosuspensions for oral bioavailability improvement of insoluble drugs: Preparation, characterization, and pharmacokinetic studies. *Journal of Pharmaceutical Innovation, 12*(3), 271–280.

77. Singh, A. K., Pandey, H., Ramteke, P. W., & Mishra, S. B., (2019). Nano-suspension of ursolic acid for improving oral bioavailability and attenuation of type II diabetes: A histopathological investigation. *Biocatalysis and Agricultural Biotechnology, 22*, 101433.

78. Oraebosi, M. I., Olurishe, T. O., & Ayanwuyi, L. O., (2019). Chrono modulated nifedipine supports concurrent glimepiride administration with subsequent amelioration of retinopathy and peripheral neuropathy in diabetic rats. *Egyptian Journal of Basic and Clinical Pharmacology, 9*.

79. Zhang, P., He, L., Zhang, J., Mei, X., Zhang, Y., Tian, H., & Chen, Z., (2019). Preparation of novel berberine nano-colloids for improving wound healing of diabetic rats by acting Sirt1/NF-κB pathway. *Colloids and Surfaces B: Biointerfaces*, 110647.

80. Nejati-Koshki, K., Mortazavi, Y., Pilehvar-Soltanahmadi, Y., Sheoran, S., & Zarghami, N., (2017). An update on application of nanotechnology and stem cells in spinal cord injury regeneration. *Biomedicine and Pharmacotherapy, 90*, 85–92.

81. Mohsen, A. M., (2019). Nanotechnology advanced strategies for the management of diabetes mellitus. *Current Drug Targets, 20*(10), 995–1007.

82. Moradkhani, M. R., Karimi, A., & Negahdari, B., (2018). Nanotechnology application for pain therapy. *Artificial Cells, Nanomedicine, and Biotechnology, 46*(2), 368–373.

83. Fangueiro, J. F. P., (2016). *Cationic Lipid Nanomedicines for the Treatment of Diabetic Retinopathy*.

84. Gonçalves, N. P., Vægter, C. B., Andersen, H., Østergaard, L., Calcutt, N. A., & Jensen, T. S., (2017). Schwann cell interactions with axons and microvessels in diabetic neuropathy. *Nature Reviews Neurology, 13*(3), 135–147.

85. Gonçalves, N. P., Vægter, C. B., & Pallesen, L. T., (2018). Peripheral glial cells in the development of diabetic neuropathy. *Frontiers in Neurology, 9*, 268.

86. Li, L., Sheng, X., Zhao, S., Zou, L., Han, X., Gong, Y., Yuan, H., et al., (2017). Nanoparticle-encapsulated emodin decreases diabetic neuropathic pain probably via a mechanism involving P2X3 receptor in the dorsal root ganglia. *Purinergic Signaling, 13*(4), 559–568.

87. Deshayes, S., Silva, N. M., Cogez, J., Baldolli, A., Fedrizzi, S., Bienvenu, B., & Aouba, A., (2016). Multiple cranial neuropathies following zoledronic acid infusion: A relationship? Clinical features and pathogenic discussion concerning a case. *Osteoporosis International, 27*(8), 2627–2629.

88. Zappavigna, S., Luce, A., Porru, M., Cossu, A. M., Ferri, C., Lusa, S., Abate, M., et al., (2017). Stealth liposomes for the delivery of zoledronic acid into tumors enhance the anticancer activity of the drug. *Translational Medicine Reports, 1*(2).

89. Yan, X., Xu, L., Bi, C., Duan, D., Chu, L., Yu, X., Wu, Z., et al., (2018). Lactoferrin-modified rotigotine nanoparticles for enhanced nose-to-brain delivery: LESA-MS/MS-based drug biodistribution, pharmacodynamics, and neuroprotective effects. *International Journal of Nanomedicine, 13,* 273.

90. Roy, M., Pal, R., & Chakraborti, A. S., (2017). *Pelargonidin-PLGA Nanoparticles: Fabrication, Characterization, and Their Effect on Streptozotocin Induced Diabetic Rats.*

91. Nair, A. S., (2019). Nefopam: Another pragmatic analgesic in managing chronic neuropathic pain. *Indian Journal of Palliative Care, 25*(3), 482.

92. Blanton, H. L., Brelsfoard, J., DeTurk, N., Pruitt, K., Narasimhan, M., Morgan, D. J., & Guindon, J., (2019). Cannabinoids: Current and future options to treat chronic and chemotherapy-induced neuropathic pain. *Drugs,* 1–27.

93. Maione, S., Rossi, F., Guy, G., Stott, C., & Kikuchi, T., (2018). *Cannabinoids for Use in the Treatment of Neuropathic Pain.* Google Patents.

94. Berde, C., & Kohane, D. S., (2019). *Neosaxitoxin Combination Formulations for Prolonged Local Anesthesia.* Google Patents.

95. Schafflick, D., Kieseier, B. C., Wiendl, H., & Zu Horste, G. M., (2017). Novel pathomechanisms in inflammatory neuropathies. *Journal of Neuroinflammation, 14*(1), 232.

96. Guo, J., Li, J., Zhao, J., Yang, S., Wang, L., Cheng, G., Liu, D., et al., (2017). MiRNA-29c regulates the expression of inflammatory cytokines in diabetic nephropathy by targeting tristetraprolin. *Scientific Reports, 7*(1), 1–13.

97. Luo, Q., Feng, Y., Xie, Y., Shao, Y., Wu, M., Deng, X., Yuan, W. E., et al., (2019). Nanoparticle-microRNA-146a-5p polyplexes ameliorate diabetic peripheral neuropathy by modulating inflammation and apoptosis. *Nanomedicine: Nanotechnology, Biology, and Medicine, 17,* 188–197.

98. Sifuentes-Franco, S., Pacheco-Moisés, F. P., Rodríguez-Carrizalez, A. D., & Miranda-Díaz, A. G., (2017). The role of oxidative stress, mitochondrial function, and autophagy in diabetic polyneuropathy. *Journal of Diabetes Research, 2017.*

99. Burns, A., & Self, W. T., (2018). Antioxidant inorganic nanoparticles and their potential applications in biomedicine. In: *Smart Nanoparticles for Biomedicine* (pp. 159–169). Elsevier.

100. Najafi, R., Hosseini, A., Ghaznavi, H., Mehrzadi, S., & Sharifi, A. M., (2017). Neuroprotective effect of cerium oxide nanoparticles in a rat model of experimental diabetic neuropathy. *Brain Research Bulletin, 131,* 117–122.

101. Khorrami, M. B., Sadeghnia, H. R., Pasdar, A., Ghayour-Mobarhan, M., Riahi-Zanjani, B., Hashemzadeh, A., Zare, M., & Darroudi, M., (2019). Antioxidant and toxicity studies of biosynthesized cerium oxide nanoparticles in rats. *International Journal of Nanomedicine, 14,* 2915.

102. Shanker, K., Naradala, J., Mohan, G. K., Kumar, G., & Pravallika, P., (2017). A sub-acute oral toxicity analysis and comparative *in vivo* anti-diabetic activity of zinc oxide, cerium oxide, silver nanoparticles, and *Momordica charantia* in streptozotocin-induced diabetic Wistar rats. *RSC Advances*, *7*(59), 37158–37167.

103. Amer, M. A. M., (2020). Protective effects of cerium oxide nanoparticles on oxaliplatin induced neurotoxicity in adult male albino rats. *Zagazig Journal of Forensic Medicine*, *18*(1), 52–67.

104. Sengani, M., (2017). Identification of potential antioxidant indices by biogenic gold nanoparticles in hyperglycemic Wistar rats. *Environmental Toxicology and Pharmacology*, *50*, 11–19.

105. Edrees, H. M., Elbehiry, A., & Elmosaad, Y. M., (2017). Hypoglycemic and anti-inflammatory effect of gold nanoparticles in streptozotocin-induced type 1 diabetes in experimental rats. *Nanotechnology*, *3*, 4.

106. Pop-Busui, R., Ang, L., Holmes, C., Gallagher, K., & Feldman, E. L., (2016). Inflammation as a therapeutic target for diabetic neuropathies. *Current Diabetes Reports*, *16*(3), 29.

107. Grote, C. W., & Wright, D. E., (2016). A role for insulin in diabetic neuropathy. *Frontiers in Neuroscience*, *10*, 581.

108. Farahani, B. V., Ghasemzadeh, H., & Afraz, S., (2016). Intelligent semi-IPN chitosan-PEG-PAAm hydrogel for closed-loop insulin delivery and kinetic modeling. *RSC Advances*, *6*(32), 26590–26598.

109. Bai, D. P., Lin, X. Y., Huang, Y. F., & Zhang, X. F., (2018). Theranostics aspects of various nanoparticles in veterinary medicine. *International Journal of Molecular Sciences*, *19*(11), 3299.

110. He, Z., Santos, J. L., Tian, H., Huang, H., Hu, Y., Liu, L., Leong, K. W., et al., (2017). Scalable fabrication of size-controlled chitosan nanoparticles for oral delivery of insulin. *Biomaterials*, *130*, 28–41.

111. Wong, C. Y., Al-Salami, H., & Dass, C. R., (2018). Microparticles, microcapsules and microspheres: A review of recent developments and prospects for oral delivery of insulin. *International Journal of Pharmaceutics*, *537*(1, 2), 223–244.

112. McNeil, S. E., (2016). Evaluation of nanomedicines: Stick to the basics. *Nature Reviews Materials*, *1*(10), 1, 2.

113. Wang, J., Liu, R., & Liu, B., (2016). Cadmium-containing quantum dots: Current perspectives on their application as nanomedicine and toxicity concerns. *Mini Reviews in Medicinal Chemistry*, *16*(11), 905–916.

114. Nagy, A., & Robbins, N. L., (2019). The hurdles of nanotoxicity in transplant nanomedicine. *Nanomedicine*, *14*(20), 2749–2762.

115. Danhier, F., (2016). To exploit the tumor microenvironment: Since the EPR effect fails in the clinic, what is the future of nanomedicine? *Journal of Controlled Release*, *244*, 108–121.

CHAPTER 5

An Overview of Preparation, Characterization, and Application of Aquasomes

SANUSHA SANTHOSH, NARENDRA KUMAR PANDEY,
SACHIN KUMAR SINGH, BIMLESH KUMAR, MONICA GULATI,
and HARDEEP

School of Pharmaceutical Sciences, Lovely Professional University,
Phagwara – 144411, Punjab, India,
E-mail: herenarendra4u@gmail.com (N. K. Pandey)

ABSTRACT

Ceramic nanoparticles (NPs) are nanosized carriers which are globule shaped. These comprise of a hydroxyapatite core, whose surface is non-covalently altered by oligosaccharide, onto which the drug of interest are further adsorbed on the surface. These ceramic NPs are also referred to as "aquasomes." Aquasomes are triple-layered self-assembled nanoparticulate drug delivery systems, which are used for the successful delivery of drugs without compromising with their conformational integrity. They establish a non-covalent link with various molecules which further promote more stability as compared to liposomes and other NPs. In the present chapter, various aspects of aquasomes are discussed such as properties, method of preparation, characterization, and various application in drug delivery.

5.1 INTRODUCTION

The potential need of nanoparticle as a carrier for drug was first stated on 1974 by Dr. Gregory Gregoriadis. He suggested that liposomes can be used as a nanoparticulate drug delivery system. Drug can be adsorbed or encapsulated in a nanocarrier system. Nowadays, nanoparticle drug delivery is the ideal choice for targeting the drug directly to the site of action, which can be then referred to targeted drug delivery. They have improved drug loading and produce very negligible adverse effects as compared to the conventional dosage form. They are also used for the poorly soluble drugs so as to improve its dissolution inside body. As very less amount of drug reaches the targeted site, toxicity produced because of large dose of drug is also reduced by this targeting method [1].

Aquasomes are the biocompatible and biodegradable nanocarrier which was invented by Nir Kossovsky in 1995 [15]. Aquasomes are defined as three-layered self-assembled nanoparticle carrier system which constitute of central solid nanocrystalline ceramic core which is further coated with poly-hydroxyl oligomer. Over this coated core, bioactive molecules are adsorbed or diffused with or without modification as shown in Figure 5.1.

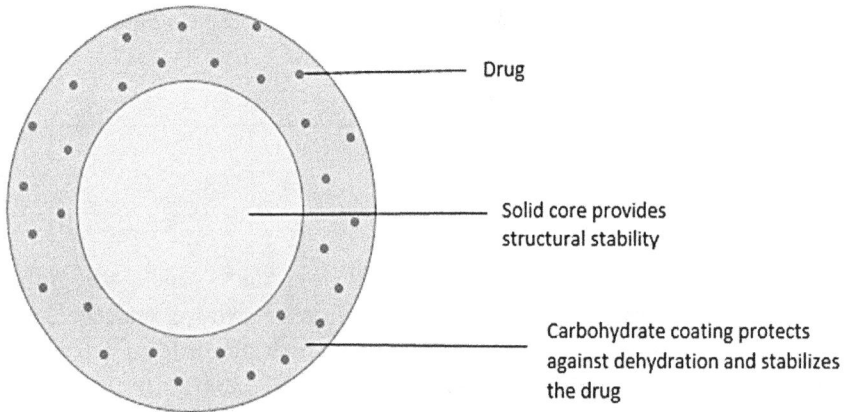

FIGURE 5.1   Structure of aquasome.

Hydroxyapatite and calcium phosphate are commonly used as ceramic core to prepare aquasomes. These aquasomes self-assemble themselves by means of ionic bonds, non-covalent bonds, or Van der Waals force [2]. Ceramic core which is coated with carbohydrate improve their cellular uptake. The main advantage of aquasomes is its non-interaction between the carrier system and the drug. It preserves the conformational integrity of the bioactive molecules which is adsorbed over the carbohydrate coated ceramic core. The structural stability is provided by the solid core while the oligomer coating helps the adsorbed bioactive molecule against dehydration and stabilizes them. It behaves like "water bodies." The water-like properties of aquasomes protect the fragile therapeutically active molecules which further help in targeting of biological molecules like peptide and protein hormones, enzymes, antigens, and genes to specific sites [3].

5.2   UNIQUE PROPERTIES OF AQUASOMES

- Aquasomes protect bioactive molecules such as proteins and peptides. There are various other carriers such as liposomes and prodrugs that can be used but they have a tendency to react between drug and carrier which may be harmful or destructive. Therefore, aquasomes are used as carrier system whose carbohydrate film protects against denaturation of proteins and peptides in the gastric environment or against the destructive pH and temperature [4].

- Aquasomes maintain optimum pharmacological activity and molecular confirmation of the drug. Usually, active molecules have unique 3D structure; molecular interaction induced internal molecular rearrangement and bulk movement. All these three properties have to be maintained for the optimum therapeutic activity, and this could be possible by formulating the drug into aquasomes [4].

- Aquasomes possess higher effective surface area and larger size. Hence, they can be loaded with considerable quantity of drug or other bioactive molecules through Van der Waals forces, ionic bonds, or non-covalent bonds [5].

- As these can be dispersed in aqueous environment due to its solid characteristic, it remains in the physical state of colloids [6].

- Aquasome's water like nature protect the bioactive molecules against dehydration and confirm to its conformational integrity and biochemical stability. Hence, aquasomes are also referred to as 'bodies of water' due to the water like behavior [7].

- Aquasomes can be used for sustained delivery of drugs by coating with certain polymers of interest which can delay the release up to extended period of time. In work done by Cherian et al. they showed that the release of insulin from the carrier was slowed down as compared to the conventional intra-venous formulation [8].

- Aquasomes can bypass the clearance by reticulo-endothelial system (RES) due to its small size and structural stability [9].

- Aquasomes are easily identified by the receptors as compared to other carrier system due to the clearance by RES. Hence, sufficient therapeutic effect can be achieved by the lower dose of dug [10].

5.3   PRINCIPLE OF SELF-ASSEMBLY

Self-assembly principle suggests that the constituent parts of final product take prescribed structural orientations in 2D or 3D space [11]. Self-assembly is an appealing approach to macromolecular synthesis because more biochemically functional products are produced in biomimetic processes. The self-assembly principle is governed by the following processes:

- Interaction between charged groups;
- Hydrogen bonding and dehydration effect;
- Structural stability.

5.3.1   INTERACTION BETWEEN CHARGED GROUPS

The interaction between the charged groups helps to stabilize the tertiary structure of folded proteins, which are amphoteric in nature. A charged polarity is produced by the chemical groups present inside the body or by the adsorbed ions from the biological fluid. The first phase of self-assembly is the interaction between the constituent subunits. In case of hydrophilic compounds, it takes place at a distance of 15 nm. For hydrophobic compounds, the range may extend up to 25 nm as well [10].

5.3.2   HYDROGEN BONDING AND DEHYDRATION EFFECT

Hydrogen bonding helps in the stabilization of secondary protein structures such as beta sheets and alpha helices. Hydrophilic compound shows hydrogen bonding which help these compounds to get arranged in the surrounding water molecules.

However, this is not the case with hydrophobic compounds. They show dehydration effect instead of hydrogen bonding due to its tendency to repel water molecule. This repulsion causes decrease in the level of entropy which helps them to get arranged in the surrounding environment. This stage is thermodynamically unstable, thus they dehydrate and get self-assembled [12].

5.3.3   STRUCTURAL STABILITY

The structural stability of bioactive molecule in therapeutic environment is measured by the interaction with charged groups and hydrogen bonding which is shown outside to the molecule and by Van der Waals forces which is shown inside of the molecule [10].

Van der Waals forces are mostly exhibited by hydrophobic compounds that are protected from aqueous environment which help to maintain internal secondary structure and conformation during self-assembly. These characteristics are responsible for sufficient hardness as well as softness of molecule, which helps to preserve the internal secondary structures of the molecule. Sugars help in molecular plasticization, in case of aquasome [4].

5.4 COMPOSITION OF AQUASOMES

5.4.1 CORE MATERIAL

Polymers (e.g., gelatin, acrylate or albumin) and ceramics (e.g., calcium phosphate, tin oxide or diamond particles) can be used for the preparation of aquasomes but ceramics are more preferred as compared to polymers due to its crystalline nature and high regularity in structure. High degree of order causes increase in the surface energy which helps in the effective binding of polyhydroxy oligomer over the ceramic core. Ceramics are biodegradable in nature. One of the commonly used cores is the calcium phosphate due its presence inside the body and is biodegraded inside body through monocytes and osteoclasts [13].

5.4.2 COATING MATERIAL

Usually, carbohydrates or polyhydroxyl oligomers are used as coating material. Commonly used carbohydrates for the preparation of aquasomes are cellobiose, trehalose, lactose, sucrose, or pyridoxal phosphate. This carbohydrate layer fulfills the main objective of aquasomes [14].

The main function of carbohydrates is to provide stabilization to the bioactive molecules as well as to protect it from dehydration. It preserves the three-dimensional structure of bioactive compound by providing aqueous like surrounding to the molecule, thus keeping it in solid dry state [15].

5.4.3 BIOACTIVE MOLECULE

They have the property of interacting with film via non-covalent and ionic interactions [8].

5.5 METHOD OF PREPARATION OF AQUASOMES

Aquasomes can be prepared by three steps as shown in Figure 5.2. These steps involve:

- Preparation of core;
- Coating of the core with carbohydrate;
- Immobilization of drug.

5.5.1 PREPARATION OF THE CORE

Fabrication of ceramic core is the first step involved in the preparation of aquasomes. These ceramic cores can be prepared by any of the following methods:

- Co-precipitation under sonication;
- Co-precipitation under reflux;
- Self-precipitation;
- Inverted magnetron sputtering.

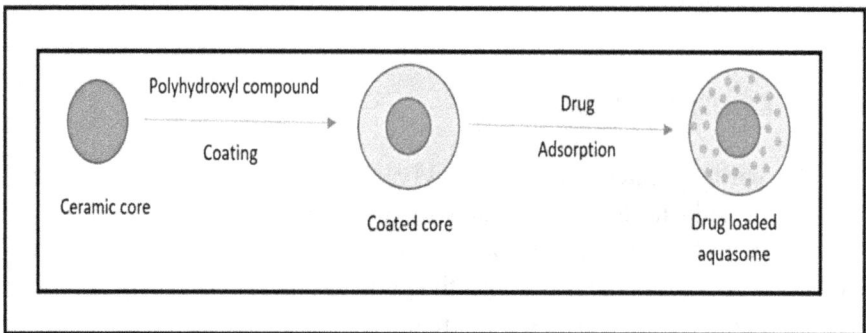

FIGURE 5.2 Preparation of aquasome.

5.5.1.1 CO-PRECIPITATION UNDER SONICATION

0.75 M disodium hydrogen phosphate was added to 0.25 M calcium chloride solution under sonication at 4°C. The formed precipitate of calcium phosphate precipitate was separated from the mixture by centrifugation for 1 hour at 15,000 rpm. Then it was washed 4–5 times with double distilled water so that it can be clear of the sodium chloride which was formed during the reaction:

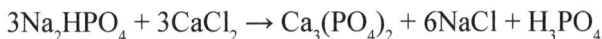

$$3Na_2HPO_4 + 3CaCl_2 \rightarrow Ca_3(PO_4)_2 + 6NaCl + H_3PO_4$$

Then the precipitate was again re-suspended in double distilled water and passed through 0.2 µm millipore filter paper so as to collect the particles below 0.2 µm [16].

5.5.1.2   CO-PRECIPITATION UNDER REFLUX

In this method, 100 mL 0.19 N diammonium hydrogen phosphate solutions was added drop wise to 100 mL 0.32 M calcium nitrate solution which was maintained at 75°C in a flask having a charged funnel, a thermometer and a reflux condenser which was fitted with a carbon dioxide trap. The synthesis should be carried out at a temperature of 75°C and in pH 8–10, which is given by the following equation:

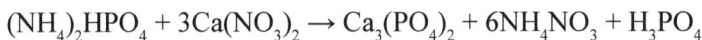

$$(NH_4)_2HPO_4 + 3Ca(NO_3)_2 \rightarrow Ca_3(PO_4)_2 + 6NH_4NO_3 + H_3PO_4$$

The resultant mixture was then magnetically stirred for 4–6 days. The precipitate thus formed was filtered, washed with double distilled water, and then finally dried at 100°C. Then sintering of powder was carried out by heating to 800–900°C in an electric furnace [17].

5.5.1.3   SELF-PRECIPITATION

The simulated body fluid of pH 7.2 which contains calcium chloride (2.5 mM), potassium chloride (5.0 mM), sodium chloride (134.8 mM), magnesium chloride (1.5 mM), disodium sulfate (0.5 mM), disodium hydrogen phosphate (1.0 mM), sodium hydrogen carbonate (4.2 mM) was used. 7.26 pH was maintained everyday using hydrochloric acid. The resultant mixture was then transferred to series of 100 ml polystyrene bottles and was properly sealed. Then it was kept for one week at 37 ± 1°C temperature. After stipulated period, precipitate will be formed at the inner side of the bottle. Then it was filtered, washed thoroughly with double distilled water, and then lastly dried at 100°C [16].

5.5.2 COATING OF THE CORE WITH CARBOHYDRATE

The next step is the coating of the ceramic core by carbohydrate layer. The process usually comprises the addition of carbohydrate, i.e., polyhydroxy oligomer to the dispersion of ceramic which is made free from contamination by cleaning in ultra-pure water. Then sonication and lyophilization are performed so that irreversible adsorption of carbohydrate is done on to the ceramic surface [13]. The carbohydrate which is not adsorbed to the surface of the core was separated by stir cell ultra-filtration or by centrifugation. Adsorption of carbohydrate over ceramic core can be done by non-solvent addition and by direct incubation method. Coating materials which are used for the preparation of aquasomes include trehalose, citrate, pyridoxal-5-phosphate, cellobiose, and sucrose [5].

5.5.3 IMMOBILIZATION OF DRUGS

The final step in the preparation of aquasomes are the adsorption of active ingredients to the carbohydrate coated core which is often done by incubating the coated core in the known concentration of drug [9]. Adsorption process involves ionic and non-covalent interaction. The factors effecting drug loading includes incubation time, incubation temperature and drug concentration [1].

5.6 EVALUATION

Table 5.1 enlists the methods and overall procedure for the evaluation of aquasomes.

TABLE 5.1 Evaluation Parameters of Aquasome

Parameters	Method	Procedure	References
Morphology	SEM and TEM	In SEM, samples are kept on the surface of a specimen coated with gold and pasted with the help of double-sided tape. In TEM, negative staining of samples is done with phosphotungistic acid.	[18]
Structural analysis	FTIR	KBr method is used and can be analyzed by observing the IR peaks in the range of 400–4000 cm^{-1}. Resultant peaks are then matched with the reference peaks for accurate results.	[19]

TABLE 5.1   *(Continued)*

Parameters	Method	Procedure	References
Crystallinity	XRD	Aquasome are studied for diffraction and are compared with standard. Calcium phosphate core and lactose showed same sharp crystalline peaks but when they are mixed together, it showed peaks of amorphous structure which is due to saturation of core with carbohydrate.	[20]
Concanavalin A induced aggregation	Concanavalin A induced aggregation method	Helps to determine how much amount of sugar is coated over the core. The reagent is added to mixture of suspension of sugar coated core and is placed in quartz cuvettes and absorbance is determines using UV visible spectrophotometer at time interval of every 5 min. The obtained results are then subtracted from blank formulation which is performed in the absence of concanavalin A.	[4]
Colorimetric analysis of sugar coating over core	Anthrone method	Anthrone test produce green colored product when sugars are subjected to hydrolysis to form simple sugar and finally to hydroxyl methyl furfural. Samples are heated with anthrone reagent in water bath and are rapidly cooled. Greenish blue solution is obtained whose absorbance is observed using glucose as standard.	[17]
Drug payload		Also called drug loading efficiency. Blank aquasome (without drug) is incubated at 4°C in a drug solution of known concentration for 24 hrs. Supernatant formed is centrifuged for 1 hr at 1500 rpm at 4°C. After loading, the drug that remains in the supernatant liquid is further analyzed by any method of analysis.	[20]
In vitro drug release		Drug release is measured in a buffer media at 37°C of suitable pH with constant stirring. Samples are taken out at regular interval of time and replacing it with same amount of fresh buffer. Samples are then centrifuged at high speed and the formed supernatant is collected and analyzed in UV spectrophotometer to determine the amount of drug released.	[1]

5.6.1   MORPHOLOGY

Vengala et al. prepared aquasome which is loaded with an antipsychotic drug pimozide. The average size of the formed aquasome was determined by the scanning electron microscope (SEM). The images of SEM showed that the particles were of uniform size with spherical nanoparticles (NPs) as shown in Figure 5.3. The size of aquasome was within the range of 60–300 nm and the average size of drug-loaded aquasome was found to be 90 nm. After determining the average size of pure drug with trinocular microscope, it was found to be 1210 nm. This proved that the fabrication of pure drug to aquasomal formulation reduced the particle size to nanometer range [16].

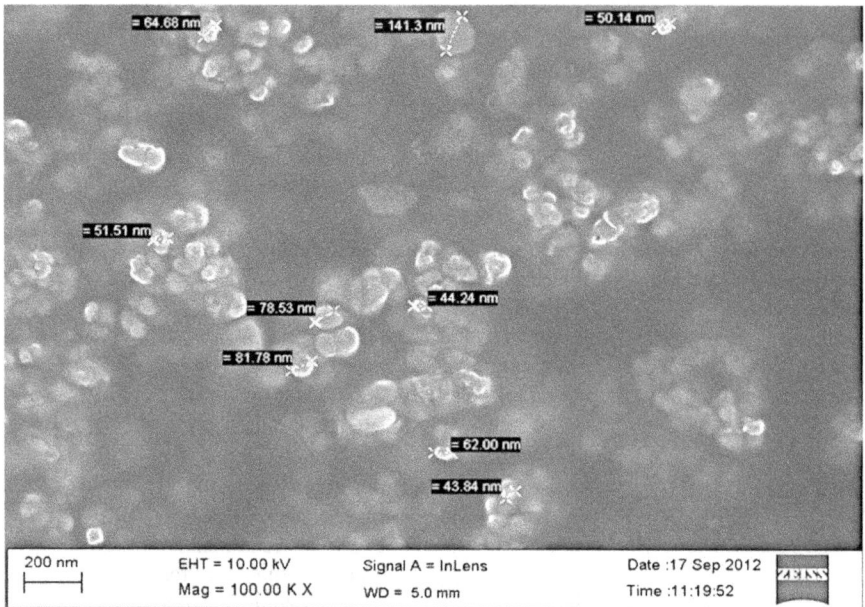

FIGURE 5.3   Scanning electron microscopy image of pimozide lactose aquasomes [16].

Source: Copyright © 2013 JPR Solutions.

Pandey et al. loaded the ovalbumin (OVA) as allergen model to the aquasome formulation which consisted of hydroxyapatite as core and trehalose as carbohydrate coating. The morphology of the hydroxyapatite alone and the final aquasome was determined by the transmission electron microscopy (TEM). The microstructure which is shown by the plain hydroxyapatite core (Figure 5.4(A)) proved that the spherical nanosized particles were dense and have rough surfaces. They also showed aggregation which may be observed due to the drying process. While the TEM images of aquasomal formulation after adsorption of OVA showed that the systems were spherical, smaller in size and were elongated in shape (Figure 5.4(B)). The nanometric architecture of formed aquasome was confirmed by the photon correlation spectroscopy (PCS). The average size of the hydroxyapatite core, blank aquasome and OVA adsorbed aquasome was found to be 39 ± 11 nm, 43 ± 14 nm and 47 ± 19 nm, respectively [2].

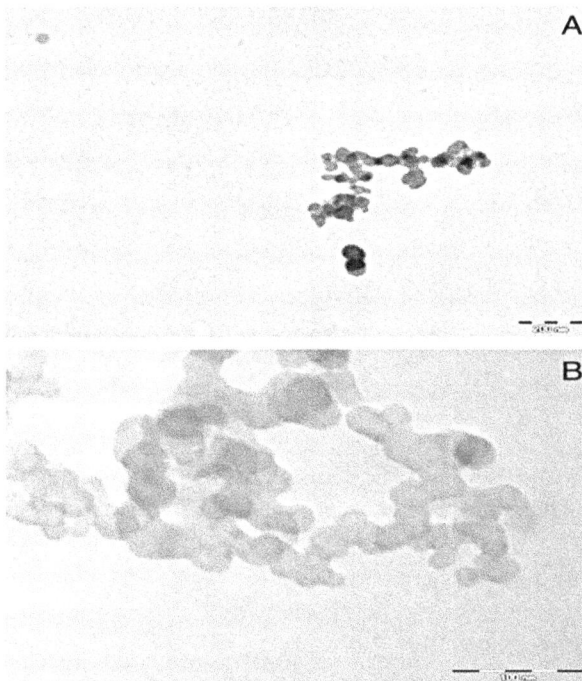

FIGURE 5.4 TEM image of antigen-loaded formulation of: (A) plain hydroxyapatite ceramic core; and (B) OVA adsorbed aquasomes [2].

Source: Copyright © 2011 Elsevier B.V.

In the work done by Kommineni et al. they prepared piroxicam-loaded aquasomes in which the negative stained TEM was used to evaluate the morphology of the aquasome. The images shown by TEM confirmed that the particles were of spherical shape as demonstrated in Figure 5.5 and the size of the particles ranged from 60–300 nm. The average particle size of carbohydrate-coated core was 56.56 ± 5.93 nm while the average particle size of drug loaded particles were 184.75 ± 13.78 nm, which comply with the nanometric dimension. Electron microscopy also showed heterogeneous population of large and small particles, which mostly existed as loose aggregates [17].

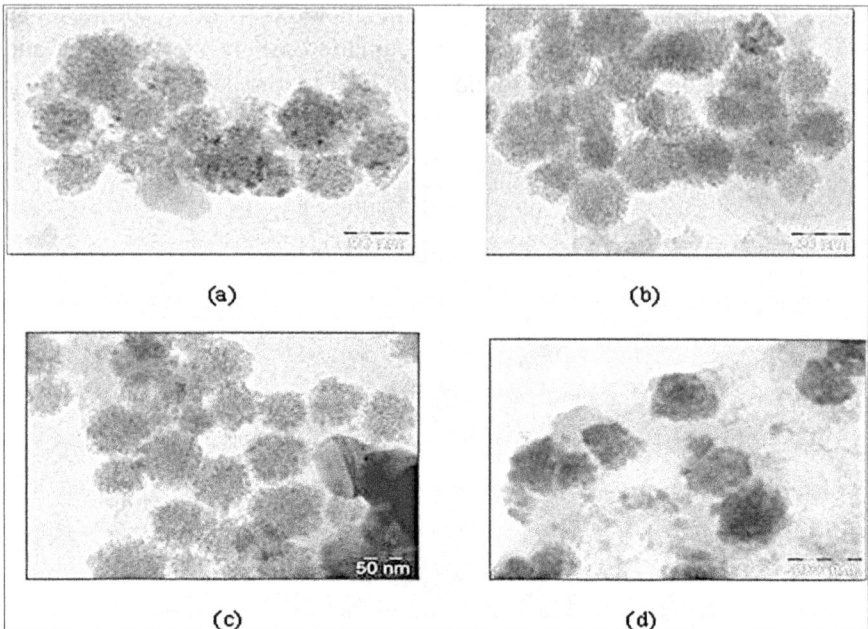

FIGURE 5.5 TEM images of ceramic core based, lactose coated piroxicam nanoparticles (a) represents lactose coated core (a scale bar of 50 nm); and (b), (c), and (d) represent piroxicam loaded aquasomes (b and c represent a scale bar of 50 nm and d represents 200 nm) [17].

Source: Copyright © 2012 Taylor and Francis.

In the work done by Khopade et al. they prepared aquasome on which hemoglobin was adsorbed. Figure 5.6 shows the SEM images which proved that the particles were of spherical shape. However, they had the tendency to aggregate which may be due to the process of drying [19].

In the work done by Goyal et al. they prepared aquasome and loaded it with bovine serum albumin (BSA) as a model antigen. Zeta sizer was used for particle size analysis and electron microscopy was used to determine the size and shape of the preparation. Hydroxyapatiite was incorporated as the ceramic core whose average particle size was found to be 150.32 nm as shown in Figure 5.7(A). In this work they have compared the coating of the core with two different types of carbohydrate coating, i.e., cellobiose, and trehalose. Hydroxyapatite core coated with cellobiose (Figure 5.7(B)) and trehalose (Figure 5.7(C)) have average particle size of 247.26 nm and 264.14 nm, respectively. After loading it with BSA, the size of cellobiose and trehalose aquasome changed to 291.24 nm and 286.56 nm, respectively. From these results, they concluded that when concentration of sugar increased, the shape and size of the aquasome also

FIGURE 5.6   Scanning electron microscopy image of hemoglobin adsorbed, sugar-coated hydroxyapatite cores before dispersing in albumin/buffer [19].

Source: Copyright © 2002 Elsevier B.V.

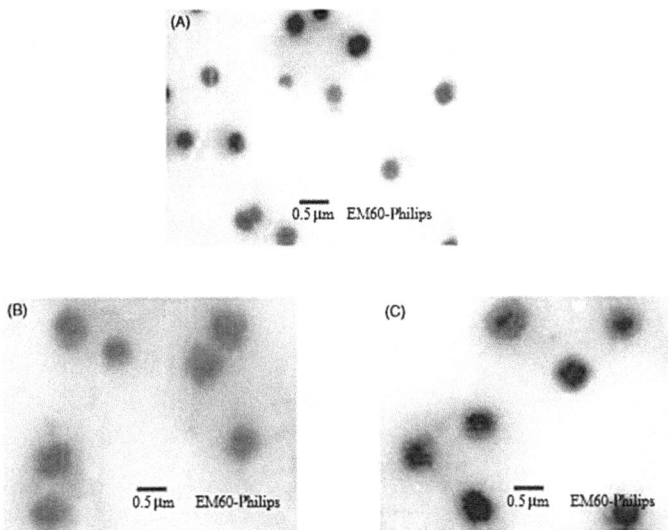

FIGURE 5.7   Electron microphotograph of different formulations: (A) transmission electron microscope (TEM) image plain hydroxyapatite; (B) TEM image of cellobiose-coated aquasomes; (C) TEM image of trehalose-coated aquasomes [14].

Source: Copyright © 2008 Taylor and Francis.

changed from small uniform particles to irregular large particles. They also observed that small particles were present as loose aggregates [14].

In the work done by Rawat et al. they prepared serratiopeptidase (STP) protein adsorbed aquasome formulation and further coated it with alginate. The average size and size distribution was determined by TEM after negative staining with phosphotungstic acid. The average size of calcium phosphate nanocore (Figure 5.8(A)) and chitosan-coated nanocore (CNC) (Figure 5.8(B)) was found to be 142 ± 4.26 nm and 415 ± 4.26 nm, respectively. They observed that as chitosan (CS) concentration increases, the size of CNC also got increased up to a particular concentration (i.e., 0.5% w/v). Beyond that optimum concentration, aggregation was observed, which may be due to the high viscosity of the CS solution. After adsorbing enzyme over the CNC, the size got increased to 524 ± 4.46 nm (Figure 5.8(C)). Further, when the CNC was encapsulated in alginate, the size again increased to 925 ± 6.81 nm (Figure 5.8(D)) [21].

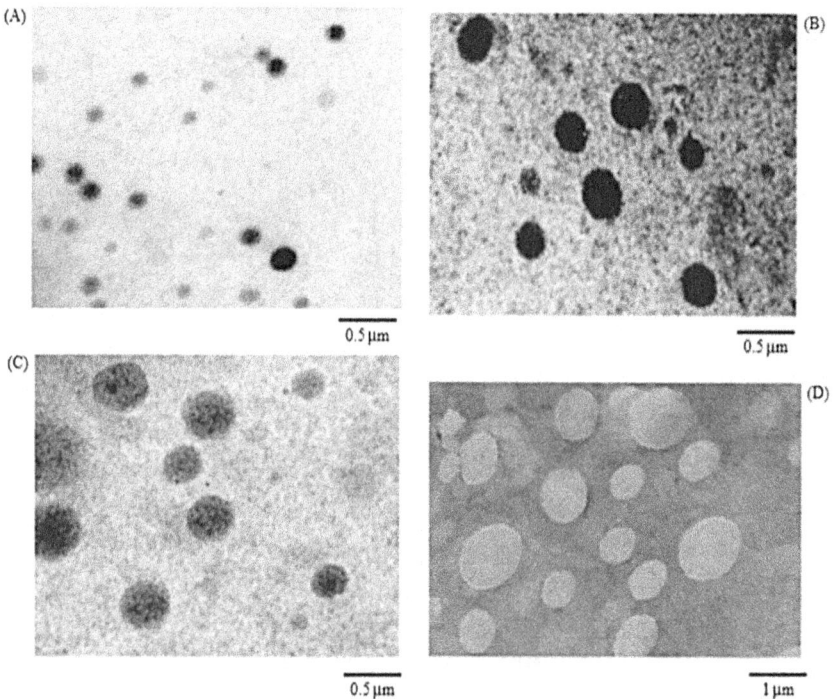

FIGURE 5.8   TEM of various stages of nanocarriers; (A) Calcium phosphate nanocores (NC); (B) Chitosan-coated nanocores (CNC); (C) STP-adsorbed CNC; (D) Alginate-encapsulated CNC (ACNC) [21].

Source: Copyright © 2008 Taylor and Francis.

In the work done by Rojas-Oviedo et al. poorly soluble indomethacin drug was formulated into aquasome using calcium phosphate as ceramic core and lactose as carbohydrate. They checked the effect of different concentration of lactose solution (0.03, 0.06, and 0.09 M) on the size of final preparation and observed that the size of aquasome increased as concentration of lactose got increased (69.76, 82.34, and 96.17 nm respectively). Figure 5.9(a–c) shows SEM images of polyhydroxylated nuclei of samples which were prepared with 0.03, 0.06, and 0.09 M lactose concentration, respectively showed that the polyhydroxylated NPs were ovoid shaped with a particle size of 200 nm. Figure 5.9(d–f) shows TEM images of aquasomal preparation in which the false color technique was used. The red color was assigned to the calcium phosphate core and the green color was assigned to the lactose layer [20].

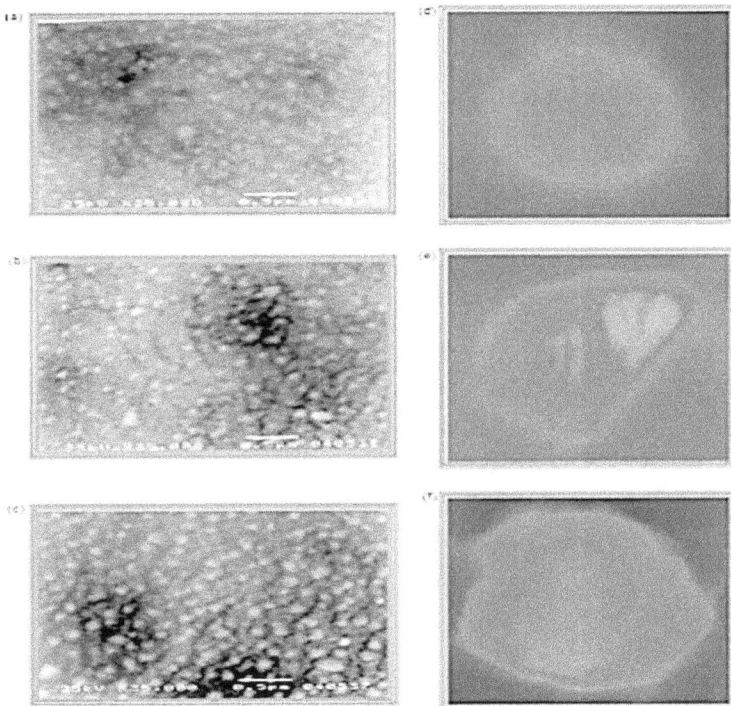

FIGURE 5.9 (a–c)-represents the SEM image of the polyhydroxylated nuclei which is prepared from 0.03, 0.06, and 0.09 M of lactose solution. The morphology was observed to be ovoid shaped with particle size of around 200 nm layer; (d–f)-shows TEM image which shows false color technique. With the help of this technique, the core and the coating can be differentiated. The red color represents the calcium phosphate core and the green color represents the lactose [20].

Source: Copyright © 2007 Elsevier B.V.

5.6.2 PHYSICOCHEMICAL INTERACTIONS

5.6.2.1 FOURIER TRANSFORM INFRARED (FTIR)

Fourier transform infrared (FTIR) spectroscopy is a technique to confirm the development of ceramic core, formation of carbohydrate coating over the ceramic core and the presence of drug on carbohydrate-coated core.

Vengala et al. confirmed the presence of pimozide drug and the interaction of this drug with lactose coated ceramic core using FTIR spectroscopy in they used the potassium bromide pellet method. The aquasomal preparation of pimozide was mixed with potassium bromide in the ratio 1:100. This formed mixture was compacted under 10 tons/cm² pressure in vacuum. This leads to the formation of a transparent pellet which is further used for the FTIR analysis. Figure 5.10 shows characteristic bands for core, lactose, and pimozide. IR bands at 891.11 cm^{-1} and 1172.72 cm^{-1} corresponds to the phosphate (P-O) and phosphate (P=O) of core. IR bands at 2875.86 cm^{-1} corresponds to the presence of symmetrical CH$_2$ stretching in lactose. IR bands at 1685.79 cm^{-1} and 1064.71 cm^{-1} corresponds to the presence of C=O bonding and C-F bending of pimozide [16].

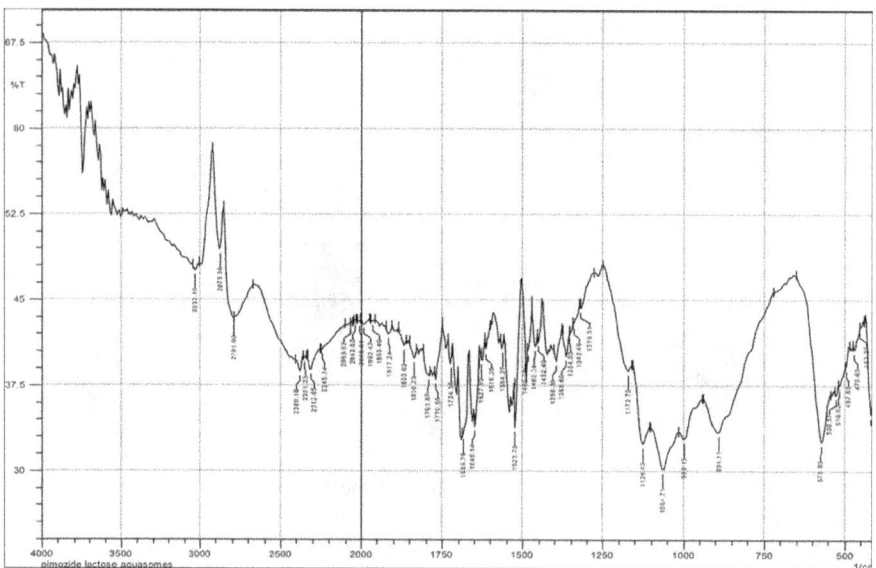

FIGURE 5.10 FTIR spectra of pimozide lactose aquasome [16].

Source: Copyright © 2013 JPR Solutions.

Study done by Rawat et al. confirmed the CS coating over the ceramic nanocore by FTIR spectroscopy. Figure 5.11 shows the characteristic band at 3,434 cm^{-1} which was due to the presence of -OH and -NH$_2$ group stretching vibration found in CS matrix. CS has excellent film-forming abilities and has the property to stabilize the core through non-covalent and ionic forces. When observing the band of CNC, it was found that peak at 1,644 cm^{-1} disappeared and, in that place, a new sharp peak at 1,632 cm^{-1} evolved. -NH$_2$ bending vibration peak shifted from 1,604 cm^{-1} to 1,536 cm^{-1}. These shifts in peaks of -NH$_2$ group can be attributed to the linkage between ammonium ion of CS and phosphoric group of calcium phosphate ceramic core [21].

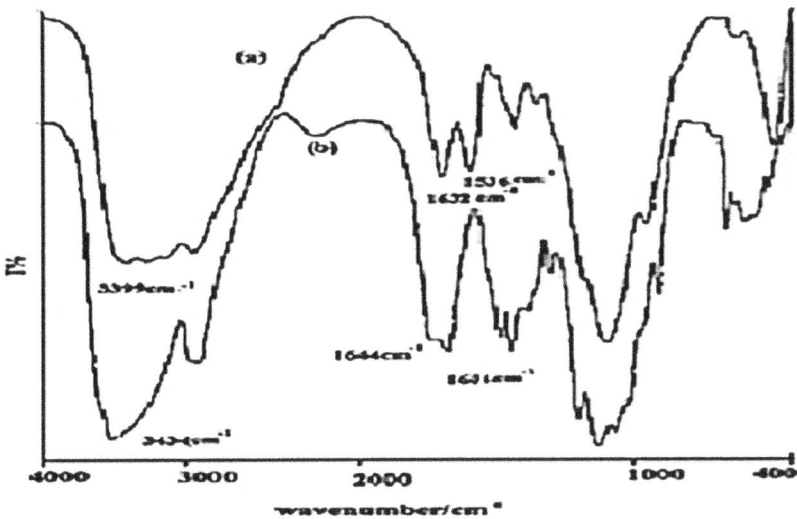

FIGURE 5.11   FTIR spectra of (A) chitosan-coated calcium phosphate nanocores and (B) pure chitosan. FTIR spectra indicate the clear shift in the peak height around 1,604/cm to 1,536/cm, indicating the linkage of phosphoric group of calcium phosphate with ammonium group of chitosan [21].

Source: Copyright © 2008 Taylor and Francis.

In the work done by Goyal et al. they prepared a potassium bromide sample disk with 1% (w/w) of hydroxyapatite powder which was compressed and dried at high temperature of 100°C. IR spectra were determined in the range of 4,000–400 cm^{-1} using FTIR spectrophotometer. From Figure 5.12, it can be observed that the characteristic bands appeared at 1,093, 1,033, 962, 602, 564, and 473 cm^{-1} which confirm the presence of PO$_4^{3-}$ in hydroxyapatite. They also performed the FTIR spectra of aquasome which is prepared with two different carbohydrates, i.e., trehalose, and cellobiose to perform a comparative study [14].

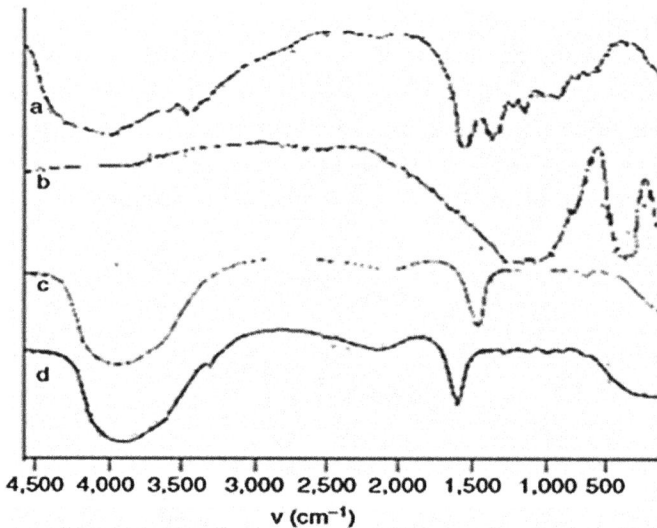

FIGURE 5.12   Fourier-transformed infrared (FTIR) spectra of different formulation at 24°C. (a) Plain BSA; (b) plain hydroxyapatite; (c) aquasomes formulation coated with cellobiose; (d) aquasomes formulation coated with trehalose [14].

Source: Copyright © 2008 Taylor and Francis.

5.6.2.2   X-RAY DIFFRACTION (XRD)

Kaur et al. performed powder x-ray diffraction (PXRD) for calcium hydrogen phosphate ($CaHPO_4$) core, trehalose aquasome, cellobiose aquasome, and pyridoxal-5-phosphate aquasome which is shown in Figure 5.13. In the study, sharp and intense peaks of core ($CaHPO_4$) indicated its crystalline nature. After coating of core with three different carbohydrate solutions separately, the sharp peaks of core converted to broad peaks. This is due to the conversion of crystalline to amorphous structure. Deformation was observed maximum in case of pyridoxal-5-phosphate aquasome which is due to the presence of charged group and electronegative atoms present in them. PO_4^{2-} group in pyridoxal-5-phosphate helps in formation of intermolecular hydrogen bond with $CaHPO_4$ and thus favor the surrounding water molecules. Generally amorphous phase requires minimal energy due to the irregular structural geometry thus leads to better solubility and bioavailability [6].

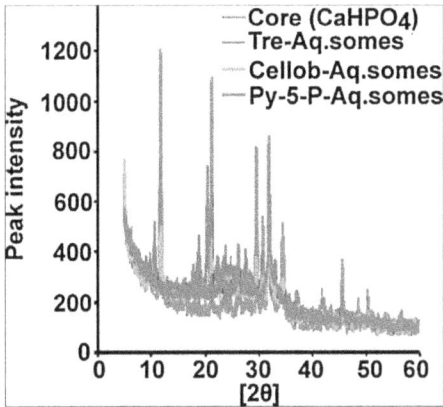

FIGURE 5.13   Powder x-ray diffraction (PXRD) pattern of core $(CaHPO_4)$ particles, trehalose aquasomes (Tre-Aquasomes), cellobiose aquasomes (Cellob-Aquasomes), and pyridoxal-5-phosphate aquasomes (Py-5-P-Aquasomes). Crystal lattice arrangement of core particles was deformed to amorphous structure upon coating with polysaccharides. Maximal amorphization was done by Py-5-P oligomer [14].

Source: Copyright © 2008 Taylor and Francis.

In the study done by Pandey et al. crystalline nature of hydroxyapatite ceramic core was determined by X-ray diffraction peaks. Figure 5.14 shows intense absorption peaks at 31–32, 49–50; 25–27 (2θ angle) which is attributed to the crystalline nature of hydroxyapatite [2].

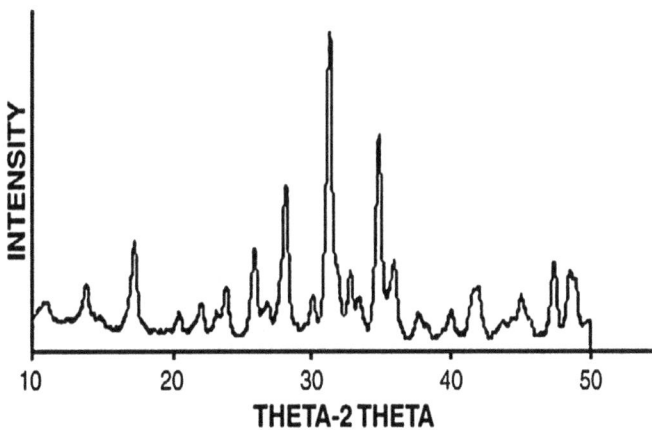

FIGURE 5.14   X-ray diffraction pattern of hydroxyapatite core particles obtained from the mixture of calcium nitrate and di-ammonium hydrogen phosphate [2].

Source: Copyright © 2011 Elsevier B.V.

5.6.3 QUANTIFICATION OF SUGAR COATING

As per Kommineni et al. quantification of carbohydrate coating over the core can be done by anthrone method. Anthrone test produces green colored product when sugars are subjected to hydrolysis using hydrochloric acid to form simple sugar and finally, they form hydroxyl methyl furfural. Samples are heated with anthrone reagent in water bath and are rapidly cooled. Absorbance is measured for the greenish-blue solution using glucose as standard [17].

The coating of carbohydrate over the core with respect to time was estimated by concanavalin an induced aggregation assay as per the studies done by Khopade et al. When concanavalin A is added to the sugarcoated core, the particles tend to aggregate causing increment in the optical density. This optical density is measured for absorbance at 450 nm. From Figure 5.15, we can observe that the absorbance is increased over the time period, which confirms the carbohydrate layer over the core [19].

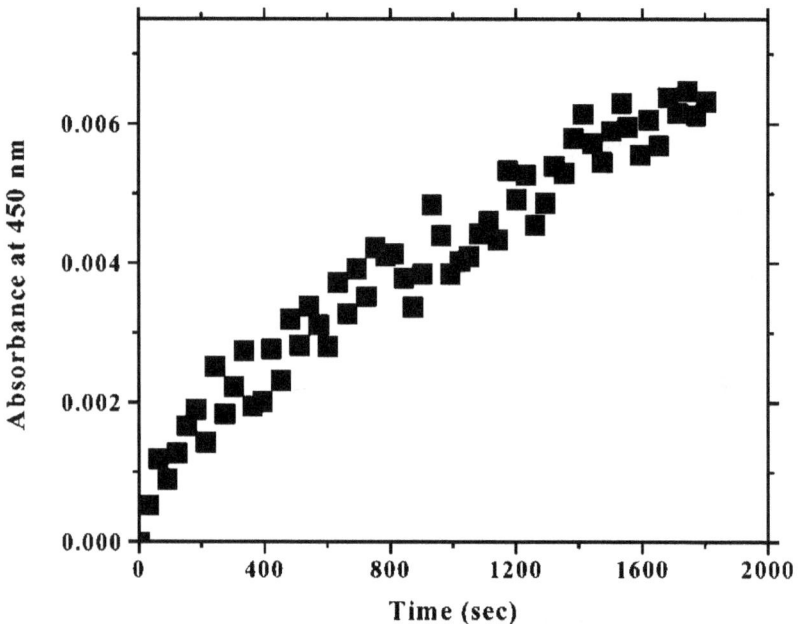

FIGURE 5.15 Concanavalin A induced aggregation of sugar-coated hydroxyapatite cores. Concanavalin A solution (100 μg ml⁻¹) was added to sugar-coated hydroxyapatite core suspension (10 μg ml⁻¹) in quartz cuvette and absorbance at 450 nm was measured as a function of time. The data was subtracted from blank experiment conducted without addition of concanavalin A [19].

Source: Copyright © 2002 Elsevier B.V.

Vengala et al. discussed the procedure to quantify the sugar coating. 50 mg of carbohydrate-coated core was weighed accurately and are dissolved in 5 ml of distilled water. From this solution, 2 mL of solution was taken and added to 5.5 ml anthrone reagent. The resultant mixture was boiled at 100°C for 10 minutes. After complete cooling of the solution, absorbance was measured at λ_{max} of 625 nm so as to estimate the amount of sugar which has been coated over the core [16].

Goyal et al. also confirmed the coating of carbohydrates (trehalose and cellobiose) over the surface of the core by concanavalin-A-induced aggregation studies. From Figure 5.16, it can be interpreted that the cellobiose-coated formulation aggregates at a slower rate as compared to the trehalose coated preparation. Thus, it can be confirmed that trehalose has more efficiency to coat the core at faster rate as compared to cellobiose [14].

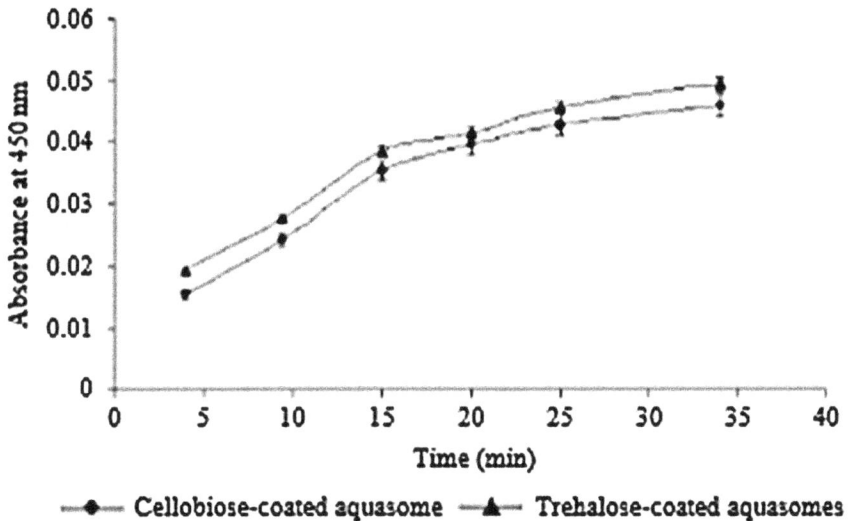

FIGURE 5.16   Concanavalin-A-induced aggregation of various aquasomal formulations [14].
Source: Copyright © 2008 Taylor and Francis.

5.6.4   DRUG LOADING

Cherian et al. estimated the amount of drug which has been coated over the sugar-coated core. Different aquasome preparation containing cellobiose, trehalose, and pyridoxal-5-phosphate as carbohydrate layer was used to study the drug loading. For this, they prepared various aquasome formulations without the drug and were incubated in known concentration of drug solution at 4°C for 24 hours. Then the drug-aquasome mixture is centrifuged for 1 hour at 15,000 rpm in a refrigerated centrifuge having temperature below 4°C. The remaining drug which is unadsorbed will be present in the supernatant is quantified spectrophotometrically at 276 nm. Aquasome containing cellobiose as sugar showed drug loading of 53.68% ± 4.64%, while aquasome containing pyridoxal-5-phosphate and trehalose as sugar showed drug loading of 59.74% ± 3.65% and 48.98% ± 3.96%, respectively. After drug loading, the size of the aquasome increased due to adsorption of drug over the sugar-coated core studied [8].

Kommineni et al. loaded the piroxicam drug over the lactose-coated core with the help of adsorption technique. Lactose coated core was incubated in different concentration of drug solution (0.5, 1.0, 1.3, 1.5 and 2.0% w/v) for 24 hours. It was found that drug loading increased linearly as the concentration of piroxicam increased. After 1.5% w/v concentration of drug solution, the drug started to get crystallize which is indicative to high amount of unadsorbed drug. Therefore, 1.5% w/v was selected as the concentration for optimum drug loading [17].

5.6.5 DRUG RELEASE

Goyal et al. performed the *in vitro* release studies of cellobiose-coated aquasome and trehalose coated aquasome in phosphate buffer saline (PBS) (pH 7.4). Samples were withdrawn at regular interval of time and concentration of bovine serum albumin (BSA) was measured. From Figure 5.17, it can be observed that faster release at the initial stages occurred which may be due to desorption of antigen from the surface. In the second stage, slower release rate was observed which is due to the sustained release of BSA from the matrix of aquasome. From the graph, it can be concluded that the release of cellobiose-coated aquasome was faster as compared to the trehalose-coated aquasome. This is due to the differences in zeta potential and difference in the adsorption patterns of antigen on the carbohydrate core which controls the movement of BSA to the external environment [14].

Kommineni et al. performed the *in vitro* release studies of marketed formulation of piroxicam (Pirox-10®) and the aquasome formulation of piroxicam in acidic buffer (0.1N HCl) and phosphate buffer (pH 6.8).

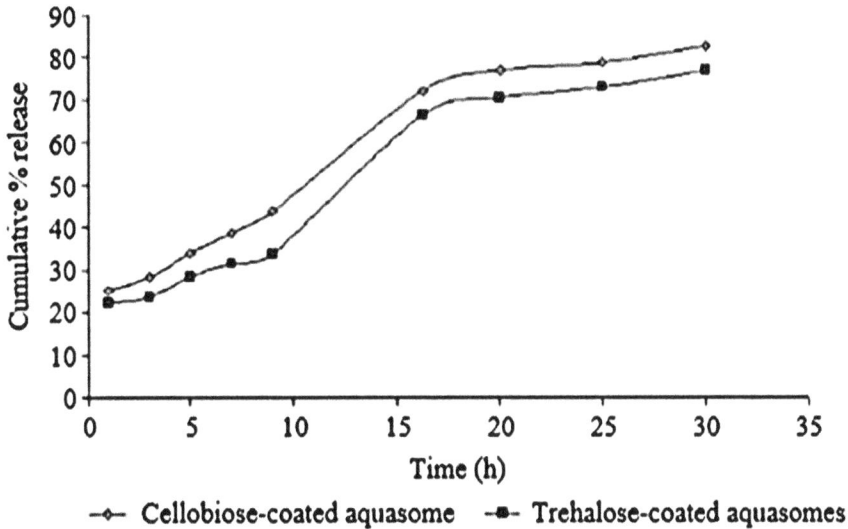

FIGURE 5.17   Cumulative percentage release rate of different aquasomal formulations [14]. *Source:* Copyright © 2008 Taylor and Francis.

From Figure 5.18, it is observed that the Pirox-10® released 77.62 ± 2.84% of the drug in 15 min but the release of drug from the matrix of aquasome was only 11.30 ± 3.73% in 45 min. From this, it can be concluded that although aquasomes are nanosized formulation, it has potential to release the drug in a controlled manner. Thus, dissolution profile indicated that gradual linear release of drug was observed without burst effect which is attributed to the interaction between the carbohydrate coating and the drug. 79.38 ± 1.53% of drug was released in acidic buffer and 68.51 ± 1.39% of drug was released in phosphate buffer over a time period of 8 hours. Since a similar release pattern was observed in both the buffers, therefore pH-independent drug release was observed [17].

Pandey et al. performed the *in vitro* release studies of OVA adsorbed aquasome. Samples were taken out every 10 minutes. From Figure 5.19, it can be observed that at initial 10 minutes, 40% of OVA were released into the phosphate buffer medium. After this initial 10 minutes, the release of OVA was found to be very small and further, it decreased over time. From the graph, it can be concluded that 90% of the drug was released at 50 minutes [2].

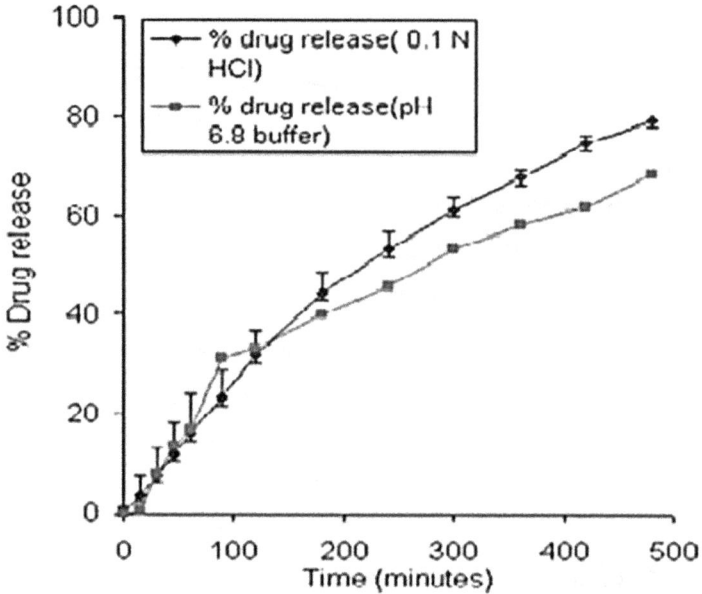

FIGURE 5.18 *In vitro* release profile of piroxicam from optimized formulation (1.5% w/v). ◆ represents dissolution kinetics in acidic buffer (0.1 N HCl) and ▪ represents dissolution kinetics in alkaline buffer (pH 6.8) [17].

Source: Copyright © 2012 Taylor and Francis.

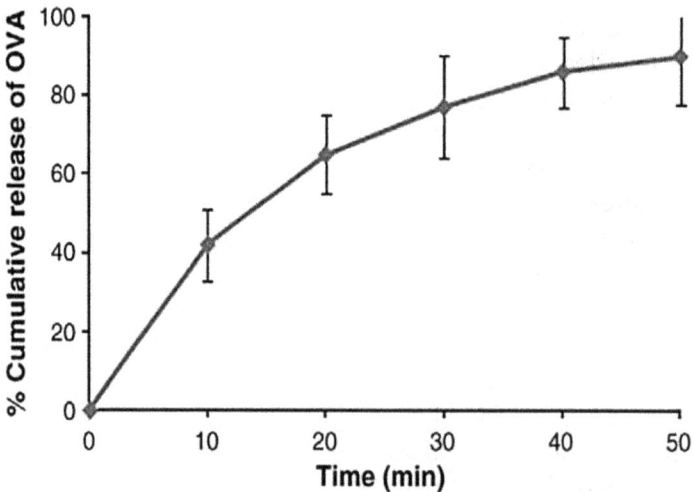

FIGURE 5.19 *In vitro* release of OVA from aquasomes in PBS. Data express the mean of the cumulative amount of OVA released vs. time (mean ± S.D, n = 3) [2].

Source: Copyright © 2011 Elsevier B.V.

Vengala et al. performed the *in vitro* release of pimozide aquasome using type-I USP-dissolution apparatus. 900 mL of 0.1 N hydrochloric acid solutions was used as the dissolution medium. 5 mL of aliquots were withdrawn at a time interval of 5 minute. From Figure 5.20, it can be observed that the aquasome of pimozide has improved dissolution rate as compared to the pure drug. This may be attributed to the aqueous environment and the nano-size of aquasome which is produced after the formulation. First order kinetics was followed by the formulation which describes the immediate release of drug. Thus, it can be concluded that the aquasome can be used for improving the solubility of less soluble drugs [16].

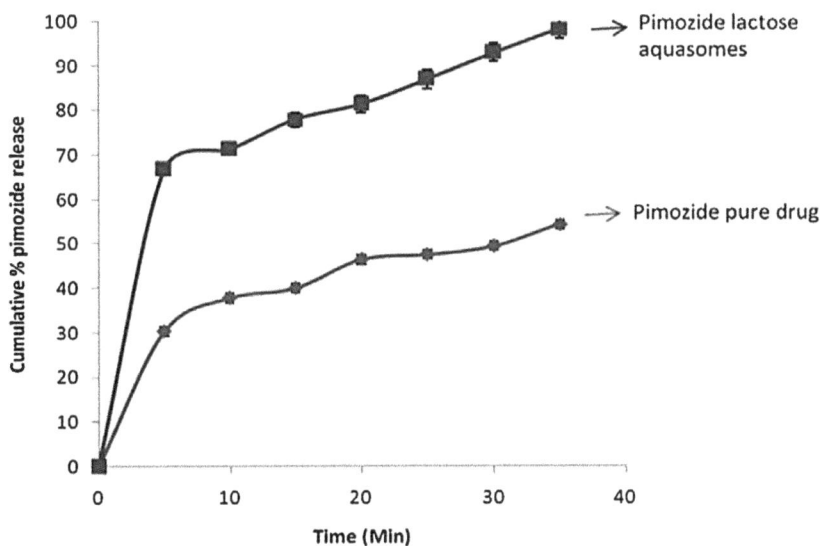

FIGURE 5.20   *In vitro* drug release of the pimozide pure drug and pimozide lactose aquasome formulation in 0.1 N hydrochloric acid solution [16].

Source: Copyright © 2013 JPR Solutions.

In vitro drug release was performed by Rawat et al. for the alginate gel-encapsulated chitosan-coated nanocores (ACNC) to check the release of encapsulated protein. The study was performed for the duration of 6 hours in pH 1.2 HCl buffer and pH 7.4-PBS. From Figure 5.21, it can be seen that in HCl buffer, only 26.58% ± 2.4% of STP protein was released in 6 hours. Whereas 99.6% ± 0.2% of protein was released in PBS solution for the same duration of time. This difference in the release pattern is due to the alginate coating which produces pH sensitivity [21].

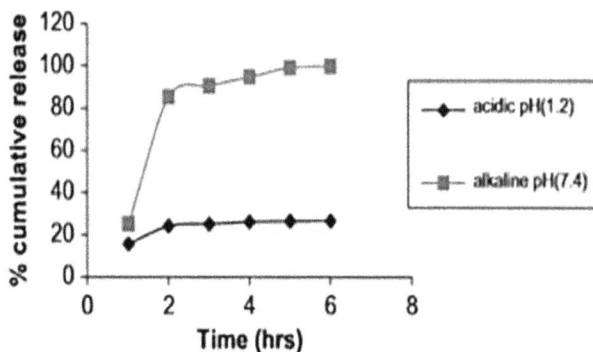

FIGURE 5.21 *In vitro* release profile of STP from optimized ACNC in acidic pH (1.2) and alkaline pH (7.4) buffer [21].

Source: Copyright © 2008 Taylor and Francis.

5.7   APPLICATION

5.7.1   INSULIN DELIVERY

In the insulin delivery, the biggest limiting factor for its use as a clinical tool is the interaction between the carrier system and the insulin. In such cases, aquasomes have proved to be a better nanocarrier as compared to other carriers because they provide conformational stability and high degree of exposure to the insulin. As a result, better therapeutic effect can be obtained. It also protects insulin from denaturation produced by the dehydration effect of insulin which is produced inside the body.

In the research done by Cherian et al. they concluded that aquasomes protected the spatial arrangement of the peptide drug. Thus providing better therapeutic response. 60% increment in the reduction of blood glucose level was seen as compared to normal porcine insulin injection. This effect is shown because drug is released at a slower rate as compared to normal porcine injection and also prevents insulin protein from dehydration or denaturation. However, aquasomes of insulin is not launched in the market because various other parameters such as pharmacokinetic, clinical, and toxicological studies are yet to be studied [8].

5.7.2   DELIVERY OF POORLY SOLUBLE DRUGS

In the research done by Vengala et al. they prepared the aquasome of pimozide. Pimozide falls in the category of anti-psychotic drug having less aqueous solubility after oral administration. In the study, they tried to increase the solubility of pimozide by formulating it into aquasome. After performing the dissolution study, they have concluded the percentage drug release was higher in case of aquasome as compared to the pure drug. This may be due to the small particle size of drug which in turn increases the surface area of the drug [16].

In another research done by Oviedo et al. they prepared aquasome and loaded it with indomethacin, a drug which shows lower aqueous solubility and shows polymorphism [20].

5.7.3   PSORIASIS

Psoriasis is a common skin disease which can be identified by plaques of skin or scaly, thickened patches. This is due to the over production of skin cells.

In the research done by Tiwari et al. they prepared aquasome of dianthranol. Dianthranol is a very promising drug against arthritis by producing anti-proliferative and anti-inflammatory actions.

They prepared topical formulation of aquasome as a cream, by dispersing 5.9 g of drug in the aquasome. The drug release study indicated that 55.93% drug was released in 7 hrs, which proved that the surface bound drugs are easily released. It also exhibits a controlled type of release. From the study, it was also concluded that the consistency of aquasome based cream as better than the plain cream [22].

5.7.4   ANTIGEN DELIVERY

Antigen formulated to aquasome provides conformational stabilization and high degree of surface exposure to antigen, thus producing specific and strong immune response by increasing the availability and activity of antigen *in vivo*. It has many advantages when used as a vaccine delivery system because when antigens are adsorbed on the aquasome surface, they produce both humoral and cellular immune response. It increases the duration for immune response and also reduce the side effects which can be produced by other nano carriers. It has potential to target special type of immune responses like cell-mediated immunity (CMI) or T-helper 1 (Th1) CD4+ T-cell response. The increased immunological activity is due to the presence of mannose-like binding lectins (MLBLs) on the aquasome which is confirmed by concanavalin-A-induced assay.

If the transport of immunogen to the antigen presenting cell (APCs) is confirmed, then it can be efficiently targeted for gene therapy. APCs possess MHC-I and MHC-II molecules. These molecules lead to the presentation and processing of immunogen via both endocytic and cytosolic pathways to exhibit both cellular and humoral response.

In the research done by Goyal et al. they prepared aquasome and loaded with bovine serum albumin (BSA) as a model antigen and checked for antigen preserving action of aquasome and its adjuvanticity. After study, they have concluded that the aquasome preserved the structural integrity of the antigen which further led to immunopotentiation activity [14].

5.7.5   ALLERGEN

In the research done by Pandey et al. they formulated aquasome and loaded with OVA as an allergen model. They compared the immunization activity of aquasome of OVA allergen with other OVA formulation like alum adsorbed OVA. After experimentation in mice, they have concluded that the aquasome of OVA are safer and more effective against an anaphylactic shock in sensitized mice. After the intradermal injection in mice, mixture of Th1 and Th2 immune response were observed. OVA adsorbed aquasome have higher survival rate and exhibited lower level of serum histamine and IgE [2].

5.7.6   HEMOGLOBIN

Khopade et al. prepared, and evaluated aquasome of hemoglobin. Hemoglobin are nowadays used as oxygen carrier as a resuscitative agent.

Calcium hydroxyapatite was used as a core which was formed by the self-precipitation method using half-generation PAMAM dendrimers from SBF. This core possessed good hemoglobin loading capacity (about 13.7 mg hemoglobin per g of core) and was stable for at least 30 days. No conversion of hemoglobin to methemoglobin was observed all through the shelf life. The formulation was tested in albino rats and the study showed that the hemoglobin aquasome can be used effectively as an oxygen carrying system [19].

5.7.7   INTERFERON

Kaur et al. formulated aquasome and loaded with recombinant human interferon-a-2b (rhINF-a-2b). rhINF-a-2b is a USFDA approved cytokine which is used to treat ovarian cancer, SKOV3 cells when co-administered with cisplatin. The main drawbacks of rhINF-a-2b include short half-life (2–6 hrs), low therapeutic index, and fluctuation in plasma level and rapid proteolytic degradation in the systemic environment. Therefore, to provide protein stability and prolonged release, aquasome was coated with PEGylated phospholipid (PEG-200). The prepared PEG coated nanoformulation of aquasome exhibited sufficient pharmacological activity of protein and escaped from the macrophage uptake. The study also revealed that the drug was released through diffusion mechanism [6].

5.7.8 ENZYME DELIVERY

Rawat et al. invented aquasome of an acid labile enzyme STP. This enzyme is categorized as proteolytic enzyme whose pharmacological activity is to help in breakdown of complex proteins to simpler amino acids and also used as an anti-inflammatory.

The enzyme was first loaded into the chitosan-coated nanocore and then it was encapsulated with alginate gel so as to provide muco-adhesiveness and protection from the acidic environment of the stomach. After observing the release study of the enzyme, it was observed that the enzyme showed sustained release in the phosphate buffer because the enzyme was released slowly from the alginate gel encapsulated chitosan-coated nanocore (ACNC) and no dehydration or degradation is observed during delivery or storage [21].

5.8 LITERATURE SURVEY

Table 5.2 summarizes the research performed till date in the field of aquasome with its use and obtained result.

TABLE 5.2 Research Performed on Aquasome

Active Ingredient	Intended Use	Results Obtained	References
Mussel adhesive protein	Antigen delivery	Stability of the antigen was increased due to presence of coating material	[15]
Non-nuclear material from HIV-I	Antigen delivery	Cellular and humoral immune response was exhibited similar to that of live HIV	[15]
Insulin	Insulin delivery	Increment in biological activity of insulin and more reduction in blood glucose level	[8]
Hemoglobin	Oxygen carrier	*In vivo* study results showed that it can be potentially used as blood substitute	[19]

TABLE 5.2   *(Continued)*

Active Ingredient	Intended Use	Results Obtained	References
Hemoglobin	Oxygen carrier	Oxygen carrying capacity was same as that of fresh blood and hemoglobin in the formulation was stable throughout 30 days	[11]
Hepatitis-B vaccine	Vaccine delivery	Combination of Th1 and Th2 immune response	[14]
Bovine serum albumin	Antigen delivery	Better uptake of antigen caused enhanced immunological response and preservation of structural integrity	[14]
Serratiopeptidase	Enzyme delivery	Preservation of biological activity with sustained delivery of enzyme	[21]
Pimozide	Poorly water-soluble drug delivery	Co-precipitation by sonication process yield more amount of core and greater dissolution as compared to pure drug	[16]
Piroxicam	Poorly water-soluble drug delivery	Diffusion controlled release of drug	[16]
Etoposide	Poorly water-soluble drug delivery	Maximum amount of the injected dose was found to be in liver and spleen.	[18]

5.9   CONCLUSION

Aquasomes is considered one of the simplest yet a novel drug carrier-based formulation which is produced on the fundamental principle of self-assembly. The drug or the bioactive molecule delivered through the aquasomes show better biological and therapeutic activity even if they belong to conformationally sensitive class, by preserving the structural integrity of protein pharmaceuticals. This is due to the presence of the unique carbohydrate coating over the ceramic core which prevents the bioactive molecule from denaturation and dehydration by providing an aqueous like environment. In addition, these formulations evoke a better immunological response and could be used as immuno-adjuvant for proteinaceous antigens. Further study of aquasomes is necessary in the field of pharmacokinetics, toxicology, and animal studies to confirm to their safety and efficacy, so as to make it a clinical useful formulation to launch them commercially in the market.

KEYWORDS

- **alginate gel encapsulated chitosan-coated nanocore**
- **antigen presenting cell**
- **aquasome**
- **serum albumin**
- **carbohydrate**
- **ceramics**

REFERENCES

1. Banerjee, S., & Sen, K. K., (2018). Aquasomes: A novel nanoparticulate drug carrier. *J. Drug Deliv. Sci. Technol., 43*, 446–452.
2. Shankar, R., Sahu, S., Sudheesh, M. S., Madan, J., Kumar, M., & Kumar, V., (2011). Carbohydrate modified ultrafine ceramic nanoparticles for allergen immunotherapy. *Int. Immunopharmacology, 11*, 925–931.
3. Parashar, A., (2017). Literature review on aquasomes: A drug carrier system. *Ind. J. Med. Res. Pharm. Sci.,* 27–30.
4. Jain, S. S., Jagtap, P. S., Dand, N. M., Jadhav, K. R., & Kadam, V. J., (2012). Aquasomes: A novel drug carrier. *J. Appl. Pharm Sci., 2*(1), 184–192.
5. Manikiran, S., Krishna, C., & Rao, R., (2010). Aquasomes: Role to deliver bioactive substances. *Res. J. Pharm. Dos. Forms Technol., 2*(6), 356–360.
6. Kaur, K., Kush, P., Pandey, R. S., Madan, J., Jain, U. K., & Katare, O. P., (2015). Stealth lipid coated aquasomes bearing recombinant human interferon-α-2b offered prolonged release and enhanced cytotoxicity in ovarian cancer cells. *Biomed Pharmacother., 69*, 267–276.
7. Inde, V. V., Jangme, C. M., Patil, S. S., Inde, G. S., Chavan, D. V., Yedale, A. D., & Makne, P. D., (2013). A review on aquasomes: A potential drug delivery carrier. *Int. Res. J. Pharm. App. Sci., 3*(2), 124–129.
8. Cherian, A. K., Rana, A. C., & Jain, S. K., (2000). Self-assembled carbohydrate-stabilized ceramic nanoparticles for the parenteral delivery of insulin. *Drug Dev. Ind. Pharm., 26*(4), 459–463.
9. Narang, N., (2012). Aquasomes: Self-assembled systems for the delivery of bioactive molecules. *Asian J. Pharm., 6*, 95–100.
10. Sutariya, V., & Patel, P., (2012). Aquasomes: A novel carrier for drug delivery. *Int. J. Pharm. Sci. Res., 3*(3), 688–694.
11. Patel, S., Aundhia, C., Seth, A., Shah, N., Pandya, K., & Patel, D., (2016). Aquasomes : A novel approach in drug carrier system. *Eur. J. Pharm. Med. Res., 3*(9), 198–201.
12. Shahabade, G. S., Bhosale, A. V., Mutha, S. S., Bhosale, N. R., Khade, P. H., Bhadane, N. P., & Shinde, S. T., (2009). An overview on nanocarrier technology: Aquasomes. *J. Phar. Res., 2*(7), 1174–1177.

13. Wadhwa, S., Singh, S. K., & Gulati, M., (2018). Aquasomes: Novel nanocarriers for drug delivery. *Pharma Times., 50*(2), 68, 69.
14. Goyal, A. K., Khatri, K., Mishra, N., Mehta, A., Vaidya, B., Tiwari, S., & Vyas, S. P., (2008). Aquasomes: A nanoparticulate approach for the delivery of antigen. *Drug Dev. Ind. Pharm., 34*(12), 1297–1305.
15. Kossovsky, N., Gelman, A., Rajguru, S., Nguyen, R., Sponsler, E., et al., (1996). Control of molecular polymorphisms by a structured carbohydrate/ceramic delivery vehicle-aquasomes. *J. Cont. Rel., 39*, 383–388.
16. Vengala, P., Dintakurthi, S., Venkata, C., & Subrahmanyam, S., (2013). Lactose coated ceramic nanoparticles for oral drug delivery. *Int. Res. J. Pharm., 7*(6), 540–545.
17. Kommineni, S., Ahmad, S., Vengala, P., & Subramanyam, C. V. S., (2012). Sugar coated ceramic nanocarriers for the oral delivery of hydrophobic drugs: Formulation, optimization and evaluation. *Drug Dev. Ind. Pharm., 38*, 577–586.
18. Nanjwade, B. K., Hiremath, G. M., Manvi, F. V., & Srichana, T., (2013). Formulation and evaluation of etoposide loaded aquasomes. *J. Nanopharm. Drug Del., 1*, 92–101.
19. Khopade, A. J., Khopade, S., & Jain, N. K., (2002). Development of hemoglobin aquasomes from spherical hydroxyapatite cores precipitated in the presence of half-generation poly (amidoamine) dendrimer. *Int. J. Pharm., 241*, 145–154.
20. Oviedo, I. R., López, R. A. S., Gasga, J. R., & Barreda, C. T. Q., (2007). Elaboration and structural analysis of aquasomes loaded with indomethacin. *Eur. J. Pharm. Sci., 32*(3), 223–230.
21. Rawat, M., Singh, D., Saraf, S., & Saraf, S., (2008). Development and *in vitro* evaluation of alginate gel-encapsulated, chitosan-coated ceramic nanocores for oral delivery of enzyme. *Drug Dev. Ind. Pharm., 34*, 181–188.
22. Tiwari, T., Khan, S., Rao, N., Josh, A., & Dubey, B. K., (2012). Preparation and characterization of aquasome based formulation of dithranol for the treatment of psoriasis. *Wor. J. Pharm. Pharmaceut. Sci., 1*(1), 250–272.

CHAPTER 6

Self-Emulsifying Drug Delivery Systems: A Strategy to Improve the Bioavailability of Hydrophobic Drugs

ANKITA BANERJEE,[1] RAJESH KUMAR,[1] MONICA GULATI,[1]
SACHIN KUMAR SINGH,[1] KAMAL DUA,[2] GURVINDER SINGH,[1]
PARDEEP KUMAR,[1] and ANKIT SHARMA[1]

[1]*School of Pharmaceutical Sciences, Lovely Professional University, Phagwara, Punjab, India, E-mail: rajksach09@gmail.com (R. Kumar)*

[2]*Discipline of Pharmacy, Graduate School of Health, University of Technology, Sydney, Australia*

ABSTRACT

Designing efficacious formulations for oral route of drug delivery has been challenging but still warranted because of its high patient compliance. However, drugs in pipeline or any new chemical entities being discovered have always faced key challenges of optimum solubility and permeability without which a product cannot show the claimed activity or even enter the trials. Though there are several intriguing choices of drug development available, lipid-based formulations are gaining popularity owing to their ease of manufacturing and efficacious results. Further, narrowing the base of lipid-formulations which have gained acceptance are micro- and nanoemulsions, whose bars have been raised higher with the introduction of "self-emulsifying drug delivery system" (SEDDS), which has swept off the bits-and-flaws of conventional emulsion. Therefore, this chapter is an attempt to cover all the aspects of SEDDS formulation including "self-micro-emulsifying drug delivery system (SMEDDS)" and "self-nano-emulsifying drug delivery system (SNEDDS)"; a deep dive in its formulation criteria, selection of suitable candidate and components, self-emulsification process, probable fate of the formulation in vivo, transformation techniques, evaluation parameters, new approaches in pipeline, dosage forms that can be prepared, challenging aspects lingering around, extensive researches done since last decade to present and patents being granted so far.

6.1 INTRODUCTION

Driven by the desire to increase the efficacy, safety, convenience of administration and costeffectiveness, pharmaceutical companies have moved towards either the development of new delivery systems, repositioning, and modification of older drugs or developing new chemical entities [1]. A number of such entities have been found or developed, majority of them being lipophilic in nature with solubility issues, fail to come up. Since lipophilicity is the key to various biopharmaceutical processes, it has the ability to influence various pharmacokinetic parameters [2]. However, possessing lipophilic characters alone is insufficient to meet the need; it has to be water soluble enough to fulfill the rest. The situation is vice versa. Therefore, many delivery systems and formulation have been considered so far in order to get an optimum product with sufficient solubility and permeability [3].

Therefore, mending all the essential aspects in a point to get optimum therapeutic efficacy, formulation scientist designs the drug delivery system which would enable introduction of therapeutic entities, knocking out the developmental deficiencies [4]. Now, if we particularly focus towards the formulation point of view, there are many preferred route of administration, but, in context to patient compliance non-invasive administration is considered as one of the easiest, painless, uncomplicated, and elementary aspect of drug delivery system [5, 6].

Among all the non-invasive techniques like buccal, sub-lingual, intranasal, transdermal, and pulmonary, oral delivery of drug is considered as the most preferred one because of better patient compliance and therefore, is favored for chronic drug therapy [7] as it is cost-effective also. Ensuring the solubility and thus, bioavailability of hydrophobic drugs is one of the major challenges to formulation scientists. As a drug inside the gastro-intestinal tract must unfailingly be in the solution form to be absorbed through the GI mucosa, a poorly soluble drug may results to unpredictable, inconsistent, and incomplete absorption [8].

The output of past research work on solubility has been focused majorly on enhancing the bioavailability using various methods like-use of cosolvents and surfactants, solid dispersions, cyclodextrin inclusion complexes, and permeation enhancers that have proved to be successful. Recent years have moved ahead of these methods and focused on delivery systems. Drug delivery carriers like nanoparticles (NPs) and microspheres are being studied with much focus and enthusiasm in last two decades, and have proved remarkably successful. Some of the other carriers for drug delivery that have been developed during last decade include dendrimers, micelles, microspheres, polymeric NPs, and inorganic NPs, etc. [4].

Apart from inorganic and polymeric carriers, present-day research also focused on to lipid-based drug delivery system (LBDDS), with special emphasis on colloidal carriers for lipophilic drugs [7]. This system has come up as a versatile one that can deliver a broad range of molecules and is the least toxic to the biological systems [8]. One such example is liposome which has the ability to accommodate both lipo- and hydrophilic drugs. In addition, by changing the selection of excipients, LBDDS makes the system suitable for both lipophilic and hydrophilic entities [9]. Other lipid carriers include solid-lipid nanoparticles (SLNs) and nano-structured lipid carriers (NLCs).

As simple-oil based systems, microemulsions, and nanoemulsions have earned lots of popularity owing to their excellent penetration abilities, wetting, and spreading properties and ease of scale-up [10, 11]. Nonetheless, there are few limitations which have led towards designing a newer delivery system [12] that would overcome the associated limitations like instability on long-term storage, structural deformity at higher temperature, hydrolysis of drug in the presence of lipid contents and polydispersity [13]. Thus, the "self-emulsifying drug delivery system (SEDDS)," has outranked other lipid-based carriers. SEDDS are thermodynamically stable formulations [9, 14–16]. They are anhydrous pre-concentrate, isotropic mixtures and a simple binary system, which comprise a unique mixture of oil, surfactant (solvent), co-surfactant, (or co-solvent) and are concentrated before analysis, hence called pre-concentrate. Therefore, this approach of overcoming the solubility-related issues associated with poorly soluble drugs can be used with a variety of drug molecules [19].

6.2   SELF-EMULSIFYING DRUG DELIVERY SYSTEMS (SEDDSS)

SEDDSs or Self-emulsifying lipid therapeutic systems are oral dosage form which comprises of natural or synthetic oils, solid or liquid surfactants and co-surfactants (co-solvent may or may not be used) [17]. The beauty of this system lies in the process of self-assembly which takes place on mild-agitation in contact with aqueous media during dilution [14]. With gastro-intestinal fluids, they form fine O/W, a clear dispersion instantly, which remains stable with dilution. These emulsions can be micronized or nanonized during preparation to get micro or nano emulsions inside gastro-intestinal tract. The basis behind producing micro and nano emulsion is the range of droplet-size which varies from several microns to less than a nanometer. The only difference which would be observed easily is their appearance [18, 19].

Their preparation involves screening of different excipients, making their combination, compatibility studies of excipients, so as to create a self-emulsification system. These formulations spread rapidly in the GIT that leads to self-emulsification with the help of agitation provided by the peristaltic movements of stomach and intestine. Since the release of drug from the SEDDS takes place in GIT, the lipophilic candidate must remain solubilized till the time relevant to gastro-intestinal absorption and therefore a co-surfactant is added as an effective lipid solubilizer/emulsifier [20]. However, before that, the key step to get a good formulation is the judicious choice of oil-surfactant mixture, which decides the fate of lipophilic drugs in which case the dissolution rate happens to be the rate-determining step [19].

"Self-micro emulsifying drug delivery systems (SMEDDS),"an improved version of SEDDS, are isotropic mixtures. The difference lies in their globule or droplet size, its physical appearance, and the concentration of excipients used [21]. These are thermodynamically stable formulations. Unlike SEDDS, these are transparent formulations having unique characteristics of forming O/W micro-emulsions upon mild agitation [22]. The micro-emulsion is formed with the help of ultra-low interfacial tension, which is achieved by two or more emulsifiers, i.e., surfactant (hydrophilic) and co-surfactant (lipophilic). Since, these are absorbed by lymphatic systems and bypass hepatic metabolism, solid SMEDDS are highly preferred for eventual formulation into tablet [23].

"Self-nano emulsifying drug delivery systems (SNEDDS)" have created a niche for themselves by overcoming the demerits of nanoemulsions [24, 25]. The major limitation of nanoemulsion is "Ostwald ripening" (change observed in homogenous solutions or liquid sols leading to redeposit ion of larger particles from the parent solution) leading to high increase in poly-dispersity of the formulation. However, colloidal dispersions have several similarities but unlike them, SNEDDS are thermodynamically unstable but kinetically stable. Another edge of SNEDDS over nanoemulsion is over-coming the process of breaking of emulsion (due to polydispersity) [26]. Another advantage of SNEDDS is the interfacial surface area that comes in contact with GI fluid. Since the droplet size is already in nano range, the surface area provided by SNEDDS positively influences the transport properties. The nano-sized droplets go through a quick digestion and undergo faster absorption; bypass the microsomal enzymes metabolism and first pass metabolism [27]. Besides this, due to the presence of pre-dissolved drugs which is key step for dissolution, they provide rapid onset of action of the drug. Thus, the rate-limiting step of dissolution is solved and it enhances the bioavailability profile. Besides, SNEDDS can be prepared with ease and hold promise of scaling-up [14].

6.3 TYPES OF SELF-EMULSIFYING LIPID FORMULATION

Their types are uniquely based on the characteristics they possess:

- Self-emulsifying formulation's globule size ranges from 200 nm–5 µm and the dispersion appears to be turbid. The HLB value of the formulation prepared using surfactant is <12 and are they are thermodynamically stable in all physiological conditions. Ternary phase diagram is essentially used for characterization [28].
- Self-micro-emulsifying formulations have oil globule size of 100–250 nm and they are translucent to optically clear dispersion. Surfactants used in the formulation should be having an HLB value greater than twelve (>12). These are thermodynamically stable in physiological conditions and the development and characterization requires pseudo-ternary phase diagram [29, 30].
- Self-nano-emulsifying formulations consist of oil-droplets having size <100 nm and the dispersion appears to be optically clear. Despite being in nano-size, no phase-separation usually observed during storage. Surfactants used in the formulation should have HLB value beyond twelve (>12). Characterization and development may require constructing pseudo-ternary diagram (Figure 6.1) [31].

6.3.1 SUITABLE DRUG CANDIDATE

The ultimate objective of any oral formulation is to have optimum drug solubility in GIT and to maximize its absorption. Suitable candidates for such delivery systems are lipophilic-drugs which express dissolution rate-limited absorption [32]. SEDDSs constitute one such delivery system that helps

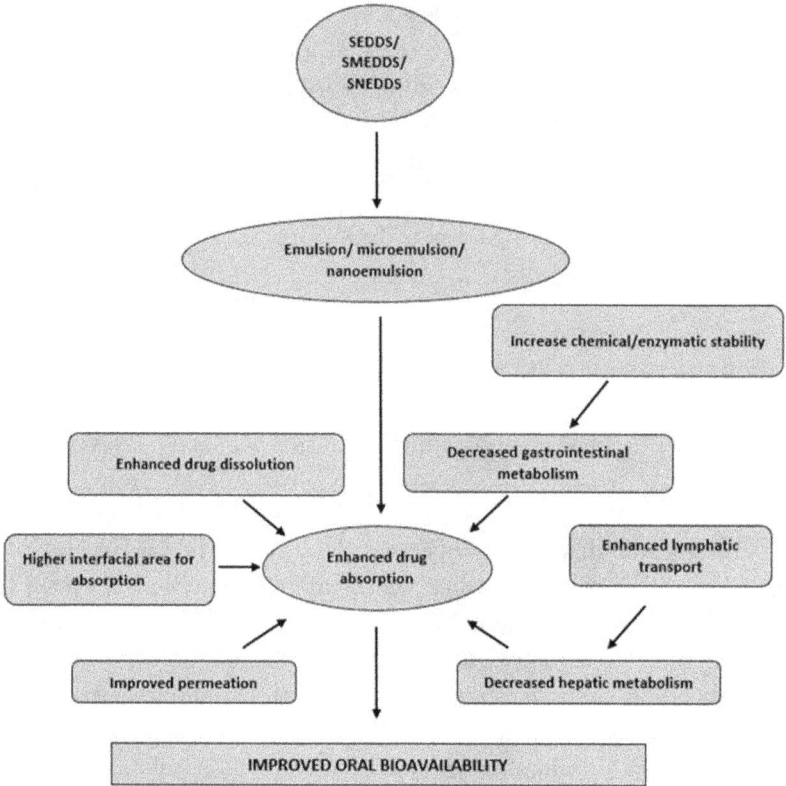

FIGURE 6.1   Factors influencing bioavailability.

Source: Adapted from Ref. [34].

improve the bioavailability followed by optimum plasma-drug concentration profile (Figure 6.2) [33].

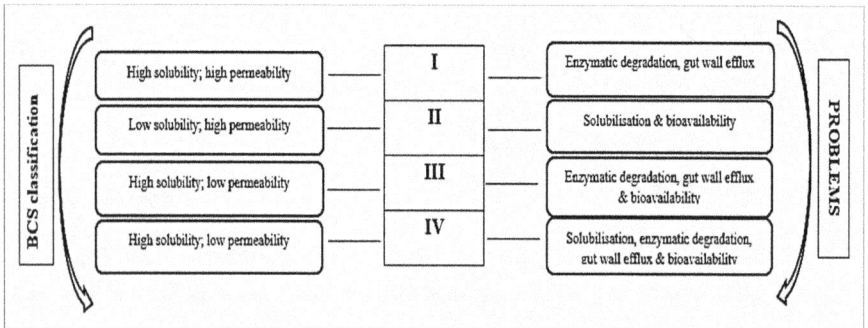

FIGURE 6.2   Problems influencing drug absorption.

6.3.2 ADVANTAGES

1. **Improved Oral Bioavailability:** Lipophilic drugs have dissolution rate-dependent drug absorption which restricts their bioavailability. Self-emulsifying systems have the ability to introduce the drug in a solubilized or micro-emulsified form (globule size – 1 to 100 nm) in GIT, ensuing increased specific surface area to transport the drug across the intestinal aqueous layer and to improve bioavailability through absorptive membrane [34].

2. **Ease of Manufacture and Scale-Up:** Self-emulsifying systems have the advantage of ease of preparation and scale-up as compared to other drug delivery systems. They require simple and economical manufacturing facilities like mixer connected with agitator and volumetric liquid filling equipment for scale-up manufacturing [35].

3. **Ability to Deliver Peptides in GIT:** Self-emulsifying systems can transport macromolecules (viz. peptides, enzyme, or hormones) and provide protection from enzymatic hydrolysis [36]. In their study prepared a prodrug whose intestinal hydrolysis was protected by cholinesterase with the help of polysorbate 20, which is an emulsifier. These systems are formed spontaneously with just mild agitation due to gastric motility, which is suitable for thermolabile drugs [24].

4. **Solid and Liquid Dosage Form:** The basic form of SEDDS are liquid and can be administered as micro or nanoemulsions. They are incorporated in soft/hard gelatin capsule or hydroxypropyl methylcellulose capsules. They can be taken orally as tablet by drying using various techniques. However, their oral bioavailability does not vary significantly on solidification [37].

5. **Quick Onset of Action:** There are several conditions, such as allergies, inflammation, and heart diseases, where facilitating faster oral absorption of drug is of prime importance. Self-emulsifying system gives quick onset of action by readily dispersing itself in the aqueous medium and larger surface area provides more absorption. Studies have demonstrated a considerable reduction in t_{max} which indicates quick release [38]. When vitamin A was given as SNEDDS in capsule and SNEDDS compressed into tablet in comparison to capsule filled with vitamin A alone, an increased bioavailability of almost twofold and 1.4 times for SNEDD-capsule and SNEED-tablet

respectively was observed in comparison to Vitamin A capsule. [39]. bioavailability.

6. **Better Drug-Loading Efficiency:** Self-emulsifying system have more drugs loading as compared to lipid solutions as they contain higher surfactant and co-surfactants concentration with less oil. It has been reported that self-emulsifying system, if formulated using medium-chain lipids, could dissolve more drug and hence, Cremophor RH40 is widely used as surfactant [40].

7. **Reduction in Drug Dose and Frequency:** Self-emulsifying systems exhibit better therapeutic effect, oral bioavailability and improved C_{max} for numerous hydrophobic drugs and hence, dose reduction is there with minimal dose-related side effects. This has been reported with some antihypertensive and anti-diabetic drugs [38].

8. **No Influence of Lipid Digestion:** Self-emulsifying systems neither get influenced by lipid breakdown (gastric reports approximately 10–25%) [41] nor bile salts emulsification or pancreatic-lipases and micelles formation. Since, these are present in micro- or nano-emulsified form; their breakdown promotes easy penetration of the mucin and aqueous layer [23].

9. **Increased Capacity of Drug Loading:** A system's drug loading capacity is ensured by the solubility of drug excipient matrices. Self-emulsifying system has added advantage of increased drug loading as it solubilizes lipophilic drugs in its oils and surfactants matrix [42, 43].

6.3.3 LIMITATIONS

Despite having several advantages, there are a few limitations associated with SEDDS. These include:

1. Due to different range of self-emulsifying systems, chemical instabilities have been observed.
2. Gastric irritation occurs due to higher concentration of surfactants. Hence, surfactants should always be selected based on their formulation as well as safety aspects. Non-ionic surfactants, though, are less toxic than ionic surfactants, lead to fluctuation or unpredictable alterations in intestinal lumen permeability.

3. Another such challenge is use of edible oils which may not be able to dissolve larger quantities of lipophilic drugs.
4. Natural emulsifiers may be preferred, being considered as safe, however, their capacity of self-emulsification is limited [40].

6.4 LIPID FORMULATION CLASSIFICATION SYSTEM (LFCS)

There are different lipid carrying delivery systems available including lipid solution, emulsion, microemulsion, and dry emulsion. To get a clear idea about different systems with available numbers of excipient combinations, "lipid formulation classification system (LFCS)" was established. This classification system was introduced as reference model for interpretation regarding *in vivo* studies readily. It also discloses the identification of particular lipid to be used looking at specific physicochemical attributes of the drugs [33]. This classification gives better idea regarding fate of the potential lipid formulation *in vivo* to avoid "trial and error" approach (Figure 6.3).

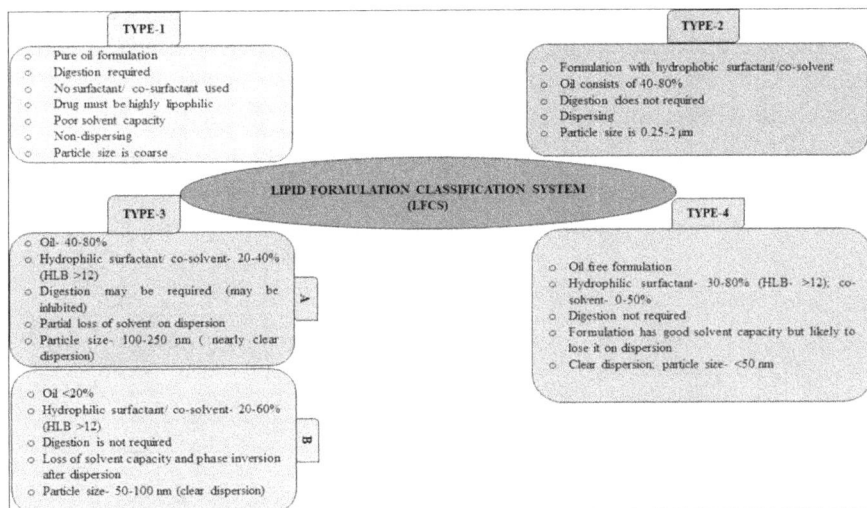

TYPE-1
- Pure oil formulation
- Digestion required
- No surfactant/ co-surfactant used
- Drug must be highly lipophilic
- Poor solvent capacity
- Non-dispersing
- Particle size is coarse

TYPE-2
- Formulation with hydrophobic surfactant/co-solvent
- Oil consists of 40-80%
- Digestion does not required
- Dispersing
- Particle size is 0.25-2 µm

LIPID FORMULATION CLASSIFICATION SYSTEM (LFCS)

TYPE-3
- Oil- 40-80%
- Hydrophilic surfactant/ co-solvent- 20-40% (HLB >12)
- Digestion may be required (may be inhibited)
- Partial loss of solvent on dispersion
- Particle size- 100-250 nm (nearly clear dispersion)

A

- Oil <20%
- Hydrophilic surfactant/ co-solvent- 20-60% (HLB >12)
- Digestion is not required
- Loss of solvent capacity and phase inversion after dispersion
- Particle size- 50-100 nm (clear dispersion)

B

TYPE-4
- Oil free formulation
- Hydrophilic surfactant- 30-80% (HLB- >12); co-solvent- 0-50%
- Digestion not required
- Formulation has good solvent capacity but likely to lose it on dispersion
- Clear dispersion; particle size- <50 nm

FIGURE 6.3 Lipid formulation classification system.

6.5 MECHANISM OF SELF-EMULSIFICATION

As the emulsification process proceeds, surfactants tend to align themselves on the oil and water interface to stabilize the formed oil droplet, thereby reduces the interfacial tension. Thus, a combination of drug, oil, and surfactant-co-surfactant altogether play an important role in the formation of a stable formulation. With a decrease in the droplet size, there is a corresponding increase in interfacial surface area. The co-solvent helps in improving emulsification by penetrating the surfactant layer of oil-water interface and thereby reduces the surface tension.

Self-emulsification is a result of the change in entropy which favors dispersion higher in comparison to the energy required for increasing the surface area, resulting in formation of free energy of emulsion which is proportional to the energy required to create a new surface between the oil and water phase. This process is governed by the equation:

$$\Delta G = \Sigma i\left(Ni4\pi ri2\sigma\right) \tag{1}$$

where; ΔG = free energy associated with the process; N = number of droplets of radius r and s is the interfacial energy; r = the radius of globules; and σ = the interfacial energy.

As the formed emulsion tends to separate for reducing the free energy of the system with time, the emulsification process requires little input of energy, involving destabilization through reduction of regional interface structure which shows no resistance against surface shearing. On the other hand, in conventional emulsification process, the agents stabilize the emulsions, forming a monolayer around the emulsion droplets thereby reducing the interfacial energy and hence providing obstruction to coalescence (Figure 6.4) [44].

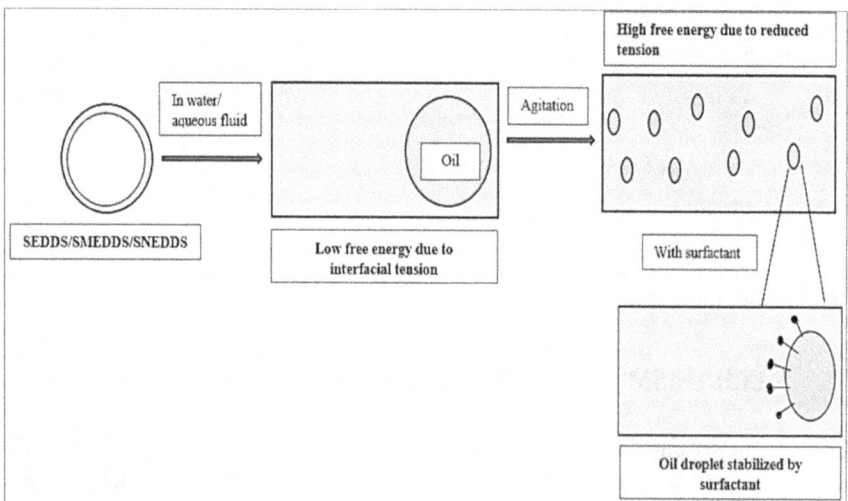

FIGURE 6.4   Emulsification mechanism.

6.6   CONSTRUCTION OF PHASE DIAGRAM

For the demonstration of self-emulsifying delivery system, the preparation of pseudo-ternary diagram is of utmost importance, as it indicates the spontaneity of the self-emulsification process. This diagram not only demonstrates the spontaneity in terms of time and phase behavior of different formulation, but also reveals the self-dispersion capability in terms of droplet size (micro- or nano-) when the formulation comes in contact with the GI luminal fluid [45, 46].

In order to prevent coalescence (due to mechanical barrier), the phase diagram optimizes the required amount of the components in the preparation. This ensures the self-emulsification capacity/potential. Generally, the greater the water penetrates the oil phase; the more it undergoes self-emulsification spontaneously, making it thermodynamically stable [47]. For specifically obtaining the nano-sized droplets *in vivo*, low energy is required (ultra-low negative interfacial tension). Phase diagram further illustrates the possible dilutions in GI luminal fluid (robustness on every trial) and probable ratios of the formulation components to be used, which reduce the chances of failure of the formulation [48].

6.7   FORMULATION OF SEDDS

Judicious selection of oils, surfactant, and co-solvent based on the solubility of the drug is an essential aspect for formulation. The design of optimal SEDDS requires preformulation, solubility studies and phase diagram studies. The essential constituents of SEDDS include:

1. **Drug Molecule:** Lipophilicity and dose of the drug are the main criteria to be considered before formulation. Drug molecules with a logP value greater than four are considered to be the most appropriate for such system. Dose of drug should be such that it should be soluble in small quantity of oil for ease of emulsification in GIT fluids after ingestion. Single dosage unit cannot be formulated if higher quantities of oil are required for the solubilization of drug.

2. **Oil:** The selection of oily phase is a major consideration which swings between the ability of oil to solubilize the drug and to promote formation of nanoemulsion having desired characteristics. Mostly the unmodified and raw forms of edible oils provide base as lipid

vehicles, but significant challenges are faced when it fails to dissolve large amounts of lipophilic drugs.

3. **Surfactant:** It is one of the essential components in the formulation, as it promotes the emulsification properties. The type and concentration of the surfactant exhibit a considerable effect on droplet size of micro- or nanoemulsions. Two important characteristics of surfactants are HLB value and concentration. For achieving high emulsification, the emulsifier involved must possess high HLB (>12) and high hydrophilicity, which ultimately promotes small o/w droplets, leading to rapid spreading of formulation in aqueous media, prolonged stay for effective absorption and eventually prevents precipitation of drug compound within GI lumen.

4. **Co-Solvent:** It lowers the surface tension and forms mixed micelle along with surfactant, which gives more surface area. In addition, it maintains the spontaneity of self-emulsification process. Ethanol, PG, and PEG are commonly used as co-solvents [13].

6.8 FACTORS AFFECTING SEDDS/SMEDDS/SNEDDS

1. **Drug Dose:** High dose of drugs are not recommended for self-emulsifying systems, since they have faster onset of action which may lead to toxicity. The pre-requisite for drug candidate is to have good solubility in at least one of the essential component of self-emulsifying system, preferably lipophilic phases [15].

2. **Solubility of the Drug in Oil Phase:** The ability of self-emulsifying system to have drug in solubilized form can be determined by the solubility of the drug in oily phase. The chances of precipitation increase if the concentration of surfactant or co-surfactant for drug solubilization is more. Further dilution of such systems will narrow the solvent capacity of surfactant or co-surfactant [49].

3. **Polarity of Oil Droplets:** Polarity reflects the affinity of the drug for oil or water. Greater affinity towards oil promotes rapid release of the drug in GI fluid, as lipid phase is the governing factors for release of drug. HLB, chain length, degree of unsaturation of the fatty acid, molecular weight of the hydrophilic portion and concentration of the emulsifier govern bipolar affinity of the droplets [50].

6.9   TECHNIQUES FOR TRANSFORMATION OF L-SEDDS TO S-SEDDS

There are a few techniques which have been used for conversion of liquid-SEDDS into solid form. Table 6.1 summarizes these techniques.

TABLE 6.1   Techniques of Preparing S-SEDDS/SMEDDS/SNEDDS

Techniques	Characteristics	Procedures	Advantages
Spray drying	• Mixing of lipids, surfactants, drug, solid carriers, and atomization of droplets	Fully solubilized liquid formulation is atomized into droplets and introduced in drying chamber. Here, the volatile phase vaporizes, forms dry particles under controlled temperature	Quality of the powder remains same; applicable for both heat-sensitive and heat-resistant products
Direct adsorption on carrier	• Microporous inorganic and colloidal adsorbents, cross-linked polymers or nanoparticle adsorbents are used	It is a simple adsorption process, involves addition of the liquid formulation along with carriers in a blender. The resulting powder may then be filled directly into capsules or with suitable excipients before compression into tablets	Cross-linked polymers are favorable to have sustain drug dissolution • Up to 70% of the SEDDS can be adsorbed with good content uniformity
Melt extrusion	• Solvent free process • High drug loading • Content uniformity • Conversion of material with plastic properties to a product of uniform shape and density	• process includes dry mixing of API and excipients, wet mass with binder followed by extrusion into thin tubular structure extrude to form spheroids of uniform size. Drying and coating is done	• Enteric-coated dry emulsion formulation preferable for the oral delivery of peptide and protein
Melt granulation/ extrusion spheronization	• Powder agglomerates are obtained through the addition of binder • Also called thermoplastic granulation technique	• Softening of agglomerate powder particles at 50–90°C followed by cooling of powder. • Consequent solidification of molten or soft binder that melts during the process. This completes the process	• One-step procedure, omitting out liquid addition and subsequent drying of conventional wet granulation. Helpful for water sensitive materials

6.9.1 FATE OF SEDDS/SMEDDS/SNEDDS AFTER ORAL ADMINISTRATION

The fate of SNEDDS after oral administration, its mechanism of drug release, enhanced bioavailability are best described by considering *in vivo* behavior of components [51]. When the formulation comes in contact with GI fluid, the SNEDDS get emulsified spontaneously in presence of bile salts which promotes formation of nano-droplets of sizeless than 100 nm [52].

For enhancement of drug absorption, several hypotheses have been put forward. Drugs get absorbed via lymphatic system (where, chylomicron synthesis the fatty components of the oil phase of the emulsion). If the drug in lipophilic, oil droplets get absorbed by micelles of bile salt along with the metabolites formed by lipid carriers [18, 53].

Time for dissolution is increased by the inhibition of gastric motility caused by the lipid phase and thus provides larger window for drug absorption. Increased bile micelles and mesenteric lymph flow are likely to be responsible for improved drug absorption (Figure 6.5) [13].

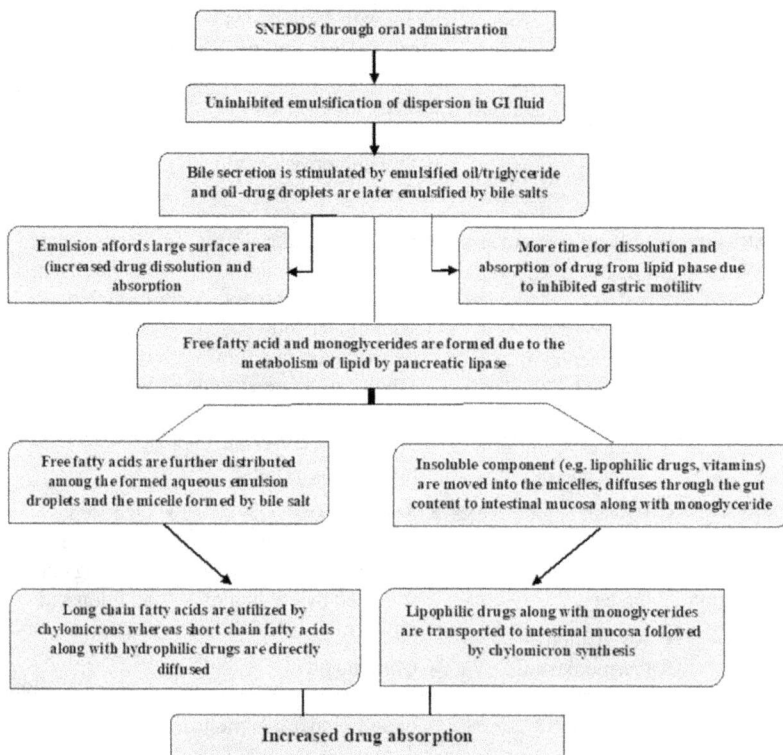

FIGURE 6.5   Fate of SEDDS/SMEDDS/SNEDDS after oral administration.

6.9.2 EVALUATION

1. **Emulsification Rate and Self-Emulsification Time:** The rate of self-emulsification is usually determined by keeping self-emulsifying formulations (pre-concentrate) in a capsule and adding it to a sufficient amount of water or bio-relevant media. The rate of dispersion is determined by visual examination [37]. Efficiency of emulsification at different compositions is tested using rotating paddle to promote emulsification and the time taken for self-emulsification is noted [49]. Light microscopy is used for evaluating the process of self-emulsification. Studies have found that the mechanism of emulsification does not progress with reduction in droplet size; but rather involves erosion of a fine cloud of small particles from the surface of large droplets [54].

2. **Turbidity Measurement:** Turbidi meters are used to establish, whether the dispersion reaches equilibrium rapidly and in a reliable time frame [55]. Hach turbidity meter and the Orbeco-Helle turbidity meter [17] are used often. The apparatus is further connected to dissolution apparatus. Every 15 sec, optical clarity of formulation is observed to determine the quality of microemulsion formed. Turbidity can also be observed by spectroscopic characterization of microemulsion at suitable dilution [54].

3. **Droplet Size:** It is critical for self-emulsification as it determines the rate and extent of drug release followed by absorption. Low dilutions are preferred for accurate droplet size evaluation. Photon correlation spectroscopy (PCS) is used for determining the droplet size of the emulsion, especially, when the properties of the emulsion do not change upon infinite aqueous dilution [40].

4. **Polarity:** Release of the drug into the aqueous phase is promoted by polarity and it reflects on the types of bonds formed and affinity of the drug towards oil or water. Polarity of the emulsion droplets characterizes the efficiency of emulsification. It is affected by the unsaturation and chain length of fatty acid and concentration of emulsifying agents [36].

5. **Zeta Potential:** The charge of the oil droplets of self-emulsifying system is also critical. It is indicative of the formulation stability and shelf life. The charge on the oil droplets should not be zero, as it signifies coagulation or precipitation, which is not suitable for

formulation [56]. Having cationic lipid in such systems is preferable for the incorporation of the drug compounds [57].

6. **Drug Precipitation and Stability on Dilution:** It indicates the ability of self-emulsifying system to retain the drug in solubilized form; hence, it is very important to evaluate the stability of the system after few dilutions. The process is carried out by diluting a single dose of self-emulsifying formulation in 250 ml of 0.1N HCl solution. The solution is then observed for drug precipitation, if any [58].

7. **Effect of pH and Robustness:** Formulation is subjected to dilution (50, 100, 1000, and 3000 time) with water, 0.1M HCl and intestinal liquid of pH 6.8. The so formed diluted emulsions are checked visually after allowing standing for 24 h, checked for any physical changes like coalescence of droplet, precipitation, or phase separation [37].

8. **Viscosity:** Brookfield viscometer and plate rheometer is used to check the viscosity of various formulations at $25 \pm 1.0°C$ [37].

9. **Dispersibility Test:** Self-emulsified oral micro/nano emulsions are tested for their efficiency using a standard USP XXII-dissolution apparatus II at 50 rpm. Each mL of different formulation is added to 500 mL of water maintaining $37 \pm 0.5°C$ (Table 6.2) [59, 60].

TABLE 6.2   Grading System According to Dispersion Time

Grades	Appearance	Dispersion Time
A	Slightly bluish or transparent	Less than 1 min
B	Bluish-white	1 min
C	Whitish milky	2 min
D	Dull grayish white	More than 5 min
E	White emulsion with large globules on surface	Less to minimal emulsification

6.9.3   NOVEL APPROACHES AND IN VITRO/IN VIVO BIOAVAILABILITY RESULTS OF SNEDDS

1. **Self-Double Nano-Emulsifying Systems (SDEDDS):** These can spontaneously emulsify to water-in-oil-in-water (w/o/w) double emulsions in GI environment which prevents enzymatic inactivation in GIT. It consists of hydrophilic surfactant and form water-in-oil emulsion, which upon further dilution with gastric fluid, forms water-in-oil-water emulsion. These are thermodynamically stable

formulation. Majority of the proteins and anti-cancer drug can be administered orally through these [61].

2. **Targeted SNEDDS:** Toxicity can be reduced with targeted delivery and SNEDDS can be considered for this purpose. The spontaneously prepared nanoemulsion can stay in the circulation for longer duration of time if PEGylation is done to surpass phagocytic system. In addition, PEG is reported to provide stealth properties to the nanoemulsions, making the surface slippery, and providing hindrance/barrier from enzymatic degradation. For further specific delivery to lymphatic system and cell targeting, ligands can be used [14].

3. **Enhanced Mucus Permeation SNEDDS:** Mucus barrier finds its importance in almost all organ targeted delivery systems like buccal cavity, nasal cavity, lungs, intestinal, and vaginal cavities. Due to faster rate of secretion and excretion rate of mucin, it gives insufficient retention time for drug carrier to stay. SNEDDS can prove to be a better strategy, as they have hydrophobic surface which can easily pass the mucus layer [14, 62].

4. **Supersaturated SNEDDS:** Selection of lipidic excipients needs to be very judicious for achieving the solubility of drug in order to avoid precipitation. To overcome this bottleneck, super-saturated SNEDDS are prepared which have the potential hydrophilic precipitation inhibitors that can stop the nucleation process, are thermodynamically stable, and exhibit longer mean absorption time (Table 6.3) [63].

TABLE 6.3 Summary of Super-Saturated SNEDDS Formulations Having Enhanced Bioavailability

Drug	Indications	Formulation	Technique/ Polymers Used for Super- Saturation	Bioavailability (in Comparison to Conventional SNEDDS)	References
Trans-resveratrol	Hypolipidemic, Anticancer	L-SNEDDS	5% hydroxyl propyl methyl cellulose (HPMC)	1.33-fold enhanced AUC and 3.92-fold increase in permeability	[64]
Ezetimibe	Hypolipidemic	L-SNEDDS	HPMC E5LV	HDL level increases by 63%; conventional SNEDDS–46.34%	[65]
Simvastatin	Anti-hyperlipidemic	SNEDDS pellets	Heating and cooling cycle	$t_{1/2}$ increased by 2.3 hrs. and relative bioavailability-180%	[66]

5. **Solid SNEDDS (Sustained/Controlled/Extended Release)**: SNEDDS have shown better pharmacokinetic parameters in comparison to any other oral dosage form. It quickly reaches to the maximum concentration (C_{max}) with shorter time length (T_{max}), which may lead to dose fluctuation. In case of potent drugs, this issue becomes very critical. Therefore, the research interest has shifted to formulate solid SNEDDS which would possess reduced C_{max} with better bioavailability and with longer mean residence time (MDT). A formulation having rate-controlled release with zero-order can be considered ideal for use (Table 6.4).

TABLE 6.4 Summary of Sustained/Controlled Release SNEDDS Formulations Having Enhanced Bioavailability

Drug	Indication	Formulation	Polymers Selected	Bioavailability (Comparison to Conventional SNEDDS)	References
Cyclosporine A	Immunosuppressant	Zero-order kinetics (controlled release)	Cellulose acetate	80–85% drug release over a period of 12 hrs.	[67]
Ziprasidone	Schizophrenia	Sustained release pellets	HPMC and MCC	95% drug release over a period of 12 hrs.	[68]
Felodipine	Anti-hypertensive	Extended release gel	SNES gel enclosed in hydrophobic gelucire coat	73% drug release over a period of 12 hrs.	[69]

6.9.4 SNEDDS FOR BIO-MACROMOLECULES

As modern therapeutics are pacing towards higher level targeting, selectivity, and specificity, regulatory bodies are approving many biopharmaceuticals like protein, peptides, and gene delivery. Focus for encouraging such development is on refined systems of delivery for such modified and recombinant products. However, there are many drawbacks relating to large molecules, e.g., poor permeation due to large size and hydrophilicity and degradation in gastric pH; hence, design of such a formulation is required which can provide better retention of molecules in gastric fluid without degradation. SNEDDS offers an attractive option for the successful delivery of biomacromolecules by providing protection against gastric lability and enhancing absorption through GIT (Table 6.5).

TABLE 6.5 Summary of Biomolecules-Loaded SNEDDS Formulations Having Enhanced Bioavailability

Drug	Formulation	Droplet Diameter	Loading Technique in SNEDDS	Bioavailability (Comparison to Conventional SNEDDS)	References
Insulin	L-SNEDDS	30–45 nm	Insulin with DMPG (Hydrophobic ion pair)	79% *in vitro* release after 180 min and 30% release after 9 hrs.	[70]
pDNA	L-SNEDDS	45–47 nm	Lipofectin-pDNA (Hydrophobic ion pair)	Resistance towards DNAase increased by 8 folds	[71]
Curcumin	L-SNEDDS	150–600 nm	Phospholipid with curcumin	>80% drug release by 15 min	[38]

6.10 DOSAGE FORM DEVELOPMENT OF S-SEDDS

1. **Dry Emulsions:** These are powdered forms of emulsions which upon coming in contact with aqueous fluid emulsify spontaneously. Dry emulsions form O/W type of emulsion and comprise solid carriers like lactose, maltodextrin, etc. They are prepared by rotary evaporation, freeze-drying, or spray drying. The use of spray drying technique is more favored for preparation of dry emulsions [15]. Recently enteric-coated dry emulsions have been reported [72] and their application extended for the oral delivery of peptide and protein drugs [16, 73].

2. **Self-Emulsifying Sustained and Controlled-Release Tablets:** Self-emulsifying (SE) tablets maintain a consistent higher level of active ingredient and its concentration in blood plasma over an equal time frame as compared to simple sustained release tablets. The recently developed self-emulsifying tablet is "SE osmotic pump tablet," where the osmotic pump system acts as the carrier. It provides a stable plasma concentrations and controlled drug release rate over a longer period [74, 75].

3. **Self-Emulsifying Sustained and Controlled-Release Pellets:** Pellets have created a niche for themselves owing to their ease in manufacturing, reducing intra-subject and inter-subject variability, better plasma profiles and minimizing GI irritation without compromising drug's bioavailability. Thus, it would be satisfactory to combine them with self-emulsifying systems. The pellets can further

be coated with water-insoluble polymer to reduce the drug release and hence give a range of release rates. SEDDS are not anticipated to influence the ability of the polymer film to control drug dissolution [76].

4. **Self-Emulsifying Suppositories:** SEDDS have been reported not only enhance GI adsorption, but also rectal/vaginal absorption. Suppositories of SEDDS of glycyrrhizin have shown satisfactory results when therapy is given through vaginal or rectal route with a mixture of a C6-C18 fatty acid glycerol ester and a C6-C18 fatty acid macrogol ester along with glycyrrhizin [16].

5. **Self-Emulsifying Implants:** These came as a boon in the wide application of S-SEDDS. One such good example is Carmustine, which is a chemotherapeutic agent for malignant brain tumors. However, the effectiveness of the drug is compromised owing its short half-life. To enhance the stability, self-emulsified carmustine is formulated with PLGA wafers of flat and smooth surface by compression molding. This self-emulsifying implant has increased the half-life of carmustine up to 130 min (*in vitro*) in comparison with 45 min of plain carmustine. *In vitro* release of carmustine from SE PLGA wafers was prolonged up to 7 days, exhibited higher antitumor activity and was found to be less susceptible to hydrolysis [77].

6.11 CHALLENGING ASPECTS AND DEVELOPMENTS

A novel formulation despite having numerous advantages and potential for scale-up remains largely unexplored due to underlying challenges which cannot be ignored. Though enormous research work has been done on self-emulsifying systems, confusion still persists in terms of evaluating the formulations visually, referred to as "phase-transition" giving a wrong interpretation. Moreover, the unavailability of sufficient *in vitro-in vivo* correlation (IVIVC), data has limited the translation of SNEDDS into more marketed products. However, there are few of the recent approaches which are suggested for developing optimized lipid formulations [78]:

1. **HLB-Response Surface Methodology:** Ternary phase diagram and commonly used trial-error methods are employed for preparing self-emulsifying systems [14]. To minimize such misleading calculation

and to emphasize alternative approach for development of self-emulsifying system, "hydrophilic-lipophilic" balance-associated response surface methodology has been introduced which would demonstrate the minimum and C_{max} of excipients required for an optimized formulation [14].

2. **IVIVC for Self-Emulsifying Systems:** The fate of self-emulsifying systems after their ingestion can be understood by focusing on *in vitro* tests like-dispersion, precipitation, and lipolysis of the formulation in well-defined media.

3. *In Vitro* **Precipitation Test and** *In Vivo* **Data Interpretation:** According to concept of absorption, a fully soluble drug would get readily absorbed whereas the precipitated drug would not be available for absorption. In case of self-emulsifying systems, with proper screening of oils and lipid, a corrective data can be obtained by plotting the amount of drug dissolved *in vitro* in comparison with the AUC or *in vivo* data. However, precipitation studies have also been reported in biorelevant media of different pH in 96-well microtiter plate [79]. To determine the level of precipitation, ordered rankings are made namely fast; slow and; no-precipitation. This ordered ranking has been observed for few drugs in higher animals to correlate well with the levels of data *in vivo*, showing strong level-A correlation. These studies have concluded that even a class-II drug can behave as class-I drug after incorporating it in self-emulsifying system [80, 81].

4. *In Vitro* **Lipolysis and** *in Vivo* **Data Interpretation:** This study is very significant in correlating the data. The study involves mimicking the exact *in vivo* condition like adding bile salts and pancreatic enzymes for initiating lipid digestion. By drawing out samples at different intervals, checking the drug content *in vitro* and comparing it with *in vivo* plasma concentration time profile, a correlation is developed [82]. A few reports are available for correlation of *in vitro* lipolysis data with *in vivo* results [83]. The study found strong correlation only when the lipid content is less, which indicates that the solubilization capability of formulation depends on the type and amount of lipid being used. It also indicates the essential ingredients and their effect on *in vivo* bioavailability of any drug [84, 85].

6.12   APPLICATIONS (TABLE 6.6)

TABLE 6.6   Research Done so far Using Various Techniques in Self-Emulsifying System Over Last Decade

Drug	Formulation	Activity	References
Docetaxel	Evaluation of a self-nano-emulsifying docetaxel delivery system	Anti-cancer	[86]
Clopidogrel	Bioavailability enhanced clopidogrel-loaded solid SNEDDS: Development and *in vitro/in vivo* characterization	Antiplatelet	[87]
Efavirenz	Fabrication and evaluation of self-nano-emulsifying oil formulations (SNEOFs) of Efavirenz	Antiretroviral	[88]
Bruceine D	Self-nano-emulsifying drug delivery system of Bruceine D: A new approach for anti-ulcerative colitis	Anti-ulcerative colitis	[89]
Orlistat	Orlistat-loaded solid SNEDDS for the enhanced solubility, dissolution, and *in vivo* performance	Gastrointestinal lipase inhibitor	[90]
Diacerein	Solid form of lipid-based self-nano emulsifying drug delivery systems for minimization of diacerein adverse effects: Development and bioequivalence evaluation in albino rabbits	Anti-osteoarthritic	[91]
Glimepiride	Impact of various solid carriers and spray drying on pre/post compression properties of solid SNEDDS loaded with glimepiride: *In vitro-ex vivo* evaluation and cytotoxicity assessment	Anti-diabetic	[92]
Mebendazole	Self-nano-emulsifying Drug delivery system of mebendazole for treatment of lymphatic filariasis	Lymphatic Filariasis	[93]
Cinnarizine	Multi-layer self-nano-emulsifying pellets: An innovative drug delivery system for the poorly water-soluble drug cinnarizine	Anti-histaminic	[94]
Candesartan citexetil	Oral solid self-nano emulsifying drug delivery systems of Candesartan citexetil: Formulation, characterization, and *in vitro* drug release studies	Angiotensin II receptor antagonist	[95]
Polypeptide-k	Solid self-nano emulsifying drug delivery systems for oral delivery of polypeptide-k: Formulation, optimization, *in vitro* and *in vivo* anti-diabetic evaluation	Anti-diabetic	[96]
Cilostazol	Development of novel cilostazol-loaded solid SNEDDS using a SPG membrane emulsification technique: Physicochemical characterization and *in vivo* evaluation	Antiplatelet	[97]

TABLE 6.6 *(Continued)*

Drug	Formulation	Activity	References
Glipizide	Design, optimization, and evaluation of glipizide solid self-nano-emulsifying drug delivery for enhanced solubility and dissolution	Anti-diabetic	[98]
Glipizide	Solid super saturable self-nano emulsifying drug delivery systems for improved dissolution, absorption, and pharmacodynamic effects of Glipizide	Anti-diabetic	[99]
Naringenin	Self-nano-emulsifying drug delivery system (SNEDDS) of the poorly water-soluble grapefruit flavonoid Naringenin: Design, characterization, *in vitro* and *in vivo* evaluation	Anti-inflammatory, hepato-protective, and anti-lipid peroxidation	[100]
Docetaxel	Formulation and evaluation of liquid loaded tablets containing docetaxel-self nano emulsifying drug delivery systems	Anti-mitotic chemotherapy agent	[101]
Tacrolimus	Solid self-nano-emulsifying drug delivery system (SNEDDS) for enhanced oral bioavailability of poorly water-soluble tacrolimus: Physicochemical characterization and pharmacokinetics	Macrolide antibiotics	[102]
Rosuvastatin calcium	Self-nano emulsifying drug delivery system (SNEDDS) of Rosuvastatin calcium: Design, formulation, bioavailability and pharmacokinetic evaluation	Anti-hyperlipidemia	[103]
Tacrolimus	Self micro-emulsifying drug delivery system of tacrolimus: Formulation, *in vitro* evaluation and stability studies	Immunosuppressant	[104]
Tacrolimus	Rice germ oil as multifunctional excipient in preparation of self-micro emulsifying drug delivery system (SMEDDS) of Tacrolimus	Macrolide antibiotic	[105]
Sulpiride	Intestinal permeability studies of sulpiride incorporated into self-micro-emulsifying drug delivery system (SMEDDS)	Antipsychotic	[106]
Lutein	Novel self-nano-emulsifying drug delivery system for enhanced solubility and dissolution of Lutein	Age related macular degeneration	[107]
Valsartan	Preparation and bioavailability assessment of SMEDDS containing Valsartan	Angiotensin-II antagonist	[108]

6.13 PATENTS (TABLE 6.7)

TABLE 6.7 Patents Granted Since 2001 to 2019

Patent number	Publication Date	Patent Title	References
US20190060300	02/28/2019	Self-Emulsifying Compositions of CB2 Receptor Modulators	[109]
US20190275006	09/12/2019	Self-emulsifying pharmaceutical compositions of hydrophilic drugs and preparation thereof	[110]
US10172838	01/08/2019	Self-emulsifying formulation of CARP-1 functional mimetics	[111]
US10220025	03/05/2019	Self-emulsifying formulation of CARP-1 functional mimetics	[112]
US10441569	10/15/2019	Self-emulsifying formulations of DIM-related indoles	[113]
WO/2018/011808A1	01/18/2018	Self-emulsifying compositions of cannabinoids	[114]
US20180036233	02/08/2018	Self-emulsifying drug delivery (SEDDS) for ophthalmic drug delivery	[115]
US9918965	03/20/2018	Self-emulsifying formulations of DIM-related indoles	[116]
WO/2017/211909A1	12/14/2017	Self-emulsifying lipid compositions	[117]
EP3045165A2	07/20/2016	Self-micro-emulsifying oral pharmaceutical composition of hydrophilic drug and preparation method thereof	[118]
WO/2016/141098A1	09/09/2016	Self-emulsifying drug delivery system (SEDDS) for ophthalmic drug delivery	[119]
US20150320864	11/12/2015	Self-micro-emulsifying drug delivery system with increased bioavailability	[120]
US20150196537	07/16/2015	Self-emulsifying pharmaceutical compositions of hydrophilic drugs and preparation thereof	[121]
WO/2015/142307A1	09/24/2015	Self-micro/nano-emulsifying drug carrying system for oral use of Rosuvastatin	[122]
WO/2015/022454A1	02/19/2015	Novel self-emulsifying instant solid system made from cyclodextrins and oil(s) for oral administration	[123]
US8728518	05/20/2014	Butylphthalide self-emulsifying drug delivery system, its preparation and method and application	[124]
US20130317117	11/28/2013	Self-micro-emulsifying drug delivery system with increased bioavailability	[125]
EP2425818A1	03/07/2012	Self-micro-emulsifying oral pharmaceutical composition of hydrophilic drug and preparation method thereof	[126]

TABLE 6.7   *(Continued)*

Patent number	Publication Date	Patent Title	References
US20110294900	12/01/2011	Self-emulsifying drug delivery system for a curcuminoid based composition	[127]
EP2425818A4	10/24/2012	Self-micro-emulsifying oral pharmaceutical composition of hydrophilic drug and preparation method thereof	[128]
WO/2010/010431A1	01/28/2010	Self-nano-emulsifying curcuminoids composition with enhanced bioavailability	[129]
WO/2008/142090A1	11/27/2008	Self-emulsifying formulation of Tipranavir for oral administration	[130]
US20080319056	12/25/2008	Butylphthalide self-emulsifying drug delivery system, its preparation method and application	[131]
WO/2008/142090A1	11/27/2008	Self-emulsifying formulation of Tipranavir for oral administration	[130]
EP1480636B1	04/18/2007	Self-emulsifying drug delivery systems for taxoids	[132]
EP1787638A1	05/23/2007	Butylbenzene phthalein self-emulsifying drug delivery system, its preparation method and application	[133]
US20070104740	05/10/2007	Self-micro-emulsifying drug delivery systems of a HIV protease inhibitor	[134]
EP1498143A1	01/19/2005	Self-emulsifying and self-microemulsifying formulations for the oral administration of toxoids	[135]
US20050232952	10/20/2005	Self-emulsifying drug delivery systems for poorly soluble drugs	[136]
WO/2005/063209A1	07/14/2005	Self-micro-emulsifying drug delivery systems of a HIV protease inhibitor	[137]
EP1340497A1	09/03/2003	Self-emulsifying drug delivery systems for poorly soluble drugs	[138]
WO/2003/074027A2	09/12/2003	Self-emulsifying drug delivery systems for poorly soluble drugs	[139]
DE10026698A1	12/06/2001	Self-emulsifying drug formulation and use of this formulation	[140]
US20010036962	11/01/2001	Self-emulsifying systems containing anticancer medicament	[141]
US20010025046	09/27/2001	Self-emulsifying systems containing anticancer medicament	[141]
US20010036962	11/01/2001	Self-emulsifying systems containing anticancer medicament	[141]

6.14 MARKETED PRODUCTS (TABLE 6.8)

TABLE 6.8 Marketed Products of SEDDS

Brand Name	Compound	Dosage Form	Company Name	Indication
Sandimmune®	Cyclosporine A/I	Soft gelatine capsule	Novartis Pharmaceuticals Corporation	Immunosuppressant (organ prophylaxis)
Neora®	Cyclosporine	Soft gelatine capsule	Novartis Pharmaceuticals Corporation	Systemic immunosuppressant
Gengraf®	Cyclosporine A/III	Hard gelatine capsule	AbbVie Inc.	Systemic immunosuppressant
Norvir®	Ritonavir	Soft gelatine capsule	AbbVie Inc.	Antiretroviral
Fortovase®	Saquinavir	Soft gelatine capsule	Roche Laboratories Inc.	Antiretroviral
Agenerase®	Amprenavir	Soft gelatine capsule	GlaxoSmithKline	Antiretroviral
Depakene®	Valproic acid	Soft gelatine capsule	AbbVie Inc.	Partial seizures
Rocaltrol®	Calcitriol	Soft gelatine capsule	Roche Products Limited	Hyperparathyroidism and hypocalcemia
Targretin®	Bexarotene	Soft gelatine capsule	Ligand Pharmaceuticals/ Eisai Ltd.	T-cell lymphoma
Vesanoid®	Tretinoin	Soft gelatine capsule	Roche Laboratories Inc.	Acne and acute promyelocytic leukemia
Accutane®	Isotretinoin	Soft gelatine capsule	Roche Laboratories Inc.	Severe recalcitrant nodular acne
Aptivus®	Tipranavir	Soft gelatine capsule	Boehringer Ingelheim Pharmaceuticals, Inc.	Antiretroviral
Convulex®	Vaproic acid	Soft gelatine capsule	Pharmacia	Antiepileptic
Lipirex®	Fenofibrate	Hard gelatine capsule	Genus	Anti-hyperlipoproteinemic

6.15 CONCLUSION

Conventionally, drugs which prove to be clinically effective are often diffi-cult to convert into pragmatic formulations, owing to their poor aqueous solubility or low permeability which lead to lower therapeutic response (causing multiple dose regimen; also may lead to toxicity). Therefore, converting a drug into such a formulation which would not only reduce the dosing frequency, but also reduce the dose with maximum efficacy is remains a challenge for formulation scientists. Despite having promising effects there still remain some challenges. In order to avoid such hurdles, a low energy-self emulsifying system is introduced which has again put the focus back on the importance of nanoemulsions. The lack of scale-up in large numbers for numerous drugs is because of the lack of established models for *in vivo* studies and incomplete IVIVC data. Newer methodologies, however, are in the pipelines for resolving these issues.

KEYWORDS

- *in vitro-in vivo* correlation
- lipid-based drug delivery system
- lipid formulation classification system
- mean residence time
- nano-structured lipid carriers
- self-double nano-emulsifying systems

REFERENCES

1. Kengar, M. D., Howal, R. S., Aundhakar, D. B., Nikam, A. V., & Hasabe, P. S., (2019). Physico-chemical properties of solid drugs: A review. *Asian Journal of Pharmacy and Technology, 9*(1), 53–59.
2. Kesarwani, P., Rastogi, S., Bhalla, V., & Arora, V., (2014). Solubility enhancement of poorly water-soluble drugs: A review. *International Journal of Pharmaceutical Sciences and Research, 5*(8), 3123.
3. Nemane, D. D., Anantwar, S. P., & Bhagat, S. S., (2019). Review on: Conventional and novel techniques for solubility enhancement. *European Journal of Biomedical, 6*(2), 202–208.
4. Asadujjaman, M., & Mishuk, A. U., (2013). Novel approaches in lipid-based drug delivery systems. *Journal of Drug Delivery and Therapeutics, 3*(4), 124–130.
5. Mathias, N. R., & Hussain, M. A., (2010). Non-invasive systemic drug delivery: Developability considerations for alternate routes of administration. *Journal of Pharmaceutical Sciences, 99*(1), 1–20.
6. Chien, Y., (1991). *Novel Drug Delivery Systems*. CRC Press.
7. Reddy, L. H. & Murthy, R. S., (2002). Floating dosage systems in drug delivery. *Crit Rev Ther Drug Carrier Syst, 19*(6), 553–585.
8. Senior, J. R., &; Isselbacher, K. J., (1962). Direct esterification of monoglycerides with palmityl coenzyme A by intestinal epithelial subcellular fractions. *J Biol Chem, 237*(5), 1454–1459.
9. Rahman, A., Harwansh, R., Mirza, A., Hussain, S., & Hussain, A., (2011). Oral lipid-based drug delivery system (LBDDS): Formulation, characterization and application: A review. *Current Drug Delivery, 8*(4), 330–345.

10. Jaiswal, M., Dudhe, R., & Sharma, P., (2015). Nanoemulsion: An advanced mode of drug delivery system. *3 Biotech*, *5*(2), 123–127.

11. Shrestha, H., Bala, R., & Arora, S., (2014). Lipid-based drug delivery systems. *Journal of Pharmaceutics*, *2014*.

12. Makadia, H. A., Bhatt, A. Y., Parmar, R. B., Paun, J. S., & Tank, H., (2013). Self-nano emulsifying drug delivery system (SNEDDS): Future aspects. *Asian Journal of Pharmaceutical Research*, *3*(1), 21–27.

13. Cherniakov, I., Domb, A. J., & Hoffman, A., (2015). Self-nano-emulsifying drug delivery systems: An update of the biopharmaceutical aspects. *Expert Opinion on Drug Delivery*, *12*(7), 1121–1133.

14. Rehman, F. U., Shah, K. U., Shah, S. U., Khan, I. U., Khan, G. M., & Khan, A., (2017). Fromnanoemulsionss to self-nanoemulsions, with recent advances in self-nanoemulsifying drug delivery systems (SNEDDS). *Expert Opinion on Drug Delivery*, *14*(11), 1325–1340.

15. Tang, J. L., Sun, J., & He, Z. G., (2007). Self-emulsifying drug delivery systems: Strategy for improving oral delivery of poorly soluble drugs. *Current Drug Therapy*, *2*(1), 85–93.

16. Tang, B., Cheng, G., Gu, J. C., & Xu, C. H., (2008). Development of solid self-emulsifying drug delivery systems: Preparation techniques and dosage forms. *Drug Discovery Today*, *13*(13/14), 606–612.

17. Nazzal, S., Smalyukh, I., Lavrentovich, O., & Khan, M. A., (2002). Preparation and *in vitro* characterization of a eutectic based semisolid self-nanoemulsified drug delivery system (SNEDDS) of ubiquinone: Mechanism and progress of emulsion formation. *International Journal of Pharmaceutics*, *235*(1/2), 247–265.

18. Gursoy, R. N., & Benita, S., (2004). Self-emulsifying drug delivery systems (SEDDS) for improved oral delivery of lipophilic drugs. *Biomedicine and Pharmacotherapy*, *58*(3), 173–182.

19. Shanmugam, S., Park, J. H., Kim, K. S., Piao, Z. Z., Yong, C. S., Choi, H. G., & Woo, J. S., (2011). Enhanced bioavailability and retinal accumulation of lutein from self-emulsifying phospholipid suspension (SEPS). *Int J Pharm*, *412*(1–2), 99–105.

20. Singh, B., Bandopadhyay, S., Kapil, R., Singh, R., & Katare, O. P., (2009). Self-emulsifying drug delivery systems (SEDDS): Formulation development, characterization, and applications. *Critical Reviews™ in Therapeutic Drug Carrier Systems*, *26*(5).

21. Dokania, S., & Joshi, A. K., (2015). Self-micro emulsifying drug delivery system (SMEDDS): Challenges and road ahead. *Drug Delivery*, *22*(6), 675–690.

22. Yadav, S. K., Parvez, N., & Sharma, P. K., (2014). An insight to self-emulsifying drug delivery systems, their applications and importance in novel drug delivery. *JSIR*, *3*(2), 273–281.

23. Kyatanwar, A. U., Jadhav, K. R., & Kadam, V. J., (2010). Self micro-emulsifying drug delivery system (SMEDDS). *Journal of Pharmacy Research*, *3*(1), 75–83.

24. Kawakami, K., Yoshikawa, T., Moroto, Y., Kanaoka, E., Takahashi, K., Nishihara, Y., & Masuda, K., (2002). Microemulsion formulation for enhanced absorption of poorly soluble drugs: I. Prescription design. *Journal of Controlled Release*, *81*(1/2), 65–74.

25. Shah, P., Bhalodia, D., & Shelat, P., (2010). Nanoemulsion: A pharmaceutical review. *Systematic Reviews in Pharmacy*, *1*(1).

26. Jain, S., Kambam, S., Thanki, K., & Jain, A. K., (2015). Cyclosporine A loaded self-nanoemulsifying drug delivery system (SNEDDS): Implication of a functional excipient-based co-encapsulation strategy on oral bioavailability and nephrotoxicity. *RSC Advances, 5*(61), 49633–49642.

27. Ruan, J., Liu, J., Zhu, D., Gong, T., Yang, F., Hao, X., & Zhang, Z., (2010). Preparation and evaluation of self-nanoemulsified drug delivery systems (SNEDDSs) of matrine based on drug–phospholipid complex technique. *International Journal of Pharmaceutics, 386*(1/2), 282–290.

28. Gupta, R., Gupta, R., & Singh, R., (2009). Enhancement of oral bioavailability of lipophilic drugs from self-micro emulsifying drug delivery systems (SMEDDS). *Int. J. Drug Dev. Res., 1*, 10–18.

29. Saritha, D., Bose, P. S. C., & Nagaraju, R., (2014). Formulation and evaluation of self-emulsifying drug delivery system (SEDDS) of Ibuprofen. *Int. J. Pharm. Sci. Res., 5*, 3511–3519.

30. Alexander, A., (2012). A review on novel therapeutic strategies for the enhancement of solubility for hydrophobic drugs through lipid and surfactant based self-micro emulsifying drug delivery system: A novel approach. *Am. J. Drug Disc. Develop, 2*(4), 143–183.

31. Wadhwa, J., Nair, A., & Kumria, R., (2011). Self-emulsifying therapeutic system: A potential approach for delivery of lipophilic drugs. *Brazilian Journal of Pharmaceutical Sciences, 47*(3), 447–465.

32. Li, R., Lim, S. J., Choi, H. G., & Lee, M. K., (2010). Solid lipid nanoparticles as drug delivery system for water-insoluble drugs. *Journal of Pharmaceutical Investigation, 40*, 63–73.

33. Kohli, K., Chopra, S., Dhar, D., Arora, S., & Khar, R. K., (2010). Self-emulsifying drug delivery systems: An approach to enhance oral bioavailability. *Drug Discovery Today, 15*(21–22), 958–965.

34. Zanchetta, B., Chaud, M., & Santana, M., (2015). Self-emulsifying drug delivery systems (SEDDS) in pharmaceutical development. *J. Adv. Chem. Eng., 5*, 1–7.

35. Betageri, G. V., (2019). Self-emulsifying drug delivery systems and their marketed products: A review. *Asian Journal of Pharmaceutics (AJP): Free Full Text Articles from Asian J. Pharm., 13*(02).

36. Khan, B. A., Bakhsh, S., Khan, H., Mahmood, T., & Rasul, A., (2012). Basics of self-micro emulsifying drug delivery system. *Journal of Pharmacy and Alternative Medicine, 1*(1), 13–19.

37. Singh, M. A., (2015). Self-emulsifying systems: A review. *Asian Journal of Pharmaceutics (AJP): Free Full Text Articles from Asian J. Pharm., 9*(1), 13–18.

38. Date, A. A., Desai, N., Dixit, R., & Nagarsenker, M., (2010). Self-nanoemulsifying drug delivery systems: Formulation insights, applications and advances. *Nanomedicine, 5*(10), 1595–1616.

39. Taha, E., Ghorab, D., & Zaghloul, A. A., (2007). Bioavailability assessment of vitamin A self-nanoemulsified drug delivery systems in rats: A comparative study. *Medical Principles and Practice, 16*(5), 355–359.

40. Khan, A. W., Kotta, S., Ansari, S. H., Sharma, R. K., & Ali, J., (2012). Potentials and challenges in self-nanoemulsifying drug delivery systems. *Expert Opinion on Drug Delivery, 9*(10), 1305–1317.

41. Thomas, N., Holm, R., Rades, T., & Müllertz, A., (2012). Characterizing lipid lipolysis and its implication in lipid-based formulation development. *The AAPS Journal*, *14*(4), 860–871.

42. Balasaheb, S. P., Balvant, P. S., & Kadam, A., (2017). Antipsoriatic activity of gel containing methylsulphonylmethane powder and seed oil of *Pongamia pinnata* Linn. *Research Journal of Pharmacology and Pharmacodynamics*, *9*(3), 122–130.

43. Singh, R., & Lillard, Jr. J. W., (2009). Nanoparticle-based targeted drug delivery. *Experimental and Molecular Pathology*, *86*(3), 215–223.

44. Li, L., Zhou, C. H., & Xu, Z. P., (2019). Self-nanoemulsifying drug-delivery system and solidified self-nanoemulsifying drug-delivery system. In: *Nanocarriers for Drug Delivery* (pp. 421–449). Elsevier.

45. Zhang, P., Liu, Y., Feng, N., & Xu, J., (2008). Preparation and evaluation of self-micro emulsifying drug delivery system of oridonin. *International Journal of Pharmaceutics*, *355*(1/2), 269–276.

46. Gui, S. Y., Wu, L., Peng, D., Liu, Q., Yin, B., & Shen, J. Z., (2008). Preparation and evaluation of a microemulsion for oral delivery of berberine. *Die Pharmazie-An International Journal of Pharmaceutical Sciences*, *63*(7), 516–519.

47. Huang, J., Wang, Q., Sun, R., Li, T., Xia, N., & Xia, Q., (2018). A novel solid self-emulsifying delivery system (SEDS) for the encapsulation of linseed oil and quercetin: Preparation and evaluation. *Journal of Food Engineering*, *226*, 22–30.

48. Syed, H. K., & Peh, K. K., (2014). Identification of phases of various oil, surfactant/co-surfactants and water system by ternary phase diagram. *Acta Pol. Pharm.*, *71*(2), 301–309.

49. Charman, S. A., Charman, W. N., Rogge, M. C., Wilson, T. D., Dutko, F. J., & Pouton, C. W., (1992). Self-emulsifying drug delivery systems: Formulation and biopharmaceutic evaluation of an investigational lipophilic compound. *Pharmaceutical Research*, *9*(1), 87–93.

50. Nigade, P. M., Patil, S. L., & Tiwari, S. S., (2012). Self-emulsifying drug delivery system (SEDDS): A review. *International Journal of Pharmacy and Biological Sciences*, *2*(2), 42–52.

51. Carey, M. C., Small, D. M., & Bliss, C. M., (1983). Lipid digestion and absorption. *Annual Review of Physiology*, *45*(1), 651–677.

52. Patel, D., & Sawant, K. K., (2009). Self-micro-emulsifying drug delivery system: Formulation development and biopharmaceutical evaluation of lipophilic drugs. *Current Drug Delivery*, *6*(4), 419–424.

53. Porter, C. J., & Charman, W. N., (2001). Intestinal lymphatic drug transport: An update. *Advanced Drug Delivery Reviews*, *50*(1/2), 61–80.

54. Pouton, C. W., (1997). Formulation of self-emulsifying drug delivery systems. *Advanced Drug Delivery Reviews*, *25*(1), 47–58.

55. Gursoy, N., Garrigue, J. S., Razafindratsita, A., Lambert, G., Benita, S., & Bloom, D. R., (2003). Excipient effects on *in vitro* cytotoxicity of a novel paclitaxel self-emulsifying drug delivery system. *Journal of Pharmaceutical Sciences*, *92*(12), 2411–2418.

56. Rios, G., Pazos, C., & Coca, J., (1998). Destabilization of cutting oil emulsions using inorganic salts as coagulants. *Colloids and Surfaces A: Physicochemical and Engineering Aspects*, *138*(2/3), 383–389.

57. Clogston, J. D., & Patri, A. K., (2011). Zeta potential measurement. In: *Characterization of Nanoparticles Intended for Drug Delivery* (pp. 63–70). Springer.

58. Hunter, R. J., (2013). *Zeta Potential in Colloid Science: Principles and Applications* (Vol. 2) Academic Press.

59. Sagar, K. S., (2015). A review-self nanoemulsifying drug delivery system (SNEDDS). *International Journal of Research in Pharmaceutical and Nano Sciences, 4*(6), 385–397.

60. Gupta, S., Kesarla, R., & Omri, A., (2013). Formulation strategies to improve the bioavailability of poorly absorbed drugs with special emphasis on self-emulsifying systems. *ISRN Pharmaceutics, 2013*.

61. Qi, X., Wang, L., Zhu, J., Hu, Z., & Zhang, J., (2011). Self-double-emulsifying drug delivery system (SDEDDS): A new way for oral delivery of drugs with high solubility and low permeability. *International Journal of Pharmaceutics, 409*(1/2), 245–251.

62. Dünnhaupt, S., Kammona, O., Waldner, C., Kiparissides, C., & Bernkop-Schnürch, A., (2015). Nano-carrier systems: Strategies to overcome the mucus gel barrier. *European Journal of Pharmaceutics and Biopharmaceutics, 96*, 447–453.

63. Thomas, N., Holm, R., Müllertz, A., & Rades, T., (2012). *In vitro* and *in vivo* performance of novel supersaturated self-nanoemulsifying drug delivery systems (super-SNEDDS). *Journal of Controlled Release, 160*(1), 25–32.

64. Singh, G., & Pai, R. S., (2016). *In vitro* and *in vivo* performance of super saturable self-nanoemulsifying system of trans-resveratrol. *Artificial Cells, Nanomedicine, and Biotechnology, 44*(2), 510–516.

65. Bandyopadhyay, S., Katare, O., & Singh, B., (2014). Development of optimized supersaturable self-nanoemulsifying systems of ezetimibe: Effect of polymers and efflux transporters. *Expert Opinion on Drug Delivery, 11*(4), 479–492.

66. Thomas, N., Holm, R., Garmer, M., Karlsson, J. J., Müllertz, A., & Rades, T., (2013). Supersaturated self-nanoemulsifying drug delivery systems (Super-SNEDDS) enhance the bioavailability of the poorly water-soluble drug simvastatin in dogs. *The AAPS Journal, 15*(1), 219–227.

67. Zhang, X., Yi, Y., Qi, J., Lu, Y., Tian, Z., Xie, Y., Yuan, H., & Wu, W., (2013). Controlled release of cyclosporine A self-nanoemulsifying systems from osmotic pump tablets: Near zero-order release and pharmacokinetics in dogs. *International Journal of Pharmaceutics, 452*(1, 2), 233–240.

68. Miao, Y., Chen, G., Ren, L., & Pingkai, O., (2016). Characterization and evaluation of self-nanoemulsifying sustained-release pellet formulation of ziprasidone with enhanced bioavailability and no food effect. *Drug Delivery, 23*(7), 2163–2172.

69. Patil, P. R., Biradar, S. V., & Paradkar, A. R., (2009). Extended release felodipine self-nanoemulsifying system. *AAPS PharmSciTech., 10*(2), 515–523.

70. Karamanidou, T., Karidi, K., Bourganis, V., Kontonikola, K., Kammona, O., & Kiparissides, C., (2015). Effective incorporation of insulin in mucus permeating self-nanoemulsifying drug delivery systems. *European Journal of Pharmaceutics and Biopharmaceutics, 97*, 223–229.

71. Mahmood, A., Prüfert, F., Efiana, N. A., Ashraf, M. I., Hermann, M., Hussain, S., & Bernkop-Schnürch, A., (2016). Cell-penetrating self-nanoemulsifying drug delivery systems (SNEDDS) for oral gene delivery. *Expert Opinion on Drug Delivery, 13*(11), 1503–1512.

72. Toorisaka, E., Hashida, M., Kamiya, N., Ono, H., Kokazu, Y., & Goto, M., (2005). An enteric-coated dry emulsion formulation for oral insulin delivery. *Journal of Controlled Release, 107*(1), 91–96.

73. Yi, T., Wan, J., Xu, H., & Yang, X., (2008). A new solid self-micro emulsifying formulation prepared by spray-drying to improve the oral bioavailability of poorly water-soluble drugs. *European Journal of Pharmaceutics and Biopharmaceutics, 70*(2), 439–444.

74. Katteboina, S., Chandrasekhar, P., & Balaji, S., (2009). Approaches for the development of solid self-emulsifying drug delivery systems and dosage forms. *Asian J. Pharm. Sci, 4*(4), 240–253.

75. Bagul, N., Patel, V., Shahiwala, A., & Misra, M., (2012). Design and development of solid self-emulsifying osmotic delivery system of nifedipine. *Archives of Pharmacy Practice, 3*(2), 128.

76. Serratoni, M., Newton, M., Booth, S., & Clarke, A., (2007). Controlled drug release from pellets containing water-insoluble drugs dissolved in a self-emulsifying system. *European Journal of Pharmaceutics and Biopharmaceutics, 65*(1), 94–98.

77. Chae, G. S., Lee, J. S., Kim, S. H., Seo, K. S., Kim, M. S., Lee, H. B., & Khang, G., (2005). Enhancement of the stability of BCNU using self-emulsifying drug delivery systems (SEDDS) and *in vitro* antitumor activity of self-emulsified BCNU-loaded PLGA wafer. *International Journal of Pharmaceutics, 301*(1/2), 6–14.

78. Yang, S. G., (2010). Biowaiver extension potential and IVIVC for BCS Class II drugs by formulation design: Case study for cyclosporine self-micro emulsifying formulation. *Archives of Pharmacal Research, 33*(11), 1835–1842.

79. Dai, W. G., Dong, L. C., Shi, X., Nguyen, J., Evans, J., Xu, Y., & Creasey, A. A., (2007). Evaluation of drug precipitation of solubility-enhancing liquid formulations using milligram quantities of a new molecular entity (NME). *Journal of Pharmaceutical Sciences, 96*(11), 2957–2969.

80. Dressman, J. B., & Reppas, C., (2000). *In vitro-in vivo* correlations for lipophilic, poorly water-soluble drugs. *European journal of Pharmaceutical Sciences, 11*, S73–S80.

81. Donato, E. M., Martins, L. A., Fröehlich, P. E., & Bergold, A. M., (2008). Development and validation of dissolution test for lopinavir, a poorly water-soluble drug, in soft gel capsules, based on *in vivo* data. *Journal of Pharmaceutical and Biomedical Analysis, 47*(3), 547–552.

82. Kollipara, S., & Gandhi, R. K., (2014). Pharmacokinetic aspects and *in vitro-in vivo* correlation potential for lipid-based formulations. *Acta Pharmaceutica Sinica B., 4*(5), 333–349.

83. Caliph, S. M., Charman, W. N., & Porter, C. J., (2000). Effect of short-, medium-, and long-chain fatty acid-based vehicles on the absolute oral bioavailability and intestinal lymphatic transport of halofantrine and assessment of mass balance in lymph-cannulated and non-cannulated rats. *Journal of Pharmaceutical Sciences, 89*(8), 1073–1084.

84. Fatouros, D. G., Nielsen, F. S., Douroumis, D., Hadjileontiadis, L. J., & Mullertz, A., (2008). *In vitro-in vivo* correlations of self-emulsifying drug delivery systems combining the dynamic lipolysis model and neuro-fuzzy networks. *European Journal of Pharmaceutics and Biopharmaceutics, 69*(3), 887–898.

85. Cuiné, J. F., Charman, W. N., Pouton, C. W., Edwards, G. A., & Porter, C. J., (2007). Increasing the proportional content of surfactant (Cremophor EL) relative to lipid in self-emulsifying lipid-based formulations of danazol reduces oral bioavailability in beagle dogs. *Pharmaceutical Research, 24*(4), 748–757.

86. Akhtartavan, S., Karimi, M., Karimian, K., Azarpira, N., Khatami, M., & Heli, H., (2019). Evaluation of a self-nanoemulsifying docetaxel delivery system. *Biomedicine and Pharmacotherapy, 109,* 2427–2433.

87. Abd-Elhakeem, E., Teaima, M. H., Abdelbary, G. A., & El Mahrouk, G. M., (2019). Bioavailability enhanced clopidogrel-loaded solid SNEDDS: Development and *in vitro/in vivo* characterization. *Journal of Drug Delivery Science and Technology, 49,* 603–614.

88. Kumar, S. S., Sankar, D. G., Biswal, S., Kumar, B. P., & Chandra, S. P., (2019). Fabrication and evaluation of self-nanoemulsifying oil formulations (SNEOFs) of Efavirenz. *Journal of Dispersion Science and Technology, 40*(3), 464–475.

89. Dou, Y. X., Zhou, J. T., Wang, T. T., Huang, Y. F., Chen, V. P., Xie, Y. L., Lin, Z. X., et al., (2018). Self-nanoemulsifying drug delivery system of bruceine D: A new approach for anti-ulcerative colitis. *International Journal of Nanomedicine, 13,* 5887.

90. Kim, D. H., Kim, J. Y., Kim, R. M., Maharjan, P., Ji, Y. G., Jang, D. J., Min, K. A., et al., (2018). Orlistat-loaded solid sNeDDs for the enhanced solubility, dissolution, and *in vivo* performance. *International Journal of Nanomedicine, 13,* 7095.

91. Naseef, M. A., Ibrahim, H. K., Nour, S. A. E. K., (2018). Solid form of lipid-based self-nanoemulsifying drug delivery systems for minimization of diacerein adverse effects: Development and bioequivalence evaluation in albino rabbits. *AAPS PharmSciTech., 19*(7), 3097–3109.

92. Rajesh, S. Y., Singh, S. K., Pandey, N. K., Sharma, P., Bawa, P., Kumar, B., Gulati, M., et al., (2018). Impact of various solid carriers and spray drying on pre/post compression properties of solid SNEDDS loaded with glimepiride: *In vitro-ex vivo* evaluation and cytotoxicity assessment. *Drug Development and Industrial Pharmacy, 44*(7), 1056–1069.

93. Rao, M. R., Raut, S. P., Shirsath, C., & Jadhav, M. B., (2018). Self-nanoemulsifying drug delivery system of mebendazole for treatment of lymphatic filariasis. *Indian Journal of Pharmaceutical Sciences, 80*(6), 1057–1068.

94. Shahba, A. A. W., Ahmed, A. R., Alanazi, F. K., Mohsin, K., & Abdel-Rahman, S. I., (2018). Multi-layer self-nanoemulsifying pellets: An innovative drug delivery system for the poorly water-soluble drug cinnarizine. *AAPS PharmSciTech., 19*(5), 2087–2102.

95. Ali, H. H., & Hussein, A. A., (2017). Oral solid self-nanoemulsifying drug delivery systems of candesartan citexetil: Formulation, characterization and *in vitro* drug release studies. *AAPS Open, 3*(1), 6.

96. Garg, V., Kaur, P., Singh, S. K., Kumar, B., Bawa, P., Gulati, M., & Yadav, A. K., (2017). Solid self-nanoemulsifying drug delivery systems for oral delivery of polypeptide-k: Formulation, optimization, *in vitro* and *in vivo* antidiabetic evaluation. *European Journal of Pharmaceutical Sciences, 109,* 297–315.

97. Mustapha, O., Kim, K. S., Shafique, S., Kim, D. S., Jin, S. G., Seo, Y. G., Youn, Y. S., et al., (2017). Development of novel cilostazol-loaded solid SNEDDS using a SPG membrane emulsification technique: Physicochemical characterization and *in vivo* evaluation. *Colloids and Surfaces B: Biointerfaces, 150,* 216–222.

98. Dash, R. N., Mohammed, H., Humaira, T., & Ramesh, D., (2015a). Design, optimization and evaluation of glipizide solid self-nanoemulsifying drug delivery for enhanced solubility and dissolution. *Saudi Pharmaceutical Journal, 23*(5), 528–540.

99. Dash, R. N., Mohammed, H., Humaira, T., & Reddy, A. V., (2015b). Solid supersaturatable self-nanoemulsifying drug delivery systems for improved dissolution,

absorption and pharmacodynamic effects of glipizide. *Journal of Drug Delivery Science and Technology, 28*, 28–36.

100. Khan, A. W., Kotta, S., Ansari, S. H., Sharma, R. K., & Ali, J., (2015). Self-nanoemulsifying drug delivery system (SNEDDS) of the poorly water-soluble grapefruit flavonoid Naringenin: Design, characterization, *in vitro* and *in vivo* evaluation. *Drug Delivery, 22*(4), 552–561.

101. Rao, B. C., Vidyadhara, S., Sasidhar, R. L., & Chowdary, Y. A., (2015). Formulation and evaluation of liquid loaded tablets containing docetaxel-self nano emulsifying drug delivery systems. *Tropical Journal of Pharmaceutical Research, 14*(4), 567–573.

102. Seo, Y. G., Kim, D. W., Yousaf, A. M., Park, J. H., Chang, P. S., Baek, H. H., Lim, S. J., et al., (2015). Solid self-nanoemulsifying drug delivery system (SNEDDS) for enhanced oral bioavailability of poorly water-soluble tacrolimus: Physicochemical characterization and pharmacokinetics. *Journal of Microencapsulation, 32*(5), 503–510.

103. Balakumar, K., Raghavan, C. V., & Abdu, S., (2013). Self-nanoemulsifying drug delivery system (SNEDDS) of rosuvastatin calcium: Design, formulation, bioavailability and pharmacokinetic evaluation. *Colloids and Surfaces B: Biointerfaces, 112*, 337–343.

104. Patel, P. V., Patel, H. K., Panchal, S. S., & Mehta, T. A., (2013). Self-micro-emulsifying drug delivery system of tacrolimus: Formulation, *in vitro* evaluation and stability studies. *International Journal of Pharmaceutical Investigation, 3*(2), 95.

105. Pawar, S. K., & Vavia, P. R., (2012). Rice germ oil as multifunctional excipient in preparation of self-micro emulsifying drug delivery system (SMEDDS) of tacrolimus. *Aaps Pharmscitech, 13*(1), 254–261.

106. Chitneni, M., Peh, K. K., Darwis, Y., Abdulkarim, M., Abdullah, G. Z., & Qureshi, M. J., (2011). Intestinal permeability studies of sulpiride incorporated into self-micro emulsifying drug delivery system (SMEDDS). *Pak. J. Pharm. Sci, 24*(2), 113–121.

107. Yoo, J. H., Shanmugam, S., Thapa, P., Lee, E. S., Balakrishnan, P., Baskaran, R., Yoon, S. K., et al., (2010). Novel self-nanoemulsifying drug delivery system for enhanced solubility and dissolution of lutein. *Archives of Pharmacal Research, 33*(3), 417–426.

108. Dixit, A. R., Rajput, S. J., & Patel, S. G., (2010). Preparation and bioavailability assessment of SMEDDS containing valsartan. *AAPS PharmSciTech., 11*(1), 314–321.

109. Anavi-Goffer, S., (2017). *Self-emulsifying compositions of cb2 receptor modulators.* WO2017149392A1.

110. Hsu, C. S., Hao, W. H., Wang, J. J., & Lin, T. H., (2010). Innopharmax Inc, *Self-emulsifying pharmaceutical compositions of hydrophilic drugs and preparation thereof.* U.S. Patent Application 12/767,293.

111. Sachdeva, M., Patel, K., & Rishi, A., (2019). Florida Agricultural and Mechanical University, *Self-emulsifying formulation of CARP-1 functional mimetics.* U.S. Patent 10,220,025.

112. Sachdeva, M., Patel, K., & Rishi, A., (2019). Florida Agricultural and Mechanical University FAMU, *Self-emulsifying formulation of CARP-1 functional mimetics.* U.S. Patent 10,172,838.

113. Zeligs, M. A., & Jacobs, I. C., (2019). BioResponse LLC, *Self-emulsifying formulations of DIM-related indoles.* U.S. Patent 10,441,569.

114. Friedman, D. A. S., & Carme-Yosef, (2018). Self-Emulsifying Compositions of Cannabinoids. WO2018011808A1

115. Shabaik, Y., Jiao, J., & Pujara, C., (2018). Allergan Inc, *Self-emulsifying drug delivery (SEDDS) for ophthalmic drug delivery.* U.S. Patent Application 15/554,983.

116. Zeligs, M. A., & Jacobs, I. C., (2018). BioResponse LLC, *Self-emulsifying formulations of DIM-related indoles*. U.S. Patent 9,918,965.

117. Derrieu, G. L., Mazzola, D. G., & Mazzola, G., (2017). Self-Emulsifying Lipid Compositions. WO2017211909A1

118. Hsu, C. S., Hao, W. H., Wang, J. J., & Lin, T. H., (2012). Self micro-emulsifying oral pharmaceutical composition of hydrophilic drug and preparation method thereof. EP2425818A4

119. Shabaik, Y. S. H. S., Jiao, J., & Pujara, C. P., (2016). Self-Emulsifying Drug Delivery System (SEDDS) for Ophthalmic Drug Delivery. WO2016141098A1

120. Hassan, E., (2017). Pharmaceutics International Inc, *Self micro-emulsifying drug delivery system with increased bioavailability*. U.S. Patent 9,770,509.

121. Hsu, C. S., Hao, W. H., Wang, J. J., & Lin, T. H., (2015). Innopharmax Inc, *Self-Emulsifying Pharmaceutical Compositions of Hydrophilic Drugs and Preparation Thereof*. U.S. Patent Application 14/669,233.

122. Karasulu, H. Y., Apaydin, S., Gundogdu, E., Yildirim Simsir, I., Turk, U. O., Karasulu, E., Yilmaz, C., & Turgay, T., (2015). Selfmicro/nanoemulsifying drug carrying system for oral use of rosuvastatin. *WO2015142307*.

123. Skiba, M., (2015). Novel Self-Emulsifying Instant Solid System Made from Cyclodextrins and Oil(S) for Oral Administration. WO2015022454A1

124. Liu, Z., Yang, L., Yang, H., Gao, Y., Shen, D., Guo, W., Feng, X., & Zheng, J., (2014). Cspc Zhongqi Pharmaceutical Technology Shijiazhuang Co Ltd, *Butylphthalide self-emulsifying drug delivery system, its preparation and method and application*. U.S. Patent 8,728,518.

125. Hassan, E., (2013). Pharmaceutics International Inc, *Self micro-emulsifying drug delivery system with increased bioavailability*. U.S. Patent Application 13/989,687.

126. Hsu, C. S., Hao, W. H., Wang, J. J., & Lin, T. H., (2012). Self micro-emulsifying oral pharmaceutical composition of hydrophilic drug and preparation method thereof. EP2425818A1

127. Kohli, K., Chopra, S., Arora, S., Khar, R. K., & Pillai, K. K., (2014). Arbro Pharmaceuticals Ltd and Jamia Hamdard (Hamdard University), *Self emulsifying drug delivery system for a curcuminoid based composition*. U.S. Patent 8,835,509.

128. Self-Micro-Emulsifying Oral Pharmaceutical Composition of Hydrophilic Drug and Preparation Method Thereof. US20100273730A1.

129. Bansal, A. K., Munjal, B., & Patel, S. B., (2010). National Institute of Pharmaceutical Education, *Self-nano-emulsifying curcuminoids composition with enhanced bioavailability*. WO2010010431A1

130. Abelaira, S., Becher, M. P., Gel, J. F., Villagra, M. F. and de Vidal, M. N. C., (2010). Boehringer Ingelheim International GmbH, *Self-emulsifying formulation of tipranavir for oral administration*. U.S. Patent Application 12/600,689.

131. Zhentao, L., Liying, Y., Hanyu, Y., Yuqing, G., Dongmin, S., Wenmin, G., Xiaolong, F., & Jia, Z., (2008). Butylphthalide self-emulsifying drug delivery system, its preparation method and application. *US Patents Application: 20080319056*.

132. Lambert, G., Razafindratsita, A., Garrigue, J. S., Yang, S. D. I. P. P., Gursoy, N., & Benita, S., (2007). Self emulsifying drug delivery systems for taxoids. *EP1480636B1*.

133. Butylbenzene Phthalein Self-Emulsifying Drug Delivery System, its Preparation Method and Application. EP1787638

134. Voorspoels, J. F., (2007). Voorspoels Jody Firmin M, *Self-microemulsifying drug delivery systems of a HIV protease inhibitor*. U.S. Patent Application 10/596,738.

135. Peracchia, M. T., Cote, S., & Gaudel, G., (2005). Aventis Pharma SA, *Self-emulsifying and self-microemulsifying formulations for the oral administration of taxoids*. U.S. Patent Application 10/894,333.

136. Lambert, G., Razafindratsita, A., Garrigue, J. S., Yang, S., Gursoy, N., Benita, S., & Novagali S. A., (2005). Yissum Research Development Company of Hebrew University of Jerusalem, *Self emulsifying drug delivery systems for poorly soluble drugs*. U.S. Patent Application 10/507,857.

137. Voorspoels, J. F., (2007). Voorspoels Jody Firmin M, *Self-microemulsifying drug delivery systems of a HIV protease inhibitor*. U.S. Patent Application 10/596,738.

138. Benita, S., Garrigue, J. S., Gursoy, N., Lambert, G., Razafindratsita, A., & Yang, S., (2003). Self-emulsifying drug delivery systems for poorly soluble drugs. *WO2003074027.*

139. Benita, S., Garrigue, J. S., Gursoy, N., Lambert, G., Razafindratsita, A., & Yang, S., (2003). Self-emulsifying drug delivery systems for poorly soluble drugs. *EP1480636 A2.*

140. Berndl, G., Breitenbach, J., Heger, R., Stadler, M., Wilke, P., & Rosenberg, J., (2013). AbbVie Deutschland GmbH and Co KG, *Self-emulsifying active substance formulation and use of this formulation*. U.S. Patent 8,470,347.

141. Liu, R. R., & Wang, Z., (2001). Abbott Laboratories, *Self-emulsifying systems containing anticancer medicament*. U.S. Patent 6,316,497.

CHAPTER 7

Solid Lipid Nanoparticles: An Overview of Production Techniques and Applications

DEEPIKA SHARMA,[1] RAJESH KUMAR,[1] RASHI MATHUR,[2]
DEEP SHIKHA SHARMA,[1] MANGESH PRADEEP KULKARNI,[1]
CHANDAN BHOGENDRA JHA,[2] GURVINDER SINGH,[1] and
PARDEEP KUMAR[1]

[1]*School of Pharmaceutical Sciences, Lovely Professional University, Phagwara, Punjab, India, E-mail: rajksach09@gmail.com (R. Kumar)*

[2]*Division of Cyclotron and Radiopharmaceutical Sciences, Institute of Nuclear Medicine and Allied Science, DRDO, New Delhi, India*

ABSTRACT

Lipids have been a matter of great interest for pharmaceuticals scientists over the last two decades especially for the effective delivery of drugs and bioactive components. Solid lipid-based nanoparticles (NPs), i.e., solid lipid nanoparticles (SLNs) are colloidal carriers which comprise lipids stabilized with a suitable surfactant. SLNs constitute a hydrophobic solid matrix core coated with a phospholipid which enables them to accommodate hydrophobic drugs in the core with greater entrapment efficiency (EE) in comparison to liposomes. SLNs have combined advantages of emulsions, liposomes, and polymeric NPs put together. Owing to their favorable technological and biological characteristics, SLNs have been able to deliver many therapeutic substances with improved efficacy in terms of drug release, site-specificity, and stability. This chapter attempts to review the methods of preparations, evaluation, and applications of SLNs in pharmaceutical and cosmetic products. In addition, the patents granted on lipid-based NPs and marketed cosmetic products thereof have been listed.

7.1 INTRODUCTION

Solid lipid nanoparticles (SLNs) came into existence in 1991. SLNs are alternate drug carriers to earlier available carriers like liposomes, niosomes, nano capsules, micelles, emulsions, and liquid crystals [1, 2]. These are novel nano-colloidal drug carrier system having sub-micron size (50–1000 nm) [2] and are made up of solid lipid mono-, di- or triglycerides, steroids, waxes, and fatty acids [1, 3] emulsifier, cryoprotectant, charge modifiers, steal thing agents, targeting ligand, and preservatives [3, 4]. SLNs are emulsions wherein the liquid lipid (oil) is replaced with a solid lipid and dispersed in aqueous solution having surfactant [5]. Owing to their technological and biological unique properties, SLNs have improved the performance of pharmaceuticals and nutraceuticals [6].

7.1.1   TECHNOLOGICAL PROPERTIES OF SLNS

SLNs have some remarkable technological characteristics viz. nano-size, narrow size distribution, more surface area, more drug loading ratio, protection of drugs from chemical and enzymatic degradation, amphililic nature, possibility of sterilization, avoidance of organic solvent and capability of delivering two active drugs simultaneously (co-delivery).

7.1.2   BIOLOGICAL PROPERTIES OF SLNS

Unique biological properties like biodegradable nature, biocompatible, site specific targeting, controlled drug release, capable of crossing of biological barriers, reduced dosing frequency, less side effects of drugs, increased bioavailability, possibility of administration via different routes (oral, parenteral, transdermal) make them a smart drug carrier [3, 6].

7.1.3   LIMITATIONS OF SLNS

Apart from their technological and biological advantages, SLNs have a few limitations as well under same set of headings:

1. **Technological Limitations:** Drug leakage after polymorphic transition during storage, agglomeration, and more polydispersity are some of the technological limitations of SLNs.
2. **Biological Limitations:** Initial bursting of loaded drug, relatively low circulation time [6].

7.2 MODELS OF DRUG LOADING TO SLNS

There are mainly three models for drug loading in SLNs as shown in Figure 7.1. These models are dependent upon the composition and method of preparation of formulation.

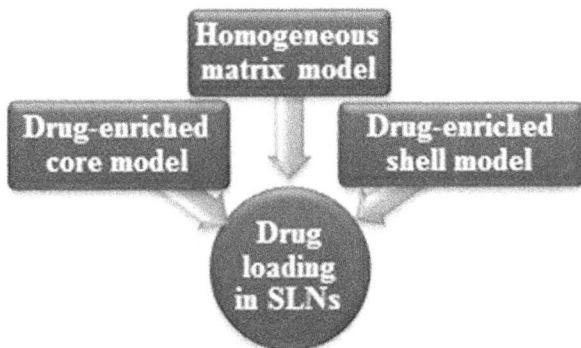

FIGURE 7.1   Different models of SLNs.

7.2.1 HOMOGENEOUS MATRIX MODEL

In this type of model, incorporation of lipophilic drugs can be done in SLNs by using cold or hot homogenization. During hot homogenization, when it is cooled the oil doesn't crystallize and there is no separation of phases between drug and lipid. In cold homogenization, the lipid has the drug in dissolved or dispersed form. High-pressure homogenization (HPH) leads to mechanical breaking of particles to nano size to form homogeneous matrix.

7.2.2 DRUG-ENRICHED CORE MODEL

In this model, precipitation of drug occurs first, Fick's law of diffusion is followed by SLN membrane, and controlled release is achieved.

7.2.3 DRUG-ENRICHED SHELL MODEL

The drug is available near the shell yielding a drug free lipid core. Separation of phases occurs during cooling of SLNs in this model and lipid gets precipitated out giving drug-free lipid core [7].

7.3   MECHANISM OF DRUG RELEASE FROM SLN

Biphasic release of drugs (initial burst release followed by prolonged release) has been observed in SLNs. Burst release has been seen in SLNs prepared by hot homogenization and at high temperature particularly. The solubility of drug increases in aqueous phase when high concentration of surfactant/high temperature is used and it lead to burst effect, whereas, if low concentration of surfactant/lower temperature is used, a little or no burst effect is seen [7].

7.4   METHODS OF PREPARATION

The main components of SLNs are lipid, emulsifier, and solvent/water. The different methods of preparation of SLNs are shown in Table 7.1.

TABLE 7.1   Methods of Preparation for SLNs

SL. No.	Preparation Methods
	High pressure homogenization:
1.	• Hot homogenization
	• Cold homogenization
2.	Solvent emulsification/evaporation method.
3.	Ultrasonication and high shear homogenization.
	Super critical fluid method:
4.	• Supper critical anti-solvent process;
	• Rapid expansion of super critical solution.
5.	Microemulsion technique.
6.	Melt-dispersion technique.
7.	Precipitation method.

7.4.1   HIGH PRESSURE HOMOGENIZATION (HPH)

This is highly applicable and consistent technique used for preparation of SLNs [8]. Through the narrow gap (ranges from few microns), the content is pushed under high pressure (100–2000 bar). At a velocity of 1000 km/h, the fluid is accelerated through a small orifice which results in breakdown of particles to sub-micron size as a result of shear stress and forces. Normally 5–10% lipid content is used but can be raised up to 40% [9]. This technique is further divided into two techniques-hot homogenization and cold homogenization [1]. These techniques are applied depending upon the lipid used, release pattern required, and stability of lipid during preparation [10].

7.4.1.1 HOT HOMOGENIZATION

Thermo stable drugs are used in this technique because of the use of heat. The process is divided into two phases [11]:

- melted lipid; and
- hot aqueous surfactant mixture.

Depending upon the nature of drug, lipophilic drug is dissolved in lipid or hydrophilic drug is dissolved in aqueous surfactant phase.

Pre-emulsion is formed by dissolving the drug in melted lipid (phase A) or hot aqueous surfactant solution (phase B). Phase A is then poured into B at same temperature under high shear mixing device for 8 minutes with rpm 2000–20,000 [2, 10]. This heated pre-emulsion is poured into high-pressure homogenizer for 3–5 cycles at 500–1500 bar which is maintained at temperature near to the melting point of lipid [12, 13]. The formed preparation is cooled at room temperature to form SLNs. Increased temperature or homogenization may also lead to particles with larger sizes (particles coalescence) due to high kinetic energy of particles or degradation of drug or the whole system [13].

7.4.1.2 COLD HOMOGENIZATION

In this technique, high-pressure milling of suspension is carried out. The thermal exposure of drug is minimized in cold technique, however, it doesn't avoid completely, since in the initial step, mixture of drug and lipid is melted [14, 15]. The cold homogenization is superior to hot homogenization because of the following reasons [1]:

- Better suited for thermolabile drugs and avoids thermal degradation of drug;
- Better drug distribution in aqueous phase during homogenization;
- Issues with hot homogenization, like drug entrapment and complex crystallization, can be overcome.

Drug is dissolved/dispensed in the melted lipid, and then it is solidified in liquid nitrogen or dry ice. Second step is involves grinding in ball mill or mortal mill to obtain the particles in micron size (50–100 μm). In a per-mix of aqueous surfactant (chilled emulsifier solution), these micro particles are dispensed, and pre-suspension is formed. This pre-suspension is passed through high-pressure homogenizer at/below the room temperature to obtained SLNs. Larger particle size and broader particle size distribution is observed in cold homogenization as compared to hot homogenization. This method consumes high energy and biomolecules get damaged at lab scale which is a major limitation of this method [10].

7.4.2   SOLVENT EMULSIFICATION/EVAPORATION METHOD

In organic solvent (cyclohexane, ether, etc.), the lipophilic drug is dissolved (A) [16]. The other phase is aqueous phase (B). The system A is added to system B slowly which leads to precipitation of the lipid. Now, this is subjected to HPH to attain nanosized SLNs [17]. To get rid of organic solvent, Rota evaporator under reduced pressure (40–60 bar) is used. The stability of emulsion is maintained by using emulsifiers (like polysorbates) and co-emulsifiers (e.g., poloxamers) [16]. This method is suitable for commercial and scale, however, it is associated with some demerits like high-energy consumption that may lead to biomolecules degradation [2, 10].

7.4.3   ULTRASONICATION AND HIGH SHEAR HOMOGENIZATION

Two methods can be used either ultra-sonication or high shear homogenization to obtain nano sized SLNs. This is only applicable for 500 mL to 1000 mL preparations. This technique is simple, easy to prepare and gives better results for less viscous preparations. Chances of metal contamination, difficulty in scale up and physically instable SLNs may result in crystal growths upon storage are major limitations of this technique [10, 15].

7.4.4   SUPER CRITICAL FLUID METHOD

There are two techniques under this method [18]:

1. **Supper Critical Anti-Solvent (SAS) Process:** Solute is dissolved in organic solvent, which is sprayed in an anti-solvent, though the organic solvent is miscible with anti-solvent, mass transfer takes place and solid particles are formed.

2. **Rapid Expansion of Super Critical Solution (RESS):** This is two-step process. The super critical fluid is saturated with substrate, and then it is passed through the nozzle. Hence decrease in temperature, leads to precipitation [19]. Carbon dioxide is used as super critical fluid mainly. Some merits of this method include collection of particles as dry powder, less temperature, and pressure stress [10].

7.4.5 MICROEMULSION TECHNIQUE

The basis of this method is the dilution of microemulsion [20]. These emulsions are of two phasic system containing inner and outer phase (e.g., o/w or w/o micro emulsion). Low melting point fatty acid (e.g., stearic acid), an emulsifier (e.g., polysorbate 20), co-emulsifiers (e.g., sodium monooctylphosphate), and water are mixed together at 60–70°C. Then, this hot micro emulsion is dispersed in cold water (2–3°C) with continuous stirring [10]. Volume ratio of hot emulsion to cold water is 1:25 to 1:50 [21, 22]. The size of SLNs formed depends upon the aqueous phase used; if water miscible solvent is used, nano-sized particles are formed, whereas, water immiscible solvent results in larger sized particles [23]. Lipid content achieved is less as compared to HPH because of dilution in former [1].

7.4.6 MELT DISPERSION TECHNIQUE

The lipid is melted (70–80°C) in organic solvent with continuous magnetic stirring. The drug is dissolved in aqueous phase and maintained at (70–80C). Same temperature for both the phases should be maintained. Now the organic phase is added drop wise into the aqueous phase under continuous stirring at 1500–2500 rpm for few hours. The system is cooled to room temperature to obtain SLNs [10, 24].

7.4.7 PRECIPITATION METHOD

Lipids are dissolved in organic phase (e.g., acetone) by sonicating for 5–10 minutes. Due to high volatility of organic phase, its evaporation is prevented by covering it properly. This pre-formed mixture is added drop wise using a syringe into aqueous phase containing surfactant and co-surfactant. The whole process takes place under continuous stirring 500–1500 rpm up to 4 hours for evaporation of organic solvent [19]. Then, it is transferred to Rota-rod evaporator under reduced pressure at 60°C for complete removal of traces of organic solvent [10].

7.5   DIFFERENT ROUTES OF ADMINISTRATION AND *IN VIVO* FATE

In vivo fate depends upon the administration route, distribution of SLNs (adsorption of biological material on the particle surface and desorption of SLN components into to biological surroundings) and the enzymatic processes. Lipase is the most important enzyme for the degradation of SLNs. Lipases split the ester linkage and form partial glycerides or glycerol and free fatty acids. Most lipases require activation by an oil/water interface, which opens the catalytic center (lid opening) [25]. Enzymatic test is performed to check the SLNs degradation (free fatty acids are measured). It was observed that longer the fatty chain in triglycerides, slower the degradation [26]. Surfactant can also increase or hinder the degradation. This is based on type of surfactant used and steric stabilization caused by it [27]. Adjusting this will help in formulating control release SLNs. SLNs prepared using waxes have slow rate of degradation as compared to SLNs prepared using triglycerides [28]. Various route of administration for SLNs are discussed in subsections [29].

7.5.1   PARENTERAL ROUTE

Proteins and peptides are administered through this route to prevent the degradation due to enzymes in GIT. Erythropoietin-SLNs were prepared and on comparison with erythropoietin alone, it was seen that erythropoietin-SLNs were non-toxic and exhibited better *in vivo* effect. Sustained release was seen in erythropoietin-SLNs with reciprocal powered time model [30].

7.5.2   ORAL ROUTE

Oral antimicrobial preparations have overcome the challenges like antimicrobial resistance, adverse effects, and the high cost of antimicrobial preparations. Anti-microbial SLNs were able to prevent the drug from the harsh environment of GIT, increase the solubility and bioavailability consequently. SLNs were also able to target the drug to specific sites [31]. Furosemide-silver complex (Ag-FSE) had broad-spectrum antibiotic activity but low solubility and bioavailability. Ag-FSE, when formulated into SLNs using hot homogenization technique, 2 and 4 fold increase in the activity of drug was noticed against *Pseudomonas aeruginosa* and *Staphylococcus aureus*, respectively [32].

7.5.3 RECTAL ROUTE

This route is mainly used for pediatric patients [2]. Bioavailability of diazepam was observed from aqueous-organic solution, submicron emulsion, and SLNs. From the study, it was concluded that submicron emulsion was comparatively a better choice for an ethanol-free drug formulation through rectal route [33]. In another study, suppositories containing diazepam-loaded SLN dispersions were used as a potential drug carrier and the prepared formulation was able to exhibit an extended *in vitro* drug release which was significantly greater than diazepam suppositories [34].

7.5.4 NASAL ROUTE

Donepezil loaded SLNs were prepared using solvent emulsification diffusion technique for delivery of the drug to brain through intra-nasal route. Pharmacokinetic and brain targeting studies (rats) revealed a significantly high drug concentration in the brain upon intra-nasal administration of donepezil loaded SLNs in comparison to donepezil solution [35].

7.5.5 OCULAR ROUTE

SLNs prepared for ocular route have many advantages like ease of sterilization by autoclaving, increasing the corneal absorption of drugs and boot ability of hydrophilic as well as lipophilic drugs [36]. The preocular retention of SLN in rabbit eyes was tested using drug-free, fluorescent SLNs (F-SLN). These were retained for longer times on the corneal surface and in the conjunctival sac when compared with an aqueous fluorescent solution. A suspension of tobramycin loaded SLN containing 0.3% w/v tobramycin was administered topically to rabbits, and the aqueous humor concentration of drug was determined up to six hours. When compared with an equal dose of tobramycin administered by standard commercial eye drops, SLN produced a significantly higher drug bioavailability in the aqueous humor [37].

Natamycin have poor corneal penetration due to which its efficacy against corneal keratitis is very poor. Natamycin-SLNs presented a promising ocular delivery system for treatment of deep corneal keratitis with sustained release and increased corneal penetration [38].

7.5.6 TOPICAL APPLICATION

Combining the nano carriers and physical penetration technologies enhances the drug penetration [39]. Tetracycline SLNs were compared to nanostructured lipid carriers (NLCs) and polymeric nanoparticle. From this study it was observed that tetracycline NLCs exhibited the most prominent *in vivo* efficiency in improving the skin permeation, analgesic time, and pain control intensity. Other experiments proved that tetracycline-PLA nanoparticles (NPs) showed advantages in serum stability and tetracycline-SLNs illustrated the best *in vitro* permeation efficiency. These three kinds of nano-systems had their own superiority in some aspects [40].

7.6 EVALUATION OF SLNS

After preparing SLNs, they need to be evaluated for particle size, shape, poly dispersity index (PDI), drug loading, entrapment efficiency (EE) and the release kinetic of drug incorporated therein. The evaluation of SLN formulations has been discussed in general as well as some specific evaluation parameters pertaining to specific dosage forms have been discussed separately.

7.6.1 GENERAL EVALUATION PARAMETERS

7.6.1.1 SURFACE CHARACTERIZATIONS: PARTICLE SIZE AND PDI

Size of SLNs significantly affects the stability and release pattern of drug from SLNs [2]. The particle size, its distribution, and PDI can be determined using dynamic light scattering (DLS) (also known as photon correlation spectroscopy or PCS) and laser diffraction at a fixed angle of 90° and at 25°C temperatures [1, 41–43]. DLS measures the differences in the intensity of scattered light which is caused due to the particle kinetics. This technique covers a size range from a few nanometers to about three microns sized particles. It doesn't depict the particle size, rather, it detects the light scattering which can be used to determine particle size; however, uncertainty is possible due to non-spherical particle shapes. Advantages of DLS include ease of sample preparation speed of analysis and sensitivity to submicron particles [1].

7.6.1.2 ELECTROPHORETIC MOBILITY

It can be determined by diluting SLNs with 0.1 Mm KCl and placing it in the electrophoretic cell of nanosizer, where electric field is applied. The test is conducted in triplicated and zeta potential is determined to interpret the outcomes as shown in Table 7.2 [41]. If both positive and negative charges are present on SLNs, it may lead to their coalescence [44].

TABLE 7.2 Zeta Potential Value and its Significance

Value	Significance	References
Zeta potential greater than ±30 mV	Aggregation is avoided due to particle-particle repulsion and Electrostatic stabilization	[1, 44]
Zeta potential ~0 mV	With the hydrophilic coating, agglomeration can be prevented and stearic stabilization	[1, 44]

7.6.1.3 SURFACE MORPHOLOGY

It can be studied by using techniques scanning electron microscopy (SEM), transmission electron microscopy (TEM), and atomic force microscopy (AFM). SEM gives 3D images of particles, whereas, TEM microscopy depicts the size, shape, and internal structure [45]. In AFM, interactions of forces between the tip and the sample surface are measured. The experiment is conducted in a non-contact mode means the space is maintained between the tip and sample surface of 10–100 Å. Temperature should be approximately 20°C with 760 mm of Hg atmospheric pressure. The samples are diluted in water (1:10v/v), prior to analyzes for getting less sticky fluid. Constant volume of SLN drops are poured on a small mica disk (diameter 1 cm) and excess of water is removed using filter paper after 2 min [41].

7.6.1.4 POWDER X-RAY DIFFRACTION (PXRD) AND DIFFERENTIAL SCANNING CALORIMETRY (DSC)

The degree of crystallinity can be assessed by the geometric scattering of radiation from crystal planes within a solid that allows the determination of presence or absence of the former. The nature of crystallinity can be determined using DSC through the determination of melting and glass point temperature [46, 47]. Thermo-analytical technique (DSC) is based on enthalpy of lipid. It determines the degree of crystallinity of lipid. Sharp peak represents crystalline compound, whereas, blunt peak represents amorphous compound [48, 49]. Powder X-ray diffraction (PXRD) can also determine the crystal nature of SLNs [49].

7.6.1.5 ACOUSTIC METHOD

It is another spectroscopy technique to determine the size and surface charge. Charges are introduced to the particles with the aid of acoustic energy, as a result of movement and generation of oscillating electric field [44]. Thus electric, field is used to determine the surface charge information [50].

7.6.2   EVALUATION OF SLNS FOR OCULAR DELIVERY

7.6.2.1 CORNEAL PERMEABILITY STUDIES

Through marginal ear vein, an IV injection of air was administered, and rabbits were sacrificed thereafter. Immediately cornea was excised and mounted on Franz's diffusion cell. It was maintained at 35 ± 1°C with a continuous stirring at 200 rpm. Baicalin SLNs (BA-SLN) and Baicalin solution (BA-sol) were pre-heated at 35°C, added to phosphate buffer pH 6.5, and placed in donor compartment. Proper oxygenation and agitation were maintained with O_2 and CO_2 (95:5) bubbled through each compartment at a rate of 3–4 bubbles/second. Samples were withdrawn from sampling port after every 40 minutes for 4 hours and fresh medium was poured to maintain the sink condition. This was repeated for three times. Below given equation was used to determine the corneal permeability coefficient (P_{app}) [51, 52]:

$$Papp = \frac{\Delta Q}{\Delta T \times Co \times A \times 60cm \,/\, second} \tag{1}$$

where; $\Delta Q/\Delta t$ is the steady-state slope of the linear portion of the plot (plotted between the amount of drug in the receiving chamber(Q) versus time (t)); A is the area of exposed corneal surface (0.5 mm²) and C_0 is the initial concentration of drug in the donor cell and 60 represents the conversion of minutes to seconds.

7.6.2.2 CORNEAL HYDRATION LEVEL

The cornea was detached from the scleral ring and weighed (w_a). The excess water on corneal ring was removed with a filter paper and then it was desiccated for 16 hrs at 60°C. The corneal sample was then reweighed to determine dry sample weight (w_b). Percentage corneal hydration was found out for both untreated corneas as well as for the cornea recovered from the permeation studies using the following equation [52–54].

$$HL\% = 1 - \left(\frac{wa}{wb}\right) \times 100 \tag{2}$$

7.6.2.3   OCULAR IRRITATION

Draize technique was used for this [55]. Baicalin-SLNs were instilled into the left eye and blank SLNs were instilled into right eye of rabbit (0.4 mL every 4 hour), four times a day for 7 days. Ocular tissue conditions were monitored 4, 12, 24, 48, and 72 hours after the end of the last installation [52, 56]. Using, Draize technique, ocular tissues were evaluated. Pathological sections were also studied.

7.6.3   EVALUATION OF TOPICAL SLNS

7.6.3.1   SPREDABILITY

One half gm of gel was weighed and placed on a glass plate which was pre-marked with diameter of 1 cm. Another glass plate was placed over it and 500 gm weight was put on the second glass plate for 5 minutes. Increase in diameter was noted [57].

7.6.3.2   RHEOLOGICAL STUDIES

Brookfield viscometer was used to perform rheological study. Thirty gm gel was weighed, placed in beaker, and waited for 5 minutes to allow its uniform distribution. The viscosity was measured using spindle at different speed like 0.5, 1, 2.5, and 5 rpm by noting down the dial readings [58].

7.6.3.3   SKIN IRRITATION STUDIES

Draize patch test was conducted. Rabbits back was clipped free of hair. Formulation was applied on the skin and spread properly. Control and SLNs without drug (blank) were also studied. All the rabbits were evaluated and noted for any redness or itching or rashes present [59].

7.6.4   EVALUATION OF NASAL SLNS

7.6.4.1   HISTOLOGICAL STUDIES

Sheep nasal mucosa was separated 1 hour before the study and washed with isotonic solution. The prepared SLN formulation was applied to it and fixed in 10% neutral bicarbonate buffered formalin solution and embedded in paraffin. The experiment was carried out in cell culture incubator and optimal conditions were provided for the viability of tissue. On glass slide, paraffin blocks (7 μm) were cut and stained with eosin and hematoxylin, and observed under microscope for any damage to the tissue [60, 61].

7.6.4.2   RADIOLABELING

With direct labeling, Technetium 99 m (^{99m}Tc) was radiolabeled on ondansetron-SLNs suspension. Stannous chloride dihydrate was added to 0.6 ml of SLNs suspension. Using 50 mM sodium bicarbonate solution, the pH was adjusted to 6.5. After the addition of sterile ^{99m}Tc-pertechnetate along with continuous stirring, the mixture was incubated at 30 ± 5°C for 30 min. Volume was made up to 1 ml with 0.9% (w/v) sodium chloride. The purity was evaluated with the help of ascending instant thin layer chromatography [60–63].

7.6.4.3   GAMMA SCINTIGRAPHY IMAGING

^{99m}Tc ondansetron-SLNs were given to rabbits in each nostril. Rabbits were slanted backward during the nasal administration. Rabbits were conscious and kept in rabbit restrainer's throughout the scintigraphy study. Images were obtained by photoemission computerized tomography [60, 61, 64].

7.7   STORAGE AND STABILITY OF SLNS

SLNs were stored at different temperature and it was observed that at different temperature conditions, different stability was noted out of which 4°C was most favorable storage temperature. At 20°C, long-term storage didn't result in drug-loaded SLN aggregation or loss of drug and at 50°C, rapid growth of particle size was observed [2, 19, 65]. Physical stability depends on the size of SLNs [66].

7.8 APPLICATIONS OF SLNS

7.8.1 BRAIN TARGETING

Blood-brain barrier (BBB) is the obstacle for drugs to go across in brain. A list of SLNs based formulations to take the drug across BBB is given in Table 7.3.

TABLE 7.3 SLN Formulation for Brain Targeting of Various Drugs

SL. No.	Drug	Observations	References
1.	Morin	• Enhanced rate and extend of drug absorbed via amplified accessibility to the enterocyte surface by reduced particle size and increased permeation through the intestinal membrane, due to sustained release properties	[77]
2.	Risperidone	• Biodistribution of risperidone SLNs was nearly 10 folds higher	[78]
3.	Resveratrol	• Treatment of tumor in brain	[79]
4.	Quercetin	• Treatment of Alzheimer's disease • Better memory retention	[80]
5.	Nevirapine	• Treatment of HIV	[81]
6.	Nutlin-3a and superparamagnetic nanoparticles	• Treatment of glioblastoma	[82]
7.	Docetaxel	• Treatment of glioblastoma multiforme	[83]
8.	Almotriptan malate	• Anti-migraine	[84]
9.	Baicalein	• Inhibition of apoptosis and a reduction in lactate dehydrogenase release • Anti-depressant effect	[85]
10.	Clonazepam	• Anticonvulsant • Antiepileptic	[86]
11.	Borneol	• Higher cellular uptake • Better blood brain barrier penetration	[87]
12.	Agomelatine	• Antidepressant activity • Prolonged intranasal retention time	[88]
13.	Rivastigmine hydrogen tartrate	• Treatment of Alzheimer's disease • sustain release of rivastigmine	[89]
14.	Olanzapine	• Treatment of acute phase schizophrenia	[90, 91]
15.	Efavirenz	• Reducing HIV • Increased permeability	[92]
16.	Methotrexate	• Treatment of glioblastoma	[93]

TABLE 7.3   SLN Formulation for Brain Targeting of Various Drugs

SL. No.	Drug	Observations	References
17.	Dexanabinol and curcumin	• Increased dopamine release	[94, 95]
18.	Rizatriptan benzoate	• Efficient management of migraine	[96]
19.	Quetiapine fumarate	• Treatment of schizophrenia	[97]
20.	Haloperidol	• Antipsychotic activity	[98]
21.	Sumatriptan succinate	• Management of migraine	[99]
22.	Rosmarinic acid	• Management of Huntington's disease	[100]
23.	Alprazolam	• Anti-depressant • Reduced the dose and dosing frequency	[101]
24.	Docetaxel and ketoconazole	• Anti-tumor activity	[102]

Risperidone-loaded SLNs were studied against risperidone I.V. and it was observed that brain/blood ratio of risperidone-loaded SLNs was 1.36 ± 0.06 as compared to 0.17 ± 0.05 from risperidone I.V. administration. The biodistribution of risperidone SLNs was nearly 10 folds higher [78].

Glycerylbehenate-based SLNs for resveratrol showed higher concentration in brain (17.28 ± 0.6344 µg/g) in comparison to free resveratrol (3.45 ± 0.3961 µg/g) and hence, to treat tumor in brain, it can be targeted by glycerylbehenate-based SLNs of resveratrol [79].

Quercetin (a potent natural antioxidant) loaded SLNs exhibited better memory retention in the treatment of Alzheimer's disease as compared to free quercetin [80].

7.8.2 CHEMOTHERAPY

Several obstacles like normal tissue toxicity, poor stability, poor specificity, and drug resistance of tumor cells were overcome by using SLNs [103]. Using emulsion solvent diffusion technique, oral SLNs of rifampicin, isoniazid, and pyrazinamide were formulated. The encapsulation efficacy was $51 \pm 5\%$ for rifampicin, $45 \pm 4\%$ for isoniazid and $41 \pm 4\%$ for pyrazinamide. It was observed that residence time for SLNs preparation was for 8 days in lungs whereas free drug was cleared within 1–2 days after administration. After 5 oral doses, the tubercle bacilli were removed and antitubercular SLNs were able to reduce the dosing frequency of drugs [104].

Docetaxel-loaded hepatoma-targeted solid lipid nanoparticles (tSLNs) were designed and prepared with galactosylated dioleoylphosphatidyl ethanolamine. Different studies were conducted and amplified accumulation of drug in tumor and higher cellular uptake by hepatoma cells were observed with no detrimental effects noticed on healthy liver. All these observations proved that tSLN enhanced the antitumor activity with little systemic toxicity for curing locally advanced and metastatic human hepatocellular carcinoma (HCC) [105].

7.8.3   LIVER TARGETING (TABLE 7.4)

TABLE 7.4   SLN Formulation for Drug Targeting to Liver

SL. No.	Drug	Observations	References
1.	Sorafenib	• Treatment of hepatocellular carcinoma	[106]
		• Better pharmacokinetic performance	
2.	Arsenic trioxide	• Treatment of hepatic carcinoma	[107]
		• Enhanced cellular uptake	
3.	Resveratrol	• Treated liver cirrhosis	[108]
4.	Silymarin	• Increased bioavailability	[109]
		• Increased peak plasma concentration	
5.	Aclacinomycin A	• Antineoplastic	[110]
6.	Quercetin	• Treated hepatocellular carcinoma	[111]
7.	Gemcitabine	• Anti-neoplastic	[112]
8.	Gene delivery	• Targeted cancer cells	[113]
		• Targeted macrophages in liver	
9.	Cucurbitacin B	• Inhibited tumor growth	[114]
		• Increased cellular uptake	

7.8.4   CARDIOVASCULAR DISEASES (TABLE 7.5)

TABLE 7.5   SLN-Based Formulations for Cardiovascular Diseases

SL. No.	Drug	Observations	References
1.	Puerarin	• To treat alcoholism • To treat myocardial ischemia	[115, 116]
2.	Paclitaxel	• Reducing atherosclerosis lesions	[117]
3.	Schisandrin B	• Treatment of myocardial infraction • Longer blood circulation time	[118]
4.	Carmustine	• Treatment of atherosclerosis lesions	[119]

7.8.5   UTERUS AND OVARY TARGETING (TABLE 7.6)

TABLE 7.6   SLN-Based Formulations for Drug Targeting to Uterus and Ovary

SL. No.	Drug	Observations	References
1.	Tamoxifen	• Increased expression of UO-44 gene	[120]
		• Treating solid tumor tissues	
2.	Chitosan	• Increased expression of vascular endothelial growth factor receptor-1	[121]
3.	β-element	• Treating ovary cancer	[122]

7.8.6   OCULAR TARGETING (TABLE 7.7)

TABLE 7.7   SLN-Based Formulations for Ocular Drug Targeting

SL. No.	Drug	Activity	References
1.	Brimonidine	• Treatment of ocular hypertension • Increased residence time and corneal penetration	[123]
2.	Tobramycin	• Active against phagocytosed Pseudomonas aeruginosa	[124]
3.	Human antigen-R	• Treatment of diabetic retinopathy	[125]

7.8.7   APPLICATION IN FOOD INDUSTRY

Bioactive lipophilic compounds like carotenoids, tocopherols, polyphenols, phytophenols, and liposoluble vitamins, all differ in their molecular properties due to which their physicochemical properties also differ. By formulating these compounds in form of SLNs, their transportation, solubility, protection, and absorption have been improved [67–69]. Carotenoids impart red and yellow color in food products and are also used as immunomodulator and reduce the risk of contracting chronic degenerative diseases such as cancer, cardiovascular disease, cataracts, and age-related macular degeneration. Dispersion of carotenoids is prone to degradation. Carotenoids have low solubility and hence low bioavailability as well. These disadvantages have been overcome by SLN-carotenoids [70–73].

The stability was found to be greatly increased in SLN formulation of alpha-tocopherol [74]. Formulating omega-3-fatty acid with high melting point soybean lecithin into SLNs overcome the problems of omega-3 fatty acid oxidation [75] which otherwise results in imparting the unpleasant flavor in food (Tables 7.8 and 7.9) [76].

TABLE 7.8 A List of Patents Granted on SLNs

SL. No.	Patent No.	Title	References
1.	US5989583A	Solid lipid compositions of lipophilic compounds for enhanced oral bioavailability	[126]
2.	WO/2000/006120A1	Lipid emulsion and solid lipid nanoparticle as a gene or drug carrier	[127–130]
3.	DE19952410A1	Solid lipid nanoparticles containing a UV filter material are useful in aqueous dispersion form as high filter content sunscreen compositions	[131]
4.	EP1378231A1	Formulation of UV absorbers by incorporation in solid lipid nanoparticles	[132]
5.	WO/2006/044660A2	Functionalized solid lipid nanoparticles and methods of making and using same	[133]
6.	WO/2006/128888A1	Use of solid lipid nanoparticles comprising cholesteryl propionate and/or cholesteryl butyrate	[134]
7.	US7147841B2	Formulation of UV absorbers by incorporation in solid lipid nanoparticles	[135]
8.	DE19952410B4	Sunscreen preparations comprising solid lipid nanoparticles	[136]
9.	WO/2008/149215A2	Method for the preparation of solid lipid micro and nanoparticles	[137]
10.	WO/2008/149215A3	Method for the preparation of solid lipid micro and nanoparticles	[138]
11.	WO/2009/111852A2	Solid lipid nanoparticle containing retinoids	[139]
12.	WO/2009/111852A3	Solid lipid nanoparticle containing retinoids	[139]
13.	WO/2010/112749A1	Solid lipid nanoparticles encapsulating minoxidil, and aqueous suspension containing same	[140]
14.	WO/2011/116963A2	Lipid nanoparticle capsule	[141]
15.	US8158601B2	Lipid formulation	[142]
16.	WO/2011/116963A3	Lipid nanoparticle capsule	[143]
17.	WO/2013/105101A1	Solid lipid nanoparticles entrapping hydrophilic/amphiphilic drug and a process for preparing the same	[144]
18.	EP2623097A1	Solid lipid nanoparticle including elastin-like polypeptide and use thereof	[145]
19.	US8663692B1	Lipid particles on the basis of mixtures of liquid and solid lipids and method for producing same	[146]
20.	US20140205722A1	Composition of solid lipid nanoparticles for the long-term conservation of fruits, vegetables, seeds, cereals, and/or fresh foodstuffs using a coating	[147]
21.	US8802644B2	Lipid formulation	[148]
22.	US8828542B2	Nanoparticles	[149]
23.	EP2777402A1	Solid lipid nanoparticles (I)	[150]
24.	WO/2014/140264A1	Solid lipid nanoparticles (I)	[151]
25.	WO/2014/140268A1	Solid lipid nanoparticles (II)	[152]

TABLE 7.8   *(Continued)*

SL. No.	Patent No.	Title	References
26.	WO/2014/191467A1	Fluorescent solid lipid nanoparticles composition and preparation thereof	[153]
27.	US9005666B2	Process for preparing lipid nanoparticles	[154]
28.	WO/2015/148483A1	Systems and methods for preparing solid lipid nanoparticles	[155]
29.	EP2983502A1	Solid lipid nanoparticles (I)	[156]
30.	EP2983490A1	Solid lipid nanoparticles (II)	[156]
31.	US9302241B2	Method of preparing lipid nanoparticles	[157]
32.	EP3017823A1	Lipid nanoparticle of polymyxin	[158]
33.	US20160199447A1	Lipid nanoparticles for wound healing	[159]
34.	WO/2017/041609A1	Solid lipid magnetic resonance nanoparticles and preparation method and use thereof	[160]
35.	US9616001B2	Solid lipid nanoparticles (I)	[161]
36.	US9662281B2	Solid lipid nanoparticles (II)	[162]
37.	US9675710B2	Lipid nanoparticles for gene therapy	[163]
38.	US9750819B2	Lipid nanoparticle compositions and methods of making and methods of using the same	[164]
39.	US9764043B2	Antithrombotic nanoparticle	[165]
40.	EP3272398A1	Lipid nanoparticle capsule	[166]
41.	US9907758B2	Process for preparing solid lipid sustained release nanoparticles for delivery of vitamins	[167]
42.	9919027	Lipid nanoparticle of polymyxin	[168]
43.	US9943846B2	Limit size lipid nanoparticles and related methods	[169]
44.	US10004695B2	GM3 functionalized nanoparticles	[170]
45.	US10039843B2	Paramagnetic solid lipid nanoparticles (pSLNs) containing metal amphiphilic complexes for MRI	[171]
46.	US10117942B2	Photoactivatable lipid-based nanoparticles as vehicles for dual agent delivery	[172]
47.	US10166298B2	Lipids and lipid nanoparticle formulations for delivery of nucleic acids	[173]
48.	US10166198B2	Solid nanoparticle with inorganic coating	[174]
49.	US10166187B2	Curcumin solid lipid particles and methods for their preparation and use	[175]
50.	US10195291B2	Compositions and methods for the manufacture of lipid nanoparticles	[176]
51.	US10206886B2	Lipid nanoparticles for wound healing	[159]
52.	US10251960B2	Fluorescent solid lipid nanoparticles composition and preparation thereof	[177]
53.	US10307490B2	Lipid nanoparticle compositions for antisense oligonucleotides delivery	[178]
54.	WO/2019/116062A1	Solid lipid nanoparticle for intracellular release of active substances and method for production the same	[179]

TABLE 7.9 Lipid Nanoparticles Based-Cosmetic Products Available in the Market [180]

SL. No.	Product Name	Main active ingredients	Producers/ Distributor
1.	Cutanova Cream Nanorepair Q10	Q 10, polypeptide, hibiscus extract, ginger extract, ketosugar	Dr. Rimpler GmbH Wedemark-Germany
2.	Intensive serum Nanorepair Q10	Q 10, polypeptide, mafane extract	
3.	Cutanova Cream Nanovital Q10	Q 10, TiO2, polypeptide, ursolic acid, oleanolic acid, sunflower seed extract	
4.	SURMER Crème Legère Nano-Protection	Kukuinut oil, Monoi Tiare Tahiti®, pseudopeptide, milk extract from coconut, wild indigo, noni extract	Isabelle Lancray
5.	SURMER Crème Riche Nano-Restructurante	Kukuinut oil, Monoi Tiare Tahiti®, pseudopeptide, milk extract from coconut, wild indigo, noni extract	
6.	SURMER Elixir du Beauté Nano-Vitalisant	Kukuinut oil, Monoi Tiare Tahiti®, pseudopeptide, milk extract from coconut, wild indigo, noni extract	
7.	SURMER Masque Crème Nano-Hydratant	Kukuinut oil, Monoi Tiare Tahiti®, pseudopeptide, milk extract from coconut, wild indigo, noni extract	
8.	Nanolipid Restore CLR	Black currant seed oil containing omega-3 and omega-6 unsaturated fatty acids	Chemisches Laboratorium Dr. Kurt Richter, (CLR) Berlin (Germany)
9.	Nanolipid Q10 CLR	Coenzyme Q10 and black currant seed oil	
10.	Nanolipid Basic CLR	Caprylic/capric triglycerides	
11.	Nanolipid Repair CLR	Black currant seed oil and manuka oil	
12.	IOPE super vital cream, Serum, Eye cream, Extra moist softener, Extra moist emulsion	Coenzyme Q10, omega-3 and omega-6 unsaturated fatty acids	Amore Pacific
13.	Regenerations crème Intensiv	Macadamia ternifolia seed oil, avocado oil, urea, black currant seed oil	Scholl, Mannheim, Germany
14.	Swiss Cellular White Illuminating Eye Essence	Glycoproteins, panax ginseng root extract, equisetum arvense extract, *Camellia sinensis* leaf extract, viola tricolor extract	Laboratoires La Prairie SA
15.	Swiss Cellular White Intensive Ampoules		Zurich, Switzerland
16.	SURMER Creme Contour DesYeux Nano-Remodelante	Kukuinut oil, Monoi Tiare Tahiti®, pseudopeptide, hydrolized wheet protein	Isabelle Lancray
17.	Olivenöl Anti Falten Pflegekonzentrat	*Olea europaea* oil, panthenol, acacia senegal, tocopheryl acetate	Dr. Theiss Naturwaren GmbH, Homburg (Germany)
18.	Olivenöl Augenpflegebalsam	Olea Europaea oil, prunus amygdalus dulcis oil, hydrolized milk protein, tocopheryl acetate, rhodiolarosea root extract, caffeine	

7.9 CHALLENGES IN SCALE-UP OF SOLID LIPID NANOPARTICLES (SLNS)

- **Safety Aspects:** Surfactants used must be from GRAS status, e.g., Lecithin, polaxemer 188, span 85, tween 80, sodium glycohalate [17].

- **For Parenterals:** Cellular binding of SLNs have to be observed. Depending upon surfactant used SLNs have distinct affinity towards RBCs. Using span 80 in SLNs increased the affinity towards RBCs and led towards aggregation of RBCs [27, 181].

- **Sterilization:** Parenteral and pulmonary SLNs face sterilization problem. SLNs melt during autoclaving and recrystallize during cooling. When control release SLNs is formulated, they can't be sterilized using autoclaving because they will be modulated in release rate, as SLNs are melted and recrystallized during sterilization. For lecithin stabalized SLNs autoclaving is possible [182]. SLNs formulated using sterically stabilizing polymer like poloxamer cannot be sterilized by autoclaving at 121°C as it may led to instability and particle aggregation because sterilization temperature is very close to the critical flocculation temperature of polymer. To overcome this problem polaxamer containing SLNs are sterilized at reduced temperature (110C) and prolonged duration of exposure [181, 182].

7.10 CONCLUSION AND FUTURE PERSPECTIVES

Lipids and fats are the natural ingredients which are easily undergo biodegradation by natural processes and produce biocompatible degradation products. SLNs are very well tolerated carrier systems for systemic as well as dermal application. Their production is feasible in laboratory and may be scaled up to large level since their production methods are simple and reproducible. Considering a huge potential of lipid-based carriers, there is a strong need of using novel lipids in combination with the safest surfactants and co-surfactants. Since some of the methods don't involve the use of organic solvents, an increasing interest has developed in these carriers to utilize their safety potential in cosmetic products as well as systemic drug delivery. In addition to various advantages, an increased benefit to risk ratio has been achieved with their use as drug delivery carriers. The introduction of lipid-based novel cosmetic products in the market has paved the way to launch topical pharmaceutical products in the coming times.

KEYWORDS

- atomic force microscopy
- Baicalin solution
- differential scanning calorimetry
- dynamic light scattering
- entrapment efficiency
- furosemide-silver complex
- high-pressure homogenization

REFERENCES

1. Mehnert, W., & Mäder, K., (2012). Solid lipid nanoparticles: Production, characterization, and applications. *Advanced Drug Delivery Reviews, 64,* 83–101.
2. Ekambaram, P., Sathali, A. A. H., & Priyanka, K., (2012). Solid lipid nanoparticles: A review. *Sci. Rev. Chem. Commun., 2*(1), 80–102.
3. Kaur, I. P., Bhandari, R., & Yakhmi, J. V., (2014). Lipids as biological materials for nanoparticulate delivery. In: *Handbook of Nanomaterials Properties* (pp. 409–455). Springer.
4. Lim, S. B., Banerjee, A., & Önyüksel, H., (2012). Improvement of drug safety by the use of lipid-based nanocarriers. *Journal of Controlled Release, 163*(1), 34–45.
5. Mukherjee, S., Ray, S., & Thakur, R., (2009). Solid lipid nanoparticles: A modern formulation approach in drug delivery system. *Indian Journal of Pharmaceutical Sciences, 71*(4), 349.
6. Geszke-Moritz, M., & Moritz, M., (2016). Solid lipid nanoparticles as attractive drug vehicles: Composition, properties and therapeutic strategies. *Materials Science and Engineering: C, 68,* 982–994.
7. Yadav, N., Khatak, S., & Sara, U. S., (2013). Solid lipid nanoparticles: A review. *Int. J. Appl. Pharm, 5*(2), 8–18.
8. Souto, E. B., Fangueiro, J. F., & Müller, R. H., (2013). Solid lipid nanoparticles (SLN™). In: *Fundamentals of Pharmaceutical Nanoscience* (pp. 91–116). Springer.
9. Müller, R. H., Mäder, K., & Gohla, S., (2000). Solid lipid nanoparticles (SLN) for controlled drug delivery: A review of the state of the art. *European Journal of Pharmaceutics and Biopharmaceutics, 50*(1), 161–177.
10. Newton, A. M., & Kaur, S., (2019). Solid lipid nanoparticles for skin and drug delivery: Methods of preparation and characterization techniques and applications. In: *Nanoarchitectonics in Biomedicine* (pp. 295–334). Elsevier.
11. Mishra, B., Patel, B. B., & Tiwari, S., (2010). Colloidal nanocarriers: A review on formulation technology, types and applications toward targeted drug delivery. *Nanomedicine: Nanotechnology, Biology and Medicine, 6*(1), 9–24.

12. Siekmann, B., & Westesen, K., (1994). Melt-homogenized solid lipid nanoparticles stabilized by the nonionic surfactant tyloxapol: I. Preparation and particle size determination. *Pharm. Pharmacol. Lett., 3*, 194–197.

13. Hu, F., Yuan, H., Zhang, H., & Fang, M., (2002). Preparation of solid lipid nanoparticles with clobetasol propionate by a novel solvent diffusion method in aqueous system and physicochemical characterization. *International Journal of Pharmaceutics, 239*(1/2), 121–128.

14. Venkateswarlu, V., & Manjunath, K., (2004). Preparation, characterization and *in vitro* release kinetics of clozapine solid lipid nanoparticles. *Journal of Controlled Release, 95*(3), 627–638.

15. Trotta, M., Debernardi, F., & Caputo, O., (2003). Preparation of solid lipid nanoparticles by a solvent emulsification-diffusion technique. *International Journal of Pharmaceutics, 257*(1/2), 153–160.

16. Yuan, H., Huang, L. F., Du, Y. Z., Ying, X. Y., You, J., Hu, F. Q., & Zeng, S., (2008). Solid lipid nanoparticles prepared by solvent diffusion method in a nanoreactor system. *Colloids and Surfaces B: Biointerfaces, 61*(2), 132–137.

17. Wissing, S., Kayser, O., & Müller, R., (2004). Solid lipid nanoparticles for parenteral drug delivery. *Advanced Drug Delivery Reviews, 56*(9), 1257–1272.

18. Freitas, C., & Müller, R. H., (1998). Spray-drying of solid lipid nanoparticles (SLNTM). *European Journal of Pharmaceutics and Biopharmaceutics, 46*(2), 145–151.

19. Lv, Q., Yu, A., Xi, Y., Li, H., Song, Z., Cui, J., Cao, F., & Zhai, G., (2009). Development and evaluation of penciclovir-loaded solid lipid nanoparticles for topical delivery. *International Journal of Pharmaceutics, 372*(1/2), 191–198.

20. Gasco, M. R., (1993). *Method for Producing Solid Lipid Microspheres Having a Narrow Size Distribution.* Google Patents. U.S. Patent No. 5,250,236. 5 October 1993.

21. Gasco, M., (1997). Solid lipid nanospheres from warm micro-emulsions: Improvements in SLN production for more efficient drug delivery. *Pharmaceutical Technology Europe, 9*, 52–58.

22. Boltri, L., Canal, T., Esposito, P., & Carli, F., (1993). In lipid nanoparticles: Evaluation of some critical formulation parameters. *Proc. Intern. Symp. Control Rel. Bioact. Mater.* (pp. 346, 347).

23. De Labouret, A., Thioune, O., Fessi, H., Devissaguet, J., & Puisieux, F., (1995). Application of an original process for obtaining colloidal dispersions of some coating polymers. Preparation, characterization, industrial scale-up. *Drug Development and Industrial Pharmacy, 21*(2), 229–241.

24. Cortesi, R., Esposito, E., Luca, G., & Nastruzzi, C., (2002). Production of lipospheres as carriers for bioactive compounds. *Biomaterials, 23*(11), 2283–2294.

25. Borgström, B., (1980). Importance of phospholipids, pancreatic phospholipase A2, and fatty acid for the digestion of dietary fat: *In vitro* experiments with the porcine enzymes. *Gastroenterology, 78*(5), 954–962.

26. Scow, R. O., & Olivecrona, T., (1977). Effect of albumin on products formed from chylomicron triacylglycerol by lipoprotein lipase *in vitro*. *Biochimica et Biophysica Acta (BBA)-Lipids and Lipid Metabolism, 487*(3), 472–486.

27. Olbrich, C., Kayser, O., & Müller, R. H., (2002). Lipase degradation of dynasan 114 and 116 solid lipid nanoparticles (SLN)-effect of surfactants, storage time and crystallinity. *International Journal of Pharmaceutics, 237*(1/2), 119–128.

28. Gordillo-Galeano, A., & Mora-Huertas, C. E., (2018). Solid lipid nanoparticles and nanostructured lipid carriers: A review emphasizing on particle structure and drug release. *European Journal of Pharmaceutics and Biopharmaceutics.*

29. Gupta, P. K., Pandit, J., Kumar, A., Swaroop, P., & Gupta, S., (2010). Pharmaceutical nanotechnology novel nanoemulsion-high energy emulsification preparation, evaluation and application. *The Pharma Research, 3,* 117–138.

30. Dara, T., Vatanara, A., Maybodi, M. N., Vakilinezhad, M. A., Malvajerd, S. S., Vakhshiteh, F., Shamsian, A., et al., (2019). Erythropoietin-loaded solid lipid nanoparticles: Preparation, optimization, and *in vivo* evaluation. *Colloids and Surfaces B: Biointerfaces.*

31. Raza, A., Sime, F. B., Cabot, P. J., Maqbool, F., Roberts, J. A., & Falconer, J. R., (2019). Solid nanoparticles for oral antimicrobial drug delivery: A review. *Drug Discovery Today.*

32. Bolla, P. K., Kalhapure, R. S., Rodriguez, V. A., Ramos, D. V., Dahl, A., & Renukuntla, J., (2019). Preparation of solid lipid nanoparticles of furosemide-silver complex and evaluation of antibacterial activity. *Journal of Drug Delivery Science and Technology, 49,* 6–13.

33. Sznitowska, M., Gajewska, M., Janicki, S., Radwanska, A., & Lukowski, G., (2001). Bioavailability of diazepam from aqueous-organic solution, submicron emulsion and solid lipid nanoparticles after rectal administration in rabbits. *European Journal of Pharmaceutics and Biopharmaceutics, 52*(2), 159–163.

34. Abdelbary, G., & Fahmy, R. H., (2009). Diazepam-loaded solid lipid nanoparticles: Design and characterization. *AAPS PharmSciTech., 10*(1), 211–219.

35. Yasir, M., Sara, U. V. S., Chauhan, I., Gaur, P. K., Singh, A. P., Puri, D., & Ameeduzzafar, (2018). Solid lipid nanoparticles for nose to brain delivery of donepezil: Formulation, optimization by Box-Behnken design, *in vitro* and *in vivo* evaluation. *Artificial Cells, Nanomedicine, and Biotechnology, 46*(8), 1838–1851.

36. More, S., Lokhande, S., & Raje, V., (2019). Review on solid lipid nanoparticles for ophthalmic delivery. *Research Journal of Pharmaceutical Dosage Forms and Technology, 11*(1), 19–34.

37. Cavalli, R., Gasco, M. R., Chetoni, P., Burgalassi, S., & Saettone, M. F., (2002). Solid lipid nanoparticles (SLN) as ocular delivery system for tobramycin. *International Journal of Pharmaceutics, 238*(1/2), 241–245.

38. Khames, A., Khaleel, M. A., El-Badawy, M. F., & El-Nezhawy, A. O., (2019). Natamycin solid lipid nanoparticles-sustained ocular delivery system of higher corneal penetration against deep fungal keratitis: Preparation and optimization. *International Journal of Nanomedicine, 14,* 2515.

39. Yang, M., Gu, Y., Tang, X., Wang, T., & Liu, J., (2019). Advancement of lipid-based nanocarriers and combination application with physical penetration technique. *Current Drug Delivery.*

40. Liu, X., & Zhao, Q., (2019). Long-term anesthetic analgesic effects: Comparison of tetracaine loaded polymeric nanoparticles, solid lipid nanoparticles, and nanostructured lipid carriers *in vitro* and *in vivo*. *Biomedicine and Pharmacotherapy, 117,* 109057.

41. Stella, B., Peira, E., Dianzani, C., Gallarate, M., Battaglia, L., Gigliotti, C., Boggio, E., et al., (2018). Development and characterization of solid lipid nanoparticles loaded with a highly active doxorubicin derivative. *Nanomaterials, 8*(2), 110.

42. Kakkar, D., Dumoga, S., Kumar, R., Chuttani, K., & Mishra, A. K., (2015). PEGylated solid lipid nanoparticles: Design, methotrexate loading and biological evaluation in animal models. *MedChemComm.*, *6*(8), 1452–1463.

43. Ud Din, F., Zeb, A., Shah, K. U., & Rehman, Z., (2019). Development, *in vitro* and *in vivo* evaluation of ezetimibe-loaded solid lipid nanoparticles and their comparison with marketed product. *Journal of Drug Delivery Science and Technology.*

44. Mishra, V., Bansal, K., Verma, A., Yadav, N., Thakur, S., Sudhakar, K., & Rosenholm, J., (2018). Solid lipid nanoparticles: Emerging colloidal nano drug delivery systems. *Pharmaceutics*, *10*(4), 191.

45. Saupe, A., Gordon, K. C., & Rades, T., (2006). Structural investigations on nanoemulsions, solid lipid nanoparticles and nanostructured lipid carriers by cryo-field emission scanning electron microscopy and Raman spectroscopy. *International Journal of Pharmaceutics*, *314*(1), 56–62.

46. Li, H., Zhao, X., Ma, Y., Zhai, G., Ling, B. L., & Hong, X. L., (2009). *J. Cont. Release*, *133*, 238–244.

47. Abdelbary, G., & Fahmy, R. H., (2009). Diazepam-loaded solid lipid nanoparticles: Design and characterization. *AAPS PharmSciTech.*, *10*(1).

48. Freitas, C., & Müller, R., (1999). Correlation between long-term stability of solid lipid nanoparticles (SLN™) and crystallinity of the lipid phase. *European Journal of Pharmaceutics and Biopharmaceutics*, *47*(2), 125–132.

49. Jenning, V., Thünemann, A. F., & Gohla, S. H., (2000). Characterization of a novel solid lipid nanoparticle carrier system based on binary mixtures of liquid and solid lipids. *International Journal of Pharmaceutics*, *199*(2), 167–177.

50. Kamble, V. A., Jagdale, D. M., & Kadam, V. J., (2010). Solid lipid nanoparticles as drug delivery system. *Int. J. Pharm. Biol. Sci*, *1*, 1–9.

51. Schoenwald, R. D., & Huang, H. S., (1983). Corneal penetration behavior of β-blocking agents I: Physicochemical factors. *Journal of Pharmaceutical Sciences*, *72*(11), 1266–1272.

52. Liu, Z., Zhang, X., Wu, H., Li, J., Shu, L., Liu, R., Li, L., & Li, N., (2011). Preparation and evaluation of solid lipid nanoparticles of baicalin for ocular drug delivery system *in vitro* and *in vivo*. *Drug Development and Industrial Pharmacy*, *37*(4), 475–481.

53. Saettone, M. F., Chetoni, P., Cerbai, R., Mazzanti, G., & Braghiroli, L., (1996). Evaluation of ocular permeation enhancers: *In vitro* effects on corneal transport of four β-blockers, and *in vitro/in vivo* toxic activity. *International Journal of Pharmaceutics*, *142*(1), 103–113.

54. Seyfoddin, A., Shaw, J., & Al-Kassas, R., (2010). Solid lipid nanoparticles for ocular drug delivery. *Drug Delivery*, *17*(7), 467–489.

55. Wilhelmus, K. R., (2001). The Draize eye test. *Survey of Ophthalmology*, *45*(6), 493–515.

56. Draize, J. H., Woodard, G., & Calvery, H. O., (1944). Methods for the study of irritation and toxicity of substances applied topically to the skin and mucous membranes. *Journal of Pharmacology and Experimental Therapeutics*, *82*(3), 377–390.

57. Shah, K. A., Date, A. A., Joshi, M. D., & Patravale, V. B., (2007). Solid lipid nanoparticles (SLN) of tretinoin: Potential in topical delivery. *International Journal of Pharmaceutics*, *345*(1/2), 163–171.

58. Madan, J. R., Khude, P. A., & Dua, K., (2014). Development and evaluation of solid lipid nanoparticles of mometasone furoate for topical delivery. *International Journal of Pharmaceutical Investigation*, *4*(2), 60.

59. Maia, C. S., Mehnert, W., & Schäfer-Korting, M., (2000). Solid lipid nanoparticles as drug carriers for topical glucocorticoids. *International Journal of Pharmaceutics, 196*(2), 165–167.

60. Joshi, A. S., Patel, H. S., Belgamwar, V. S., Agrawal, A., & Tekade, A. R., (2012). Solid lipid nanoparticles of ondansetron HCl for intranasal delivery: Development, optimization and evaluation. *Journal of Materials Science: Materials in Medicine, 23*(9), 2163–2175.

61. Jain, S. A., Chauk, D. S., Mahajan, H. S., Tekade, A. R., & Gattani, S. G., (2009). Formulation and evaluation of nasal mucoadhesive microspheres of Sumatriptan succinate. *Journal of Microencapsulation, 26*(8), 711–721.

62. Kumar, M., Misra, A., Mishra, A., Mishra, P., & Pathak, K., (2008). Mucoadhesive nanoemulsion-based intranasal drug delivery system of olanzapine for brain targeting. *Journal of Drug Targeting, 16*(10), 806–814.

63. Jogani, V. V., Shah, P. J., Mishra, P., Mishra, A. K., & Misra, A. R., (2008). Intranasal mucoadhesive microemulsion of tacrine to improve brain targeting. *Alzheimer Disease and Associated Disorders, 22*(2), 116–124.

64. Kumar, M., Misra, A., Babbar, A., Mishra, A., Mishra, P., & Pathak, K., (2008). Intranasal nanoemulsion based brain targeting drug delivery system of risperidone. *International Journal of Pharmaceutics, 358*(1/2), 285–291.

65. Paliwal, R., Rai, S., Vaidya, B., Khatri, K., Goyal, A. K., Mishra, N., Mehta, A., & Vyas, S. P., (2009). Effect of lipid core material on characteristics of solid lipid nanoparticles designed for oral lymphatic delivery. *Nanomedicine: Nanotechnology, Biology and Medicine, 5*(2), 184–191.

66. Obeidat, W. M., Schwabe, K., Müller, R. H., & Keck, C. M., (2010). Preservation of nanostructured lipid carriers (NLC). *European Journal of Pharmaceutics and Biopharmaceutics, 76*(1), 56–67.

67. McClements, D. J., Decker, E. A., Park, Y., & Weiss, J., (2009). Structural design principles for delivery of bioactive components in nutraceuticals and functional foods. *Critical Reviews in Food Science and Nutrition, 49*(6), 577–606.

68. Nakajima, M., Wang, Z., Chaudhry, Q., Park, H. J., & Juneja, L. R., (2015). Nano-science-engineering-technology applications to food and nutrition. *Journal of Nutritional Science and Vitaminology, 61*, S180–S182.

69. Badea, G., Lăcătuşu, I., Badea, N., Ott, C., & Meghea, A., (2015). Use of various vegetable oils in designing photoprotective nanostructured formulations for UV protection and antioxidant activity. *Industrial Crops and Products, 67*, 18–24.

70. Mozafari, M. R., Flanagan, J., Matia-Merino, L., Awati, A., Omri, A., Suntres, Z. E., & Singh, H., (2006). Recent trends in the lipid-based nanoencapsulation of antioxidants and their role in foods. *Journal of the Science of Food and Agriculture, 86*(13), 2038–2045.

71. Weiss, J., Takhistov, P., & McClements, D. J., (2006). Functional materials in food nanotechnology. *Journal of Food Science, 71*(9), R107–R116.

72. Hentschel, A., Gramdorf, S., Müller, R., & Kurz, T., (2008). β-Carotene-loaded nanostructured lipid carriers. *Journal of Food Science, 73*(2), N1–N6.

73. Da Silva, S. V., Ribeiro, A. P. B., & Santana, M. H. A., (2019). Solid lipid nanoparticles as carriers for lipophilic compounds for applications in foods. *Food Research International*.

74. De Carvalho, S. M., Noronha, C. M., Floriani, C. L., Lino, R. C., Rocha, G., Bellettini, I. C., Ogliari, P. J., & Barreto, P. L. M., (2013). Optimization of α-tocopherol loaded

solid lipid nanoparticles by central composite design. *Industrial Crops and Products*, *49*, 278–285.

75. Lacatusu, I., Mitrea, E., Badea, N., Stan, R., Oprea, O., & Meghea, A., (2013). Lipid nanoparticles based on omega-3 fatty acids as effective carriers for lutein delivery. Preparation and *in vitro* characterization studies. *Journal of functional Foods, 5*(3), 1260–1269.

76. Cho, Y., Shim, H., & Park, J., (2003). Encapsulation of fish oil by an enzymatic gelation process using transglutaminase cross-linked proteins. *Journal of Food Science, 68*(9), 2717–2723.

77. Zhao, C., Zhang, J., Hu, H., Qiao, M., Chen, D., Zhao, X., & Yang, C., (2018). Design of lactoferrin modified lipid nano-carriers for efficient brain-targeted delivery of nimodipine. *Materials Science and Engineering: C, 92*, 1031–1040.

78. Patel, S., Chavhan, S., Soni, H., Babbar, A., Mathur, R., Mishra, A., & Sawant, K., (2011). Brain targeting of risperidone-loaded solid lipid nanoparticles by intranasal route. *Journal of Drug Targeting, 19*(6), 468–474.

79. Jose, S., Anju, S., Cinu, T., Aleykutty, N., Thomas, S., & Souto, E., (2014). *In vivo* pharmacokinetics and biodistribution of resveratrol-loaded solid lipid nanoparticles for brain delivery. *International Journal of Pharmaceutics, 474*(1, 2), 6–13.

80. Dhawan, S., Kapil, R., & Singh, B., (2011). Formulation development and systematic optimization of solid lipid nanoparticles of quercetin for improved brain delivery. *Journal of Pharmacy and Pharmacology, 63*(3), 342–351.

81. Lahkar, S., & Kumar, D. M., (2018). Surface modified kokum butter lipid nanoparticles for the brain targeted delivery of nevirapine. *Journal of Microencapsulation, 35*(7/8), 680–694.

82. Grillone, A., Battaglini, M., Moscato, S., Mattii, L., De Julián, F. C., Scarpellini, A., Giorgi, M., et al., (2018). Nutlin-loaded magnetic solid lipid nanoparticles for targeted glioblastoma treatment. *Nanomedicine, 8*(3), 727–752.

83. Kadari, A., Pooja, D., Gora, R. H., Gudem, S., Kolapalli, V. R. M., Kulhari, H., & Sistla, R., (2018). Design of multifunctional peptide collaborated and docetaxel loaded lipid nanoparticles for antiglioma therapy. *European Journal of Pharmaceutics and Biopharmaceutics, 132*, 168–179.

84. Youssef, N. A. H. A., Kassem, A. A., Farid, R. M., Ismail, F. A., EL-Massik, M. A. E., & Boraie, N. A., (2018). A novel nasal almotriptan loaded solid lipid nanoparticles in mucoadhesive *in situ* gel formulation for brain targeting: Preparation, characterization and *in vivo* evaluation. *International Journal of Pharmaceutics, 548*(1), 609–624.

85. Chen, B., Luo, M., Liang, J., Zhang, C., Gao, C., Wang, J., Wang, J., Li, Y., Xu, D., & Liu, L., (2018). Surface modification of PGP for a neutrophil–nanoparticle co-vehicle to enhance the anti-depressant effect of baicalein. *Acta Pharmaceutica Sinica B, 8*(1), 64–73.

86. Abbas, H., Refai, H., & El Sayed, N., (2018). Superparamagnetic iron oxide–loaded lipid nanocarriers incorporated in Thermosensitive *in situ* gel for magnetic brain targeting of clonazepam. *Journal of Pharmaceutical Sciences, 107*(8), 2119–2127.

87. Song, H., Wei, M., Zhang, N., Li, H., Tan, X., Zhang, Y., & Zheng, W., (2018). Enhanced permeability of blood-brain barrier and targeting function of brain via borneol-modified chemically solid lipid nanoparticle. *International Journal of Nanomedicine, 13*, 1869.

88. Fatouh, A. M., Elshafeey, A. H., & Abdelbary, A., (2017). Agomelatine-based *in situ* gels for brain targeting via the nasal route: Statistical optimization, *in vitro*, and *in vivo* evaluation. *Drug Delivery, 24*(1), 1077–1085.

89. Malekpour-Galogahi, F., Hatamian-Zarmi, A., Ganji, F., Ebrahimi-Hosseinzadeh, B., Nojoki, F., Sahraeian, R., & Mokhtari-Hosseini, Z. B., (2018). Preparation and optimization of rivastigmine-loaded tocopherol succinate-based solid lipid nanoparticles. *Journal of Liposome Research, 28*(3), 226–235.

90. Joseph, E., Reddi, S., Rinwa, V., Balwani, G., & Saha, R., (2017). Design and *in vivo* evaluation of solid lipid nanoparticulate systems of olanzapine for acute phase schizophrenia treatment: Investigations on antipsychotic potential and adverse effects. *European Journal of Pharmaceutical Sciences, 104*, 315–325.

91. Natarajan, J., Baskaran, M., Humtsoe, L. C., Vadivelan, R., & Justin, A., (2017). Enhanced brain targeting efficacy of olanzapine through solid lipid nanoparticles. *Artificial Cells, Nanomedicine, and Biotechnology, 45*(2), 364–371.

92. Gupta, S., Kesarla, R., Chotai, N., Misra, A., & Omri, A., (2017). Systematic approach for the formulation and optimization of solid lipid nanoparticles of efavirenz by high pressure homogenization using design of experiments for brain targeting and enhanced bioavailability. *Biomed Research International, 2017*.

93. Battaglia, L., Muntoni, E., Chirio, D., Peira, E., Annovazzi, L., Schiffer, D., Mellai, M., et al., (2017). Solid lipid nanoparticles by coacervation loaded with a methotrexate prodrug: Preliminary study for glioma treatment. *Nanomedicine, 12*(6), 639–656.

94. He, X., Zhu, Y., Wang, M., Jing, G., Zhu, R., & Wang, S., (2016). Antidepressant effects of curcumin and HU-211 coencapsulated solid lipid nanoparticles against corticosterone-induced cellular and animal models of major depression. *International Journal of Nanomedicine, 11*, 4975.

95. Kakkar, V., Mishra, A. K., Chuttani, K., & Kaur, I. P., (2013). Proof of concept studies to confirm the delivery of curcumin loaded solid lipid nanoparticles (C-SLNs) to brain. *International Journal of Pharmaceutics, 448*(2), 354–359.

96. Girotra, P., & Singh, S. K., (2017). Multivariate optimization of rizatriptan benzoate-loaded solid lipid nanoparticles for brain targeting and migraine management. *AAPS PharmSciTech., 18*(2), 517–528.

97. Li, J. C., Zhang, W. J., Zhu, J. X., Zhu, N., Zhang, H. M., Wang, X., Zhang, J., & Wang, Q. Q., (2015). Preparation and brain delivery of nasal solid lipid nanoparticles of quetiapine fumarate *in situ* gel in rat model of schizophrenia. *International Journal of Clinical and Experimental Medicine, 8*(10), 17590.

98. Yasir, M., & Sara, U. V. S., (2014). Solid lipid nanoparticles for nose to brain delivery of haloperidol: *In vitro* drug release and pharmacokinetics evaluation. *Acta Pharmaceutica Sinica B, 4*(6), 454–463.

99. Hansraj, G. P., Singh, S. K., & Kumar, P., (2015). Sumatriptan succinate loaded chitosan solid lipid nanoparticles for enhanced anti-migraine potential. *International Journal of Biological Macromolecules, 81*, 467–476.

100. Bhatt, R., Singh, D., Prakash, A., & Mishra, N., (2015). Development, characterization and nasal delivery of rosmarinic acid-loaded solid lipid nanoparticles for the effective management of Huntington's disease. *Drug Delivery, 22*(7), 931–939.

101. Singh, A. P., Saraf, S. K., & Saraf, S. A., (2012). SLN approach for nose-to-brain delivery of alprazolam. *Drug Delivery and Translational Research, 2*(6), 498–507.

102. Venishetty, V. K., Komuravelli, R., Kuncha, M., Sistla, R., & Diwan, P. V., (2013). Increased brain uptake of docetaxel and ketoconazole loaded folate-grafted solid lipid nanoparticles. *Nanomedicine: Nanotechnology, Biology and Medicine*, *9*(1), 111–121.

103. Wong, H. L., Bendayan, R., Rauth, A. M., Li, Y., & Wu, X. Y., (2007). Chemotherapy with anticancer drugs encapsulated in solid lipid nanoparticles. *Advanced Drug Delivery Reviews*, *59*(6), 491–504.

104. Pandey, R., Sharma, S., & Khuller, G., (2005). Oral solid lipid nanoparticle-based antitubercular chemotherapy. *Tuberculosis*, *85*(5, 6), 415–420.

105. Selvamuthukumar, S., & Velmurugan, R., (2012). Nanostructured lipid carriers: A potential drug carrier for cancer chemotherapy. *Lipids in Health and Disease*, *11*(1), 159.

106. Tunki, L., Kulhari, H., Vadithe, L. N., Kuncha, M., Bhargava, S., Pooja, D., & Sistla, R., (2019). Modulating the site-specific oral delivery of sorafenib using sugar-grafted nanoparticles for hepatocellular carcinoma treatment. *European Journal of Pharmaceutical Sciences*, 104978.

107. Fei, W., Zhang, Y., Han, S., Tao, J., Zheng, H., Wei, Y., Zhu, J., Li, F., & Wang, X., (2017). RGD conjugated liposome-hollow silica hybrid nanovehicles for targeted and controlled delivery of arsenic trioxide against hepatic carcinoma. *International Journal of Pharmaceutics*, *519*(1/2), 250–262.

108. Singh, A., Ahmad, I., Ahmad, S., Iqbal, Z., & Ahmad, F. J., (2016). A novel monolithic controlled delivery system of resveratrol for enhanced hepatoprotection: Nanoformulation development, pharmacokinetics and pharmacodynamics. *Drug Development and Industrial Pharmacy*, *42*(9), 1524–1536.

109. Chaudhary, S., Garg, T., Murthy, R., Rath, G., & Goyal, A. K., (2015). Development, optimization and evaluation of long chain nanolipid carrier for hepatic delivery of silymarin through lymphatic transport pathway. *International Journal of Pharmaceutics*, *485*(1/2), 108–121.

110. Jia, Y., Ji, J., Wang, F., Shi, L., Yu, J., & Wang, D., (2016). Formulation, characterization, and *in vitro*/vivo studies of aclacinomycin A-loaded solid lipid nanoparticles. *Drug Delivery*, *23*(4), 1317–1325.

111. Varshosaz, J., Jafarian, A., Salehi, G., & Zolfaghari, B., (2014). Comparing different sterol containing solid lipid nanoparticles for targeted delivery of quercetin in hepatocellular carcinoma. *Journal of Liposome Research*, *24*(3), 191–203.

112. Nandini, P., Doijad, R., Shivakumar, H., & Dandagi, P., (2015). Formulation and evaluation of gemcitabine-loaded solid lipid nanoparticles. *Drug Delivery*, *22*(5), 647–651.

113. Jing, F., Li, J., Liu, D., Wang, C., & Sui, Z., (2013). Dual ligands modified double targeted nano-system for liver targeted gene delivery. *Pharmaceutical Biology*, *51*(5), 643–649.

114. Hu, H., Liu, D., Zhao, X., Qiao, M., & Chen, D., (2013). Preparation, characterization, cellular uptake and evaluation *in vivo* of solid lipid nanoparticles loaded with cucurbitacin B. *Drug Development and Industrial Pharmacy*, *39*(5), 770–779.

115. Luo, C. F., Yuan, M., Chen, M. S., Liu, S. M., Zhu, L., Huang, B. Y., Liu, X. W., & Xiong, W., (2011). Pharmacokinetics, tissue distribution and relative bioavailability of puerarin solid lipid nanoparticles following oral administration. *International Journal of Pharmaceutics*, *410*(1/2), 138–144.

116. Dong, Z., Guo, J., Xing, X., Zhang, X., Du, Y., & Lu, Q., (2017). RGD modified and PEGylated lipid nanoparticles loaded with puerarin: Formulation, characterization and protective effects on acute myocardial ischemia model. *Biomedicine and Pharmacotherapy*, *89*, 297–304.

117. Gomes, F. L., Maranhão, R. C., Tavares, E. R., Carvalho, P. O., Higuchi, M. L., Mattos, F. R., Pitta, F. G., et al., (2018). Regression of atherosclerotic plaques of cholesterol-fed rabbits by combined chemotherapy with paclitaxel and methotrexate carried in lipid core nanoparticles. *Journal of Cardiovascular Pharmacology and Therapeutics*, *23*(6), 561–569.

118. Shao, M., Yang, W., & Han, G., (2017). Protective effects on myocardial infarction model: Delivery of schisandrin B using matrix metalloproteinase-sensitive peptide-modified, PEGylated lipid nanoparticles. *International Journal of Nanomedicine*, *12*, 7121.

119. Daminelli, E. N., Martinelli, A. E., Bulgarelli, A., Freitas, F. R., & Maranhão, R. C., (2016). Reduction of atherosclerotic lesions by the chemotherapeutic agent carmustine associated to lipid nanoparticles. *Cardiovascular Drugs and Therapy*, *30*(5), 433–443.

120. Foroughi, S., Ziamajidi, N., Javid, S., Abbasalipourkabir, R., Aflatoonian, R., Ashrafi, M., & Nourian, A., (2018). Study of telomerase reverse transcriptase and uterine-ovarian-specific genes expression in the endometrial tissue of ovariectomized female Sprague-Dawley rats. *International Journal of Biological Macromolecules*, *113*, 1302–1307.

121. Farnia, P., Ghanavi, J., Bahrami, A., Bandehpour, M., Kazemi, B., & Velayati, A. A., (2015). Increased production of soluble vascular endothelial growth factors receptor-1 in CHO-cell line by using new combination of chitosan-protein lipid nanoparticles. *International Journal of Clinical and Experimental Medicine*, *8*(1), 1526.

122. Zhai, B., Zeng, Y., Zeng, Z., Zhang, N., Li, C., Zeng, Y., You, Y., et al., (2018). Drug delivery systems for elemen, its main active ingredient β-elemene, and its derivatives in cancer therapy. *International Journal of Nanomedicine*, *13*, 6279.

123. El-Salamouni, N. S., Farid, R. M., El-Kamel, A. H., & El-Gamal, S. S., (2018). Nanostructured lipid carriers for intraocular brimonidine localization: Development, *in vitro* and *in vivo* evaluation. *Journal of Microencapsulation*, *35*(1), 102–113.

124. Chetoni, P., Burgalassi, S., Monti, D., Tampucci, S., Tullio, V., Cuffini, A. M., Muntoni, E., et al., (2016). Solid lipid nanoparticles as promising tool for intraocular tobramycin delivery: Pharmacokinetic studies on rabbits. *European Journal of Pharmaceutics and Biopharmaceutics*, *109*, 214–223.

125. Amadio, M., Pascale, A., Cupri, S., Pignatello, R., Osera, C., Leggio, G. M., Ruozi, B., et al., (2016). Nanosystems based on siRNA silencing HuR expression counteract diabetic retinopathy in rat. *Pharmacological Research*, *111*, 713–720.

126. Amselem, S., (1999). *Solid Lipid Compositions of Lipophilic Compounds for Enhanced Oral Bioavailability*. U.S. Patent No. 5,989,583. 23 November 1999.

127. Jeong, S. Y., Kwon, I. C., & Chung. H. (2000). "Lipid emulsion and solid lipid nanoparticle as a gene or drug carrier." *PCT Int. Appl, WO* 2000006120.

128. Seoul 130–230, Chung, K., & Hesson, (2000). *Lipid Emulsion and Solid Lipid Nanoparticle as a Gene or Drug Carrier*.

129. Jeong, S. Y., Kwon, I. C., & Chung, H. (2000). "Lipid emulsion and solid lipid nanoparticle as a gene or drug carrier." *PCT Int. Appl, WO* 2000006120.

130. Lee, Y. J., (2005). Review of the Argynnis adippe species group (Lepidoptera, Nymphalidae, Heliconiinae) in Korea. *Lucanus 5*, 1–8.

131. Heppner, A., Hansen, P., & Schumann, C., (2001). Solid Lipid Nanoparticles Containing a UV Filter Material Are Useful in Aqueous Dispersion Form as High Filter Content Sunscreen Compositions. *Stada Arzneimittel Ag*. DE19952410A1.

132. Herzog, B., (2006). BASF Performance Products LLC, *Formulation of UV absorbers by incorporation in solid lipid nanoparticles*. U.S. Patent 7,147,841.

133. Shastri, V., Sussman, E., & Jayagopal, A., (2006). Vanderbilt University, *Functionalized solid lipid nanoparticles and methods of making and using same*. U.S. Patent Application 11/251,109.

134. Gasco, M. R., Nanovector, S. R. L., (2009). *Use of solid lipid nanoparticles comprising cholesteryl propionate and/or cholesteryl butyrate*. U.S. Patent Application 11/921,634.

135. Herzog, B., (2006). BASF Performance Products LLC, *Formulation of UV absorbers by incorporation in solid lipid nanoparticles*. U.S. Patent 7,147,841.

136. Heppner, A. F., (61197, DE), Hansen, Peter, Dr. (Dreieich, 63303, DE), Schumann, Christof (Breitscheid, 35767, DE) Sunscreen preparations comprising solid lipid nanoparticles. 2007.

137. Trotta, M., Cavalli, R., & Battaglia, L. S., (2008). Method for the preparation of solid lipid micro and nanoparticles. WO2008149215.

138. Trotta, M., Cavalli, R., & Battaglia, L. S., (2008). Method for the preparation of solid lipid micro and nanoparticles. WO2008149215A2.

139. Miranda, F. L. A., Assis, C. G. G., Lambert, O. R., Tadeu, L. B. V., Aprecida, D. O. C., & Bohórquez, M. G. A., (2009). Solid Lipid Nanoparticles Containing Retinoids. WO2009111852.

140. Padois, K., Pirot, F., & Falson, F., (2010). Solid lipid nanoparticles encapsulating minoxidil and aqueous suspension containing same. *Patent WO2010112749*.

141. Petit, J. L.V., Gonzalez, R. D., & Botello, A. F., Lipotec, S. A., (2013). *Lipid nanoparticle capsules*. U.S. Patent Application 13/636,909.

142. Chen, J., Ansell, S., Akinc, A., Dorkin, J. R., Qin, X., Cantley, W., Manoharan, M., Rajeev, K. G., Narayanannair, J. K., & Jayaraman, M., (2012). Alnylam Pharmaceuticals Inc, *Lipid formulation*. U.S. Patent 8,158,601.

143. Petit, J. L.V., Gonzalez, R. D., & Botello, A. F., & Lipotec, S. A., (2013). *Lipid nanoparticle capsules*. WO2011116963A3.

144. Kaur, I. P., & Bhandari, R., (2012). Solid lipid nanoparticles entrapping hydrophilic/ amphiphilic drug and a process for preparing the same. *PCT application number: PCT/IN2012/000154 dated*, 5(3), p.2012.

145. Park, S. M., Kim, H. R., Park, J. C., & Chae, S. Y., (2013). Samsung Electronics Co Ltd, *Solid lipid nanoparticles including elastin-like polypeptides and use thereof*. U.S. Patent Application 13/757,528.

146. Müller, R. H., Jenning, V., Mader, K., & Lippacher, A., (2014). PharmaSol GmbH, *Lipid particles on the basis of mixtures of liquid and solid lipids and method for producing same*. U.S. Patent 8,663,692.

147. Guerrero, D. Q., Zaragoza, M. D.L. L.Z., Cardenas, A. A., & Silva, E. M., (2014). Universidad Nacional Autonoma de Mexico, *Composition of solid lipid nanoparticles for the long-term conservation of fruits, vegetables, seeds, cereals and/or fresh foodstuffs using a coating*. U.S. Patent Application 14/111,439.

148. Chen, J., Ansell, S., Akinc, A., Dorkin, J. R., Qin, X., Cantley, W., Manoharan, M., Rajeev, K. G., Narayanannair, J. K., & Jayaraman, M., (2014). Tekmira Pharmaceuticals Corp, *Lipid formulation*. U.S. Patent 8,802,644.

149. Weiss, J., Schweiggert, C., Leuenberger, B., Novotny, M., Tedeschi, C., & Kessler, A., (2017). DSM IP Assets BV, *Solid lipid nanoparticles (I)*. U.S. Patent 9,616,001.

150. Weiss, J., Schweiggert, C., Leuenberger, B., Novotny, M., Tedeschi, C., & Kessler, A., (2017). DSM IP Assets BV, *Solid lipid nanoparticles (I)*. EP2777402A1

151. Weiss, J., Maier, C., Leuenberger, B., Novotny, M., Tedeschi, C., & Kessler, A., (2014). Solid Lipid Nanoparticles (II). US20160030305A1

152. Ghiani, S., Maiocchi, A., Caminiti, L., & Miragoli, L., (2014). Fluorescent Solid Lipid Nanoparticles Composition and Preparation Thereof. EP3003395A1

153. Saulnier, P., Benoit, J. P., & Anton, N., (2015). Inserm Transfert and Universite dAngers, *Process for preparing lipid nanoparticles*. U.S. Patent 9,005,666.

154. Repka, M. A., Patil, H. G., Majumdar, S., Park, J. B., & Kulkarni, V. I., (2017). University of Mississippi, *Systems and methods for preparing solid lipid nanoparticles*. U.S. Patent Application 15/128,536.

155. Weiss, J., Maier, C., Kessler, A., Tedeschi, C., Leuenberger, B., & Novotny, M., (2016). Solid Lipid Nanoparticles (II). US20160030305A1

156. Benoit, J. P., Thomas, O., Saulnier, P., & Ramadan, A. A., (2016). Method of Preparing Lipid Nanoparticles. US20120027825A1

157. Gainza, L. E., Del, P. P. A., Gainza, L. G., Ibarrola, M. O., Villullas, R. S., Fernandez, P. R., Bachiller, P. D., et al., (2016). Lipid Nanoparticle of Polymyxin. EP3017823A1

158. Lafuente, E. G., Lucea, G. G., Rincón, S. V., Navarro, M. P., Moreno, O. I., Hornes, G. A., Pérez, A. D.P., Martín, R. M.H., Olaechea, M. I., & Muñoz, J. L.P., (2019). Praxis Biopharma Research Institute, *Lipid nanoparticles for wound healing*. U.S. Patent 10,206,886.

159. Sun, J. H., Yuan, H., Liu, F., & Chen, S. Q., (2017). Solid lipid magnetic resonance nanoparticles and preparation method and use thereof. *Patent WO2017041609*.

160. Weiss, J., Schweiggert, C., Leuenberger, B., Novotny, M., Tedeschi, C., & Kessler, A., (2017). DSM IP Assets BV, *Solid lipid nanoparticles (I)*. EP2777402A1

161. Weiss, J., Maier, C., Kessler, A., Tedeschi, C., Leuenberger, B., & Novotny, M., (2017). Solid Lipid Nanoparticles (II). US20160030305A2

162. Rodriguez, G. A., Solinís, A. M. A., Del, P. R. A., Delgado, S., Vicente, D., & Pedraz, M. J. L., (2017). Lipid Nanoparticles for Gene Therapy. US9675710B2

163. Lee, R. J., (2017). Ohio State Innovation Foundation, *Lipid nanoparticle compositions and methods of making and methods of using the same*. U.S. Patent 9,750,819.

164. Myerson, J., & Wickline, S. A., (2017). Washington University in St Louis, *Antithrombotic nanoparticle*. U.S. Patent 9,764,043.

165. Petit, J. L.V., Gonzalez, R. D., & Botello, A. F., Lipotec, S. A., (2018). *Lipid nanoparticle capsules*. US20130017239A1.

166. Kaur, I. P., & Verma, M. K., (2018). Panjab University Department of Biotechnology (dbt), *Process for preparing solid lipid sustained release nanoparticles for delivery of vitamins*. U.S. Patent 9,907,758.

167. Gainza, L. E., Del, P. P. A., Gainza, L. G., Ibarrola, M. O., Villullas, R. S., Fernandez, P. R., Bachiller, P. D., et al., (2016). Lipid Nanoparticle of Polymyxin. EP3017823A1

168. Reinhard, B. M., & Gummuluru, S., (2018). GM3 Functionalized Nanoparticles. WO2015160597A1

169. Brioschi, C., Cabella, C., Ghiani, S., Maiocchi, A., Miragoli, L., & Visigalli, M., Bracco Imaging SpA, (2018). *Paramagnetic solid lipid nanoparticles (pSLNs) containing metal amphiphilic complexes for MRI.* U.S. Patent 10,039,843.

170. Puri, A., Blumenthal, R. P., Joshi, A., Tata, D. B., & Viard, M., (2018). Baylor College of Medicine and US Department of Health, *Photoactivatable lipid-based nanoparticles as vehicles for dual agent delivery.* U.S. Patent 10,117,942.

171. Du, X., & Ansell, S. M., (2019). Acuitas Therapeutics Inc, *Lipids and lipid nanoparticle formulations for delivery of nucleic acids.* U.S. Patent 10,221,127.

172. Carlsson, J. O., Johansson, A., Rooth, M.,& Nanexa Ab, (2019). *Solid nanoparticle with inorganic coating.* U.S. Patent 10,166,198.

173. Diorio, C., & Lokhnauth, J., Capsugel Belgium NV, (2019). *Curcumin solid lipid particles and methods for their preparation and use.* U.S. Patent 10,166,187.

174. Nechev, L., & Price, S., (2019). Alnylam Pharmaceuticals Inc, *Compositions and methods for the manufacture of lipid nanoparticles.* U.S. Patent 10,195,291.

175. Ghiani, S., Maiocchi, A., Caminiti, L., & Miragoli, L., (2019). Bracco Imaging SpA, *Fluorescent solid lipid nanoparticles composition and preparation thereof.* U.S. Patent 10,251,960.

176. Lee, R. J., Lee, Y. B., Kim, D. J., & Ahn, C. H., Ohio State University, (2019). *Lipid nanoparticle compositions for antisense oligonucleotides delivery.* U.S. Patent 10,307,490.

177. Tzachev, C. T., (2019). Solid Lipid Nanoparticle for Intracellular Release of Active Substances and Method for Production the Same. WO2019116062.

178. Puglia, C., & Bonina, F., (2012). Lipid nanoparticles as novel delivery systems for cosmetics and dermal pharmaceuticals. *Expert Opinion on Drug Delivery, 9*(4), 429–441.

179. Olbrich, C., Gessner, A., Kayser, O., & Müller, R. H., (2002). Lipid-drug-conjugate (LDC) nanoparticles as novel carrier system for the hydrophilic antitrypanosomal drug diminazene diaceturate. *Journal of Drug Targeting, 10*(5), 387–396.

180. Schwarz, C., Freitas, C., Mehnert, W., & Muller, R., (1995). In Sterilization and physical stability of drug-free and etomidate-loaded solid lipid nanoparticles. *Proc. Int. Symp. Control. Release Bioact. Mater., 766,* 767.

181. Cavalli, R., Marengo, E., Rodriguez, L., & Gasco, M. R., (1996). *Effect of Some Experimental Factors on the Production Process of Solid Lipid Nanoparticles.*

182. Schwarz, C., & Mehnert, W., (1995). In: *Sterilization of Drug-Free and Tetracaine-Loaded solid Lipid Nanoparticles (SLN)* (pp. 485–486). Proc. First World Meeting APGI/APV, Budapest.

CHAPTER 8

Evaluation of Polymer Electrolytes for Application in DSSC

SHIVANI ARORA ABROL[1] and CHERRY BHARGAVA[2]

[1]*Research Scholar, School of Electronics and Electrical Engineering*
Lovely Professional University, Phagwara, Punjab-144411
Email: Shivani.ar80@gmail.com

[2]*Associate Professor, Department of Computer Science and Engineering*
Symbiosis Institute of Technology, Pune, Maharashtra-412115
E-mail: cherry.bhargava@sitpune.edu.in (C. Bhargava)

ABSTRACT

Today's world is much dependent on conventional resources of energies such as fossil fuels, coal, petroleum, etc., which are exhausting very fast. Global warming is the consequence of using these resources to lay on the atmosphere with the emission of harmful pollutants, mainly carbon dioxide in the atmosphere, which leads to greenhouse effects. This being the main area of apprehension, researchers have started looking for replacements for these non-renewable energy resources. With the current global energy transition under the way, mainly two primary renewable energy resource drivers have emerged: solar energy and wind energy. Much available resource amongst the two is solar, which can be easily converted into a useful form of energy by means of photovoltaic cells. In this context, the increased efficiency of Dye-Sensitized Solar Cells with the help of the selection of an appropriate polymer electrolyte will be observed.

8.1 INTRODUCTION

The various types of solar cells available are:

    i.   Silicon-based solar cells;
    ii. Thin-layer solar cells;
    iii. Dye-sensitized solar cells.

Silicon-based solar cells, also known as first-generation solar cells, are most widely used due to their good ionic photon conversion efficiency of 25% regardless of the very high cost of production and large number of photons being wasted in heat dissipation. [3–5]. The second-generation solar cells, i.e., the thin layer solar cells have low photon conversion efficiencies but have a low production cost [6]. Moving further, in the year 1991, Brain'O Regan and Michael Gratzel projected a photovoltaic cell called Dye-Sensitized Solar Cell, also known as third-generation solar cells and which seemed to have a reasonable conversion efficiency of about 7.1% to 7.9% and a moderate current density of 12 mA cm^{-2}. The processes and the materials involved in the fabrications of these were less as compared to the other two generations. It is fabricated by semiconductor materials assembled between electrolyte and photosensitized anode thus producing a photo electrochemical effect and also have high environment congeniality and good performance under diffused light [39]. Other advantages included transparency, flexibility in shape and color.

There are several factors that determine the performance of DSSC like fill factor (FF), short circuit current density (J_{sc} in mAcm^{-2}), open circuit voltage (V_{oc}) and energy conversion efficiency (n%). Further, there are several components that a DSSC is comprised of. These are counter electrode (e.g., platinum), semiconductor electrode, dye (e.g., N719, N3, Organic dyes), transparent conducting substrate (e.g., indium doped tin oxide (ITO), fluorine doped tin oxide) and electrolyte (e.g., polymer electrolytes, I_3^-, I^-) [6]. The voltage and current dependencies of these factors are expressed in the relationships as below:

$$FF = \frac{Vmax \times Jmax}{Voc \times Jsc}$$

(1)

where; J_{max} = maximum current density in mA cm^{-2}.

$$\eta\% = \frac{Voc \times Jsc \times FF}{Pin} \times 100$$

(2)

where; P_{in} = incident power line density in mW cm^{-2}.

As is observed from equations, the improvement in form factor, efficiencies, and all other parameters can be increased with increasing current densities (J_{sc} and J_{max}), which is due to the presence of electrolytes hence they play a vital role in the fabrications of DSSC.

8.2   ELECTROLYTES IN DSSC

An electrolyte is one of the most important ingredients in fabrication of DSSC. In between the electrodes, the inner charge carrier transportation is done through electrolyte. Electrolyte also continuously renews itself as well as the dye. They further improve the stability of DSSCs and solar to electric conversion efficiencies by maintaining the performance parameters.

Figures 8.1 and 8.2 classify the electrolytes and depict an insight to the various types of polymer electrolytes present. Polymer is also called gelator. Gelator is used to solidify liquid electrolytes and the solution become viscous. For this purpose, gelator makes use of a solvent which are also known as plasticizer. Plasticizer help in reducing crystallization, glass temperature of electrolyte and polymer-polymer chain interaction and they increase meta-meric mobilities. Most commonly used gelator's are polyethylene oxide, polyacrylonitrile, polyvinylpyrrolidone, polystyrene, polyvinyl chloride, polyvinylidene ester, poly methyl methacrylate, etc.

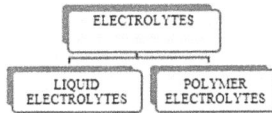

FIGURE 8.1   Types of electrolytes.

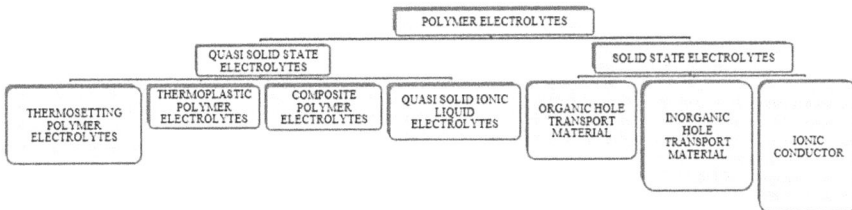

FIGURE 8.2   Types of polymer electrolyte based on physical state and composition.

8.3   REVIEW OF ELECTROLYTES

Liquid electrolytes provide an efficiency of 7.1 to 7.9%. DSSC composed of liquid electrolytes have a maximum efficiency of 13%. These electrolytes have characteristics like good conductivity, low viscosity, better interfacial contact channel between electrodes, ease of preparation and improved conversion efficiency.

However, Polymer electrolytes have a polar functional group present in them which provides intermolecular interaction for changing the ionic state and with the advent of high ionic conductivity polymeric materials (like polyethylene oxide complexes), energy storage in electrochemical devices have improved. Commonly available polymer electrolytes are ionic rubber polymer electrolytes, plasticize gel and ion conducting polymer electrolytes. Some examples of polymer electrolyte are poly methyl meth acrylate (PMMA), poly ethylene glycol (PEG), poly methyl ether methacrylate (PMEM), poly acrylonitrile (PAN), polyvinylidene fluoride-co-hexafluoropropylene, etc., [4, 5].

Liquid electrolytes have complications like leakage, evaporation of solvent, deterioration of dye and decay of counter electrodes. These problems are eliminated in quasi solid-state polymer electrolytes [22]. However, DSSC based on these electrolytes have lower efficiencies as compared to liquid electrolytes; nonetheless Quasi solid-state provides improved long-term stability and enhanced sealing capability. They show good ionic conductivity; higher than 10^{-7} S/cm and a better electrode interfacial contact and these parameters are achieved due to the amalgamation of cohesive properties of solid and diffusive properties of liquid. These electrolytes are generally found useful in batteries, sensors, and actuators, supercapacitors, fuel cells and electrochromic displays. [1] Examples of thermosetting polymer electrolyte is polyethylene oxide + I2, 4 di-o-benzylidene-d-sorbitol + methoxy propionitrile.

Comparative study of various polymer electrolytes (10 Solid State Polymer Electrolytes and 18 Quasi Solid-State Polymer Electrolytes) with an effective dye is shown tabularly in Tables 8.1 and 8.2.

8.4 PERFORMANCE OF DIFFERENT POLYMER ELECTROLYTES

Performance of different polymer electrolytes in DSSC can be calculated on the basis of ionic conductivity, Open-circuit photo-voltage, short-circuit photocurrent density, efficiency, and FF enlisted in table above can be shown graphically as under:

   i.   **For Solid-State Electrolytes Polymer Electrolytes (I–X as shown in Table 8.1):** Figures 8.3–8.5 show the open circuit photovoltage, short circuit photocurrent density and ionic conductivity of different solid-state electrolyte based DSSC.

TABLE 8.1 Performance of Different Solid-State Polymer Electrolytes

Electrolyte	Dye	Ionic Conductivity	O-C. Photo-Voltage	S-C. Photo-Current Density (mA cm^{-2})	Efficiency	Life Span
I. 1-methyl-3-propylimidazolium iodide + iodine + potassium iodide + tertbutylpyridene + polyethylene glycol dimethyllether/polyethylene oxide + GuSCN+LiClO$_4$	N/A	N/A	0.68	21	8.9 (FF = 0.62)	N/A
II. Poly N-alkyl-4-vinyl pyridine iodide + N-methyl pyridine iodide + I$_2$	N719	4.55	0.68	13.44	5.64 (FF = 0.62)	N/A
III. Polyvinyl alcohol + Polymethyl methacrylate	N/A	N/A	0.65	14.1	N/A 5.5 (FF = 0.6)	2000 hrs
IV. Polyvinyl butyral + gamma butyrolacetone + N-methyl-2-pyrrolidene + LiI+I$_2$	N719	1	0.79	10.17	5.2	2000 hrs
V. Polyethylene glycol + Al$_2$O$_3$	N719	4.1	0.667	12.42	5.1	N/A
VI. Polyethylene oxide + polyethylene glycol + potassium iodide + I$_2$	N719	0.145	0.72	11.2	3.84	N/A
VII. Phenothiazene + Polyvinylidene Fluoride + potassium iodide + I$_2$	N/A	0.745	0.616	5 ± 0.17	2.92 (FF = 0.57)	N/A
VIII. 2-(1-(anthracen-2-yl)-4,5-Diphenyl-1H-imidazol-2-yl)	N/A	N/A	0.72	4	1.21 (FF = 0.42)	N/A
IX. Poly(3-hexilthiophene)	N/A	N/A	0.89	6.2	2	N/A
X. Polyurethane + Lithium iodide	N/A	0.76	0.14	0.06	0.003 (FF = 0.26)	N/A

TABLE 8.2 Performance of Different Quasi-Solid Polymer Electrolytes

Electrolyte	Dye	Ionic Conductivity	O. C. Photo-Voltage (V)	S. C. Photo-Current Density	Efficiency (%)	Life Span
A. N-Phthaloylchitosan + polyethyleneoxide + tetrapropylammoniumiodide + dimethylformamide + tetrapropylammoniumiodide + ethylenecarbonate + BMII	N3	13.5 ± 1.2	0.7	19.68	9.61	N/A
B. Polyethyleneimine + polyethylene glycol + KI + I_2 + 4,4'((((oxybis(ethane-2,1-diyl)bis(oxy))bis(ethane-2,1diyl))bis(sulfanediyl))dipyridine	N719	1.65	0.89	12.9	9.2 (FF = 0.56)	N/A
C. Polymer gel electrolyte + propylene carbonate	N/A	10.96	0.76	17.35	8.31 (FF = 0.63)	Tested up to 1000 hrs
D. Polymethyl Methacrylate + 1-butyl-3-methylimidazoliumiodide/I_2	N719	N/A	0.75	15.5	8 (FF = 0.69)	Tested up to 1000 hrs
E. Polystyrene beads + 1-Butyl-3-methlimidazoliumiodide+I_2	N719	N/A	0.77	15.3	7.5 (FF = 0.64)	N/A
F. Electrospun polyvinylidene fluoride-co-hexafluoropropylene (e PVDF-co-HPF) + I_2 + Tetrabutylammoniumiodide + PMII + PC or EC + CN biphenyl	N719	2.9	0.72	14.62	6.82	N/A
G. Polyvinylidene fluoride-hexafluoropropyle + PMII + I_2 + NMBI in MPN	Z907	10.4–12	0.74	13.1	6.7	N/A
H. Polyacrylonnitrile: polyethylene glycol/carbon + ethylene carbonate + 1-N-butyl-3-hexyl imidazolium iodide + I_2	N719	80.6	0.76	13	6.55	Tested up to 1000 hrs
I. PMII + BMII + Polyaniline loaded carbon black	N719	N/A	0.73	12.2	5.81 (FF = 0.65)	Tested up to 1000 hrs

TABLE 8.2 *(Continued)*

Electrolyte	Dye	Ionic Conductivity	O. C. Photo-Voltage (V)	S. C. Photo-Current Density	Efficiency (%)	Life Span
J. PVDF–HPF + MPII	Z907	N/A	0.66	11.69	5.3	N/A
K. Agarose + 1-allyl-3 ethylimidadoliumiodide + propylenecarbonate + guanidiniumthiocynate + N-methlbenzimidazole/I_2	N719	N/A	0.72	11.71	5.45 (FF = 0.65)	N/A
L. Carboxymethyl cellulose sodium salt (cmc) + polyethylene oxide	N/A	N/A	0.75	10.03	N/A	N/A
M. Agarose + MPII	N719	N/A	0.7	11.73	5.25 (FF = 0.64)	N/A
N. Carboxymethyl cellulose/polyethylene oxide + sodium iodide + MPII + TBP + I_2	N/A	N/A	0.75	10.03	5.18 (FF = 0.69)	N/A
O. 1-ethyl-3-methylimidazolium bis trifluoromethylsulphonylimdide + EMIm-I + lithium iodide + I_2 + Tert-butylpyridene + carbon black composite	N3	3	0.67	11.19	4.83	N/A
P. PVA-PEO-glass fiber-mat	N/A	10	N/A	N/A	N/A	N/A
Q. Nanofiber cellulose hydrogel + gelatin + KOH	N/A	97	N/A	N/A	N/A	N/A
R. Polyacrylic acid + potassium hydroxide	N/A	456–540	N/A	N/A	N/A	N/A

OPEN CIRCUIT PHOTOVOLTAGE AND FILL FACTOR OF DIFFERENT SOLID STATE ELECTROLYTES BASED DSSC

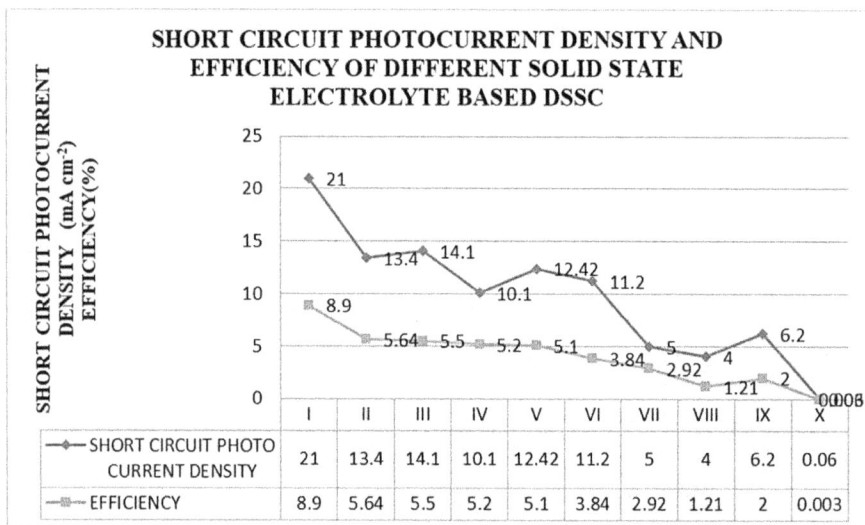

	I	II	III	IV	V	VI	VII	VIII	IX	X
OPEN CIRCUIT PHOTOVOLTAGE	0.68	0.68	0.65	0.79	0.66	0.72	0.61	0.72	0.89	0.14
FILL FACTOR	0.62	0.62	0.6				0.57	0.42		0.26

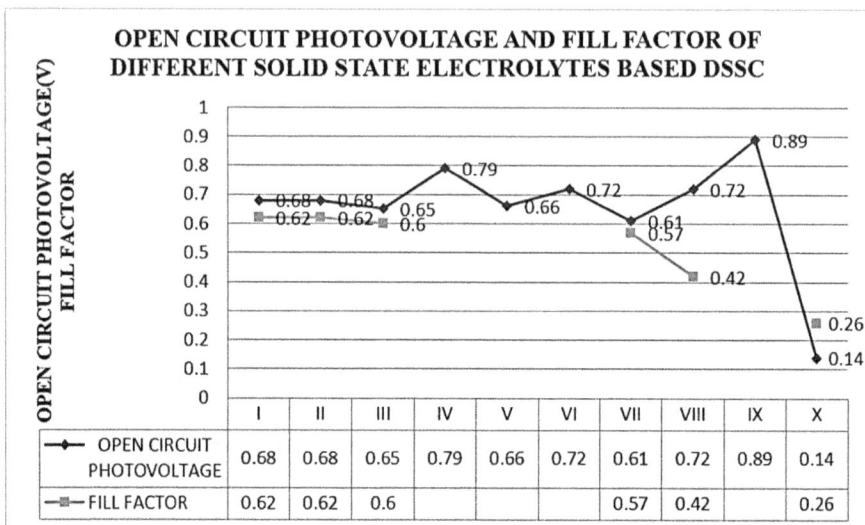

FIGURE 8.3 Open circuit photovoltage and fill factor of different solid-state electrolyte based DSSC.

SHORT CIRCUIT PHOTOCURRENT DENSITY AND EFFICIENCY OF DIFFERENT SOLID STATE ELECTROLYTE BASED DSSC

	I	II	III	IV	V	VI	VII	VIII	IX	X
SHORT CIRCUIT PHOTO CURRENT DENSITY	21	13.4	14.1	10.1	12.42	11.2	5	4	6.2	0.06
EFFICIENCY	8.9	5.64	5.5	5.2	5.1	3.84	2.92	1.21	2	0.003

FIGURE 8.4 Short circuit photocurrent density and efficiency of different solid-state electrolyte based DSSC.

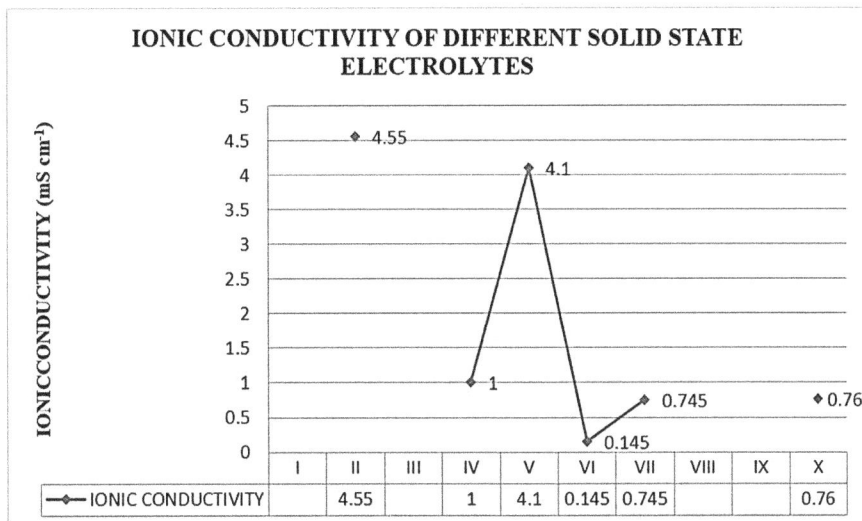

FIGURE 8.5 Ionic conductivity of different solid-state electrolytes.

ii. **For Quasi Solid Electrolytes Polymer Electrolytes, A–O (as in Table 8.2):** Figures 8.6–8.8 show the open circuit photovoltage, short circuit photocurrent density, and ionic conductivity of different quasi solid-state electrolyte based DSSC.

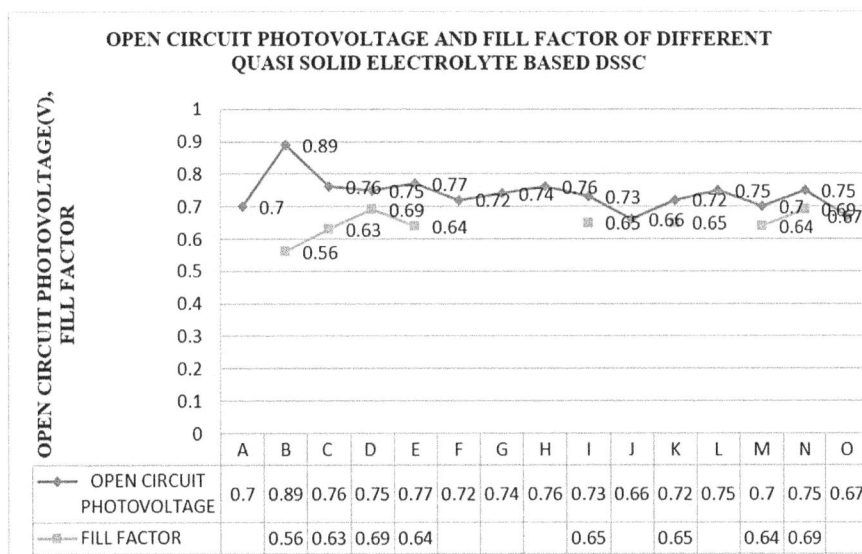

FIGURE 8.6 Open circuit photovoltage and fill factor of different quasi solid-state electrolyte based DSSC.

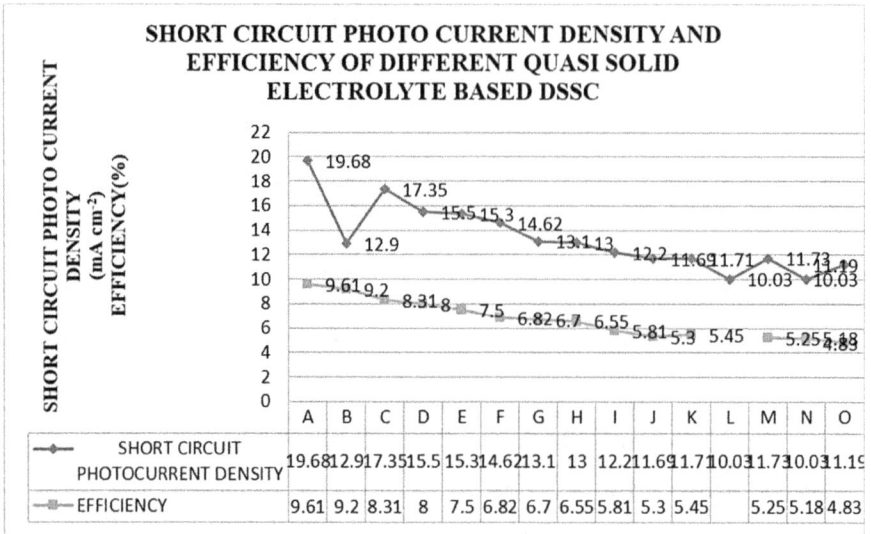

FIGURE 8.7   Short circuit photocurrent density and efficiency of different quasi solid-state electrolyte based DSSC.

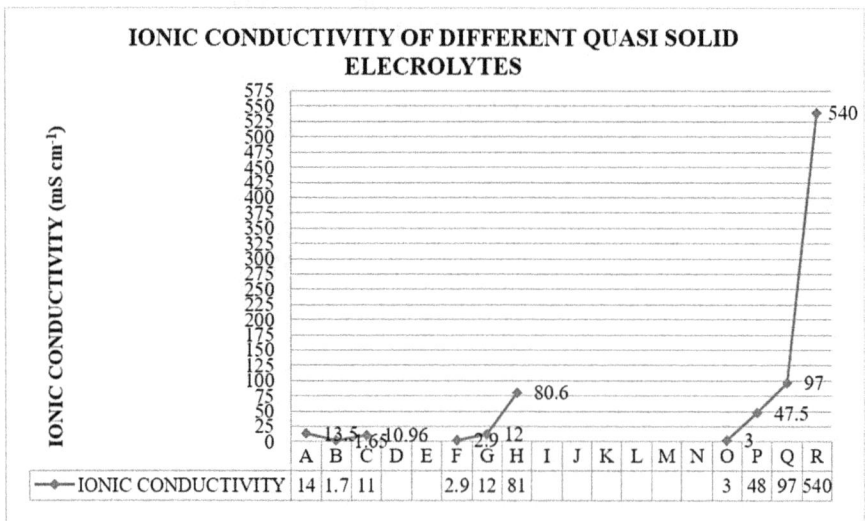

FIGURE 8.8   Ionic conductivity of different quasi solid electrolytes.

8.5 CONCLUSION

According to above analytics, best electrolyte among all is A(N-phthaloyl-chitosan + polyethylene oxide + tetrapropylammonium iodide + dimethyl-formamide + tetrapropylammonium iodide + ethylene carbonate + BMII) having efficiency 9.61%, but it can't be used here because of carbon component used in it which behaves unstable at high temperatures.

Second most efficient electrolyte shown in table is B (Polyethyleneimine +polyethylene glycol +KI + I_2 + dopant I) having efficiency of 9.2% due to desirable parameters like having ionic conductivity 1.65 mS cm^{-2}, open circuit photovoltage 0.89 V, short circuit photocurrent density 12.9 mA cm^{-2}, FF 0.56 comes under best dye class.

8.6 FUTURE SCOPE

In the year 2009, scientists discovered an organic metal Hallide ($CH_3NH_3PbI_3$) constituting Perovskite Cell, having efficiency 3.8% [43]. Perovskite solar cell is an enhanced branch of DSSC. It is a next generation low cost printable cell [42]. Its high-power conversion efficiency of around 26% has led to an increase in its efficacy in recent recently.

Its power conversion efficiency has shown an increase from 3.7% to 25.2% during the past decade [42, 45]. Perovskite cell can be created on a glass or polyethylene terephthalate with electrode (ITO/FTO) by printing techniques [41] Free carriers produce by lead halides $CH_3NH_3PbI_3$ (act as photoabsorber) collected by electrodes(Au or Ag) through p-type (organic hole transport material) and n-type(compact TiO_2) buffer layer [42] further depicting distinguishing optical properties, excitonic properties, and electrical conductivity. Metal halide can be sensitized from sources like Plumbu (Pb), C, N, and halogen [46]. Primarily design of the Perovskite Solar Cell looks similar to the DSSC, which uses a mesoporous TiO_2 layer for electron transfer but later, it was found that the Perovskite absorber layer can transfer the electrons, which increase its efficiency. It gave innovation to the development of two more revolutionary Perovskite cells, i.e., heterojunction perovskite solar cells and ultra-thin perovskite cell with absorber layers (it uses extremely thin layer as light-harvesting agents) [47, 48]. Different techniques used for the fabrication of Perovskite Cell are screen-printing, electrodeposition, vacuum thermal deposition, spin coating, and chemical vapor deposition. There are few drawbacks of Perovskite Solar Cell as well, such as unstable nature when exposed to moisture, high temperature, and continuous illumination [49, 50] due to which it has not come into use frequently. Research is done to overcome these limitations and an introduction to such cells is sure to revolutionize the solar energy arena.

KEYWORDS

- **fill factor**
- **Plumbu**
- **polyacrylonitrile**
- **polyethylene glycol**
- **polymethyl ether methacrylate**
- **polymethyl methacrylate**

REFERENCES

1. O'regan, B., & Gratzel, M., (1991). A low-cost, high efficiency solar cell based in dye sensitized colloidal TiO$_2$ films, *Nature, 353*, 737–740.
2. Jihuai, W., Zhang, L., Jianming, L., Miaoliang, H., Yunfang, H., Leqing, F., & Genggeng, L., (2015). Electrolytes in dye-sensitized solar cells. *ACS Chem. Rev.,* 2136–2173.
3. Jamalullail, N., Mohamad, I. S., Norizan, M. N., Baharum, N. A., & Mahmed, N., (2017). Short review: Natural pigments photosensitizer for dye-sensitized solar cell (DSSC). *IEEE 15th Student Conference on Research and Development (SCOReD)* (pp. 344–353).
4. Carbó-Vela, P. C., Rocha-Rangel, E., Rodríguez-García, J. A., Martínez-Peña, E., López-Hernández, J., & Armendáriz-Mireles, E. N., (2015). Construction of dye-sensitized solar cells (DSSC) with natural pigments. *Advances in Functional Materials (Conference 2015), AFM,* (pp. 194–200).
5. Xiaoru, G., Ganhua, L., & Junhong, C. (2015). Graphene-based materials for photoanodes in dye-sensitized solar cells. *Front. Energy Res., 3,* 1–15, Article: 50.
6. Mazalan, M., Mohd, N. M., Wahab, Y., Norizan, M. N., & Mohamad, I. S., (2016). Development of dye-sensitized solar cell (DSSC) using patterned indium tin oxide (ITO) glass fabrication and testing of DSSC. In: *IEEE Conference on Clean Energy and Technology (CEAT), 2013* (pp. 187–191).
7. Su'ait, M. S., Rahman, M. Y. A., & Ahmad, A., (2015). Review on polymer electrolytes in DSSC. *Solar Energy, 115,* 452–470. Elsevier.
8. Zhang, D. W., Li, X. D., Sun, Z., Chen, Y. W., & Huang, S. M., (2008). Quasi solid state DSSC prepared with a D102 sensitizer and a polymer electrolyte. In: *2nd IEEE International Nanoelectronics Conference (INEC 2008)* (pp. 761–765).

9. Sung-Hae, P., Jongchul, L., In Young, S., Jae-Ryung, L., & Taiho, P., (2013). Physically stable polymer-membrane electrolytes for highly efficient solid-state dye-sensitized solar cells with long-term stability. *Adv. Energy Mater.*

10. Federico, B., & Roberta, B., (2013). Photoinduced polymerization: An innovative, powerful and environment friendly technique for the preparation of polymer electrolyte for DSSC. *Journal of Photochemistry and Photobiology C: Photochemistry Reviews, 16*, 1–21. Elsevier.

11. Anteneh, A., & Solomon, D., (2018). Review on dye-sensitized solar cells (DSSCs). *J. Heterocyclics, 1*, 29–34. Edelweiss Publications.

12. Federico, B., & Roberta, B., (2013). *Journal of Photochemistry and Photobiology C: Photochemistry Reviews, 16*, 1–21. Elsevier.

13. Fenton, D. E., Parker, D. M., & Wright, P. V., (1973). Complexes of alkali metal ions with poly(ethylene oxide). *Polymer, 14*.

14. De Frietas, J. N., Benedetti, J. E., Frietas, F. S., Nogueira, A. F., & De Paoli, M. A., (2010). *Polymer Electrolytes for Dye-Sensitized Solar Cells* (pp. 381–430). Woodhead Publishing Limited.

15. Koh, S. N., Ramesh, S., Ramesh, K., & Joon, C. J., (2016). *A Review of Polymer Electrolytes: Fundamental, Approaches and Applications.* Springer-Verlag Berlin Heidelberg.

16. Buraidah, M. H., Shahan, S., Teo, L. P., Faisal, I. C., Careem, M. A., Albinsson, I., Mellander, B. E., & Arof, A. K., (2017). High efficient dye sensitized solar cells using phthaloyl chitosan based gel polymer electrolytes. *Electrochimica Acta.*

17. Karthikaa, P., Ganesana, S., & Arthanareeswaria, M., (2018). Low-cost synthesized organic compounds in solvent free quasi-solid state polyethyleneimine, polyethylene glycol-based polymer electrolyte for dye sensitized solar cells with high photovoltaic conversion effciencies. *Solar Energy, 160*, 225–250. Elsevier.

18. Sheng-Yen, S., Rui-Xuan, D., Po-Ta, S., Vittal, R., Jiang-Jen, L., & Kuo-Chuan, H. (2014). Novel polymer gel electrolyte with organic solvents for quasi-solid-state dye-sensitized solar cells. *ACS Appl. Mater. Interfaces.*

19. Iacopo, B., Hannes, M., & Marina, F., (2018). Solid-state dye-sensitized solar cells. *Mater. Chem. C, 1*–123.

20. Sung, K. A., Taewon, B., Sakthivel, P., Jae, W. L., Yeong-Soon, G., Jin-Kook, L., Mi-Ra, K., & Sung-Ho, J., (2012). Development of dye-sensitized solar cells composed of liquid crystal embedded, electro spun poly(vinylidene fluoride-co-hexafluoropropylene) nanofibers as polymer gel electrolytes. *ACS Appl. Mater. Interfaces, 4*, 2096−2100.

21. Karuppannan, R., Sambandam, A., & Kandasamy, J., (2015). Polymer electrolytes in dye sensitized solar cells. *Materials Focus, 4*, 262–271.

22. Chuan-Pei, L., Po-Yen, C., Vittala, R., & Kuo-Chuan, H., (2010). Iodine-free high efficient quasi solid-state dye-sensitized solar cell containing ionic liquid and polyaniline-loaded carbon black. *J. Mater. Chem., 20*, 2356–2361.

23. Hsin-Ling, H., Cheng-Fang, T., Ya-Ting, Y., & Jihperng, L., (2013). Dye-sensitized solar cells based on agarose gel electrolytes using allyl imidazolium iodides and environmentally benign solvents. *Electrochimica Acta, 91*, 208–213.

24. Federico, B., Jijeesh, R. N., & Claudio, G., (2013). Towards green, efficient and durable quasi-solid dye sensitized solar cells integrated with a cellulose-based gel-polymer electrolyte optimized by a chemometric DoE approach. *RSC Adv., 3*, 15993–16001.

25. Aswani, P., Karla, N., Nathaniel, J., Liangbing, H., Yucheng, L., & Deepa, M., (2019). Wood cellulose-based thin gel electrolyte with enhanced ionic conductivity. *MRS Communications*, 1–7. Materials Research Society.

26. Masanobu, C., Shoji, T., Eiji, H., & Hiroshi, I., (2011). Preparation and characterization of organic-inorganic hybrid hydrogel electrolyte using alkaline solution. *Polymers, 3*, 1600–1606.

27. Aswani, P., Eunhwa, J., Deepa, M., Nathaniel, J., Liangbing, H., & Yucheng, L., (2019). Cellulose hydrogel as a flexible gel electrolyte layer. *MRS Communications*, 1–7. Materials Research Society.

28. Chun-Chen, Y., & Sheng-Jen, L., (2002). Alkaline composite PEO-PVA-glass-fiber-mat polymer electrolyte for Zn-air battery. *Journal of Power Sources, 112*, 497–503. Elsevier.

29. Zahra, S., Rasoul, M., Hui-Ping, W., Jia-Wei, S., & Wei-Guang, D. E., (2015). High-performance and stable gel-state dye-sensitized solar cells using anodic TiO_2 nanotube arrays and polymer-based gel electrolytes. *ACS Appl. Mater. Interfaces.*

30. Hiroki, U., Hiroshi, M., Nobuo, T., & Shozo, Y., (2004). Improved dye-sensitized solar cells using ionic nanocomposite gel electrolytes. *Journal of Photochemistry and Photobiology A: Chemistry, 164*, 97–101.

31. Woohyung, C., Young, R. K., Donghoon, S., Hyung, W. C., & Yong, S. K., (2014). High-efficiency solid-state polymer electrolyte dye-sensitized solar cells with a bi-functional porous layer. **J. Mater. Chem. A**, **2**, 17746–17750.

32. Jihuai, W., Sanchun, H., Zhang, L., Jianming, L., Miaoliang, H., Yunfang, H., Pingjiang, L., et al., (2008). An all-solid-state dye-sensitized solar cell-based poly(N-alkyl-4-vinyl-pyridine iodide) electrolyte with efficiency of 5.64%. *J. Am. Chem. Soc., 130*, 11568, 11569.

33. Chen, K. F., Liou, C. H., Lee, C. H., & Chen, F. R., (2010). *Development of Solid Polymeric Electrolyte for DSSC Device* (pp. 003288–003290). IEEE.

34. Shaheer, A. M., Zhen, Y. L., Woojin, L., & O-Bong, Y., (2013). *Effective Inorganic-Organic Composite Electrolytes for Efficient Solid-State Dye Sensitized Solar Cells* (pp. 2414–2416). IEEE.

35. Senthil, R. A., Theerthagiri, J., & Madhavan, J., (2016). Organic dopant added polyvinylidene fluoride based solid polymer electrolytes for dye-sensitized solar cells. *Journal of Physics and Chemistry of Solids, 89*, 78–83. Elsevier.

36. Ramanpreet, K. A., Sana, S., Tanvi, Sandeep, K., Aman, M., Bedi, R. K., & Subodh, K., (2015). Designing and synthesis of imidazole-based hole transporting material for solid state dye sensitized solar cells. *Synthetic Metals, 205*, 92–97. Elsevier.

37. Matteocci, F., Casaluci, S., Razza, S., Guidobaldi, A., Brown, T. M., Reale, A., & Di Carlo, A., (2013). Solid state dye solar cell modules. *Journal of Power Sources.*

38. Su'ait, M. S., Ahmad, A., Badria, K. H., Mohamed, N. S., Rahman, M. Y. A., Azanza, R. C. L., & Scard, P., (2013). The potential of polyurethane bio-based solid polymer electrolyte for photoelectrochemical cell application. *International Journal of Hydrogen Energy*, 1–13. Elsevier.

39. Gomesh, N., Ibrahim, A. H., Syafinar, R., Irwanto, M., Mamat, M. R., Irwan, Y. M., Hashim, U., & Mariun, N., (2015). Fabrication of dye sensitized solar cell using various counter electrode thickness. *International Journal Series in Engineering Science (IJSES), 1*(1), 49–56.

40. Hubert, H., Michael, B., Peter, M., & Thilo, G., (2014). Biophotovoltaics: Natural pigments in dye-sensitized solar cells. *Applied Energy, 115*, 216–225.

41. Castro-Hermosa, S., Janardan, D., Andrea, M., & Thomas, M. B., (2017). *Perovskite Solar Cells on Paper and the Role of Substrates and Electrodes on Performance,* 0741–3106. IEEE.
42. Atsushi, W. (2016). Recent progress on perovskite solar cells and our materials science. *AM-FPD, 16,* 5–8.
43. Zhou, D., Tiantian, Z., Yu, T., Xiaolong, Z., & Yafang, T. (2018). Perovskite-based solar cells: Materials, methods, and future perspectives. *Hindawi Journal of Nanomaterials Volume 2018.*
44. Nam-Gyu, P., (2014). Perovskite solar cell an emerging photovoltaic technology. *Materials Today.* Elsevier.
45. NREL cell efficiency chart, N. E. Best Research Cell Efficiencies, 2019.
46. Akihiro, K., Kenjiro, T., Yasuo, S., & Tsutomu, M., (2009). Organometal halide perovskites as visible-light sensitizers for photovoltaic cells, *J. Am. Chem. Soc., 131,* 6050–6051.
47. Yang, L., Felix, L., Thomas, D., Alexander, S., Christian-Herbert, F., Tristan, K., Paul, P., et al., (2017). Enhancement of photocurrent in an ultra-thin perovskite solar cell by Ag nanoparticles deposited at low temperature. *RSC Adv., 7,* 1206–1214.
48. Hui-Seon, K., Chang-Ryul, L., Jeong-Hyeok, Ki-Beom, L., Thomas, M., Arianna, M., Soo-Jin, M., et al. (2012). Lead iodide perovskite sensitized all-solid-state submicron thin film mesoscopic solar cell with efficiency exceeding 9%. *Scientific Reports, 2,* 591.
49. Sheikh, I. R., Sakib, F., Sarwar, A., & Tanvir, I. D., (2017). A comparative study on different HTMs in perovskite solar cell with ZnOS electron transport layer. In: *2017 IEEE Region 10 Humanitarian Technology Conference* (pp. 546–550).
50. Dian, W., Matthew, W., Naveen, K., Elumalai, & Ashraf, U., (2016). Stability of perovskite solar cells. *Solar Energy Materials and Solar Cell, 147,* 255–275.

CHAPTER 9

Performance of Nano-Structured Thermal Spray Coatings in the Renewable Energy Sector

GAURAV PRASHAR and HITESH VASUDEV

School of Mechanical Engineering, Lovely Professional University,
Phagwara – 144411, Punjab, India,
E-mail: hiteshvasudev@yahoo.in (H. Vasudev)

ABSTRACT

The demand for renewable energy sources for power generation has attracted the attention of the research community in the last decade owing to the depletion of fossil fuels and environmental issues. Extensive experimental investigations have been conducted to extract maximum energy from natural resources in the dams and power plants. The coating technologies play a significant role in protecting the components used in power plants. The thermal spraying technique is considered as an effective and economical hard facing method for applying coatings consisting of multilayer stacks on the surface of structural components. However, the latest material development in the form of nano-size scale enhances the material properties like hardness, toughness, and resistance to sliding wear. This is mainly due to reduction in inherent defects and dislocation of grain boundaries. Hence, the deposition of nanostructured material leads to an engineered surface with outstanding coating efficiency. The current study focuses on the high-temperature oxidation behavior of nanostructured coatings deposited by thermal spraying concerning the role of main alloying elements at high temperatures towards oxidation, oxide scale growth, and spallation tendency of coatings.

9.1 INTRODUCTION

In the last few years, the use of waste and ecological biomass as a fuel is increased gradually in coal-based power plants instead of conventional fossil fuels, moving a step ahead in the direction for clean environment and to meet the recently up-graded energy targets set by EU for the year 2030 [1]. Currently, over 80% of the energy demand is fulfilled by fossil fuels but their reserves would not last long and get exhausted in coming decade. Biomass is a carbon neutral fuel, whereas use of the waste is important to reduce the number of waste landfills. However, despite of several advantages, there is also a technical challenge to use them in power plants. Biomass/waste on combustion releases a large amount of corrosive agents such as Cl_2, NaCl/KCl, HCl, and H_2O that accelerates the corrosion phenomena. Alkali chlorides (NaCl/KCl) are deposited on the structural components of the boilers and reduce the heat transfer rate, which directly affects the efficiency of power plants [2]. Hot corrosion/high-temperature oxidation is a complex and serious problem in power plants, which leads to unplanned shutdowns, increased downtimes, and material degradation [3]. Therefore, coatings that provide corrosion resistance at high temperatures are increasingly being sought. Thermally sprayed nanostructured coatings are examined by various researchers, and considerable improvement was noticed in performance due to their nano-sized variants as compared to conventional grain size [4–12]. $Ni60-TiB_2$ nanostructured composite and conventional coating was subjected to cyclic oxidation studies at 800°C by [13]. HVOF (high velocity oxygen fuel) thermal spraying technique was used to spray the coatings. Authors concluded that nanostructured composite coating display better oxidation resistance than its counterpart [13].

Hence, the focus of this article is to examine the high temperature oxidation behavior of nanostructured coatings deposited by thermal spraying (i.e., the role of main alloying elements at high temperatures towards oxidation, oxide scale growth and spallation tendency of coatings) because of their broad application to enhance the service life of structural components at high temperatures.

9.1.1 NANOCOMPOSITE COATINGS

Nanotechnologies are regarded as key technologies for innovation worldwide in almost every branch of economy. Nanocomposite coatings made up of two substances, out of which one must have a nanoscale dimension of less than 100 nm. Worldwide growth of nanocoatings can be evident from (a) the number of patent applications (inventions) that reaches to 1200 in 2012 as compared to 142 in 2000 and all these patents were filed in top 5 patent offices, and (b) number of SCI publications that shows research activities in area of nanocomposites. In the year 2000 the number was 1319 and in 2012 it was 31000 as shown in Figure 9.1.

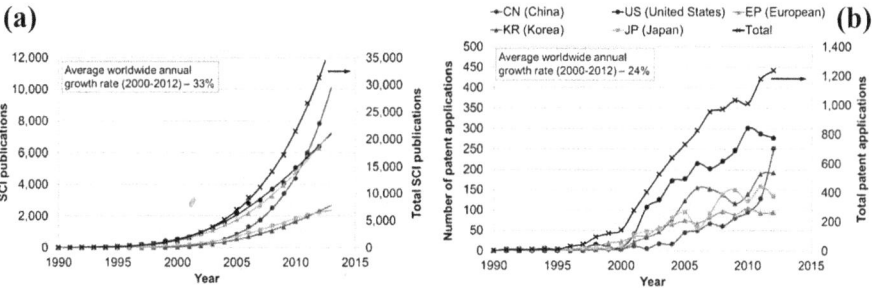

FIGURE 9.1 Number of SCI publications and total number of patents on nanocomposite coatings [14].

Source: Reprinted with permission from Ref. [14]. © 2015 Taylor & Francis.

Nanocoatings are one of the alternatives due to its excellent abilities to combat corrosion/oxidation as compared to coarse-grained coating at elevated temperatures. Nanocomposite can be broadly classified into two types (a) metal matrix composite, within the nanoscale (b) single-layered or multi-layered coating [14] as shown in Figure 9.2.

Nano coatings can be produced by three methods (a) physical, (b) chemical and, (c) mechanical. Bonding, condensation, and sputtering all comes under physical methods. Physical diffusion bonding was done by applying medium pressure and temperatures whereas in braze bonding high temperature is used with addition of lubricants. SAB (surface additive bonding) use low temperatures and pressures for getting good flat polished and clean surfaces. However, SLS (selective layer sintering) makes use of new technology named 3-D printing in which material is manufactured layer by layer. Second method of physical bonding is condensation [15]. PVD is conducted under vacuum, whereas continues pressure conditions were use in LPE (liquid phase epitaxy). Sputtering process is costly but produces good bond strengths. It is performed by MBE (molecular beam epitaxy), radio frequency (RF) magnetron, and PLD. At last, chemical method is cheaper but required costly precursors like Langmuir, sol-gel, etc. However, mechanical methods are the cheapest one and it includes painting, spraying, and dip coatings. Thin films deposited by all these methods on the surface of substrates effects their mechanical properties like fracture, strength, and ductility. Each method has its own positives and negatives, choosing the right technique depend upon the studying of all process elements in detail. Various methods of thin film deposition techniques are shown in Figure 9.3.

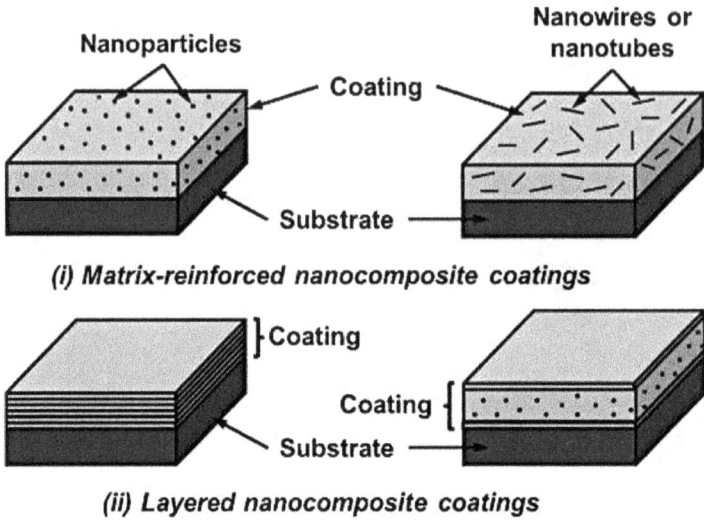

FIGURE 9.2   Schematic representation of nanocomposite coatings: (i) matrix reinforced; (ii) multi-layered.

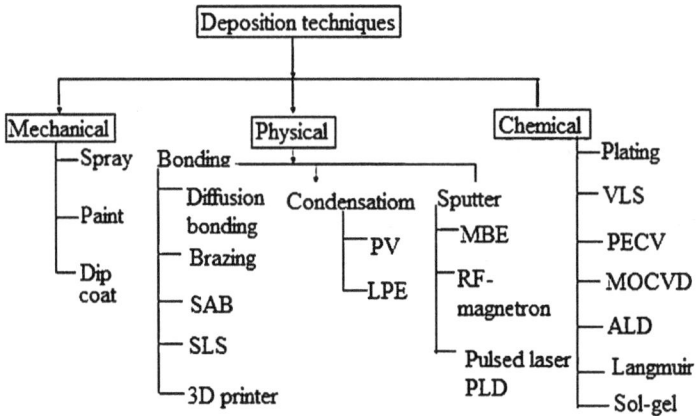

FIGURE 9.3   Thin-film deposition techniques.

9.2   STUDIES RELATED TO NANOSTRUCTURED COATINGS

9.2.1   STUDIES RELATED TO TI ALLOYS AND TI COMPOSITES

Titanium alloys and titanium-aluminum intermetallics have multiple set of properties like lightweight, high melting point, and high strength. Furthermore, they also have excellent mechanical properties which make them candidate for the applications where high temperature corrosion resistance is needed. However, oxidation resistance of Ti alloys and Ti composites at elevated temperature is poor. Reason being the development of TiO_2 fails to provide better protection against further oxidation. Hence, their use can be limited up-to 650°C ranges. An effort has been made to improve the oxidation resistance of these materials. One alternate is the addition of costly alloying additions like Nb, but the results obtained were not satisfactory.

To overcome this drawback, a special composite consisting of $Ti_3Al(O)$-Al_2O_3 was developed by ball milling TiO_2 with Al [16]. Thermal spraying was used to deposit the developed composite onto the pure titanium substrate. Oxidation studies were conducted at temperatures of 700°C and 800°C for duration of 400 hr. Coatings exhibit superior resistance against oxidation and scale spallation as shown in Figure 9.4. This may be attributed to the addition of Al_2O_3 that plays major role during oxidation by acting as diffusion barrier and inhibitor to the formation of cracks in oxide scale, Figure 9.5(a and c). Whereas, thick and dense TiO_2 scale appears onto the substrate leading to cracks formation along the scale, Figure 9.5(b and d). Hence, the temperature range for titanium alloys and titanium-aluminum intermetallics can be increased from 650°C to 800°C by adding aluminum oxide [31].

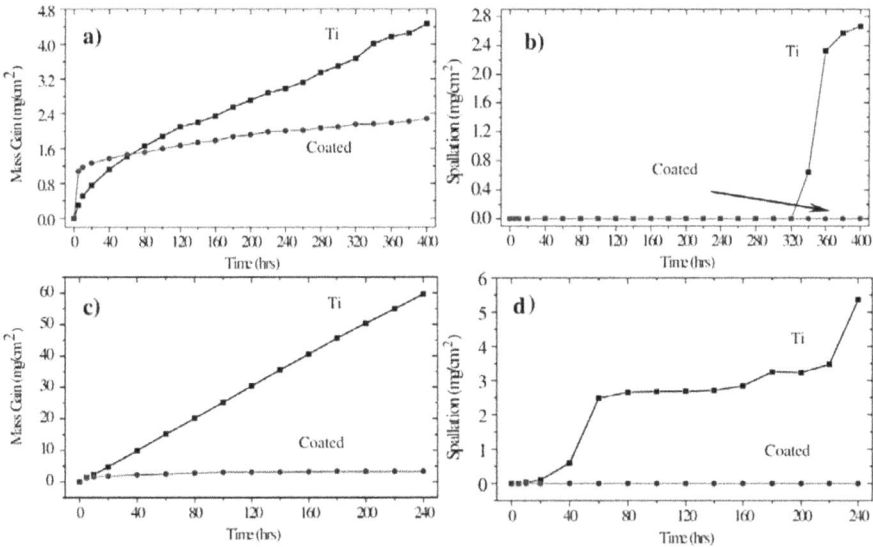

FIGURE 9.4 Oxidation kinetics and scale spallation of Ti with and without composite coating: (a–b) 700°C; (c–d) 800°C [31].

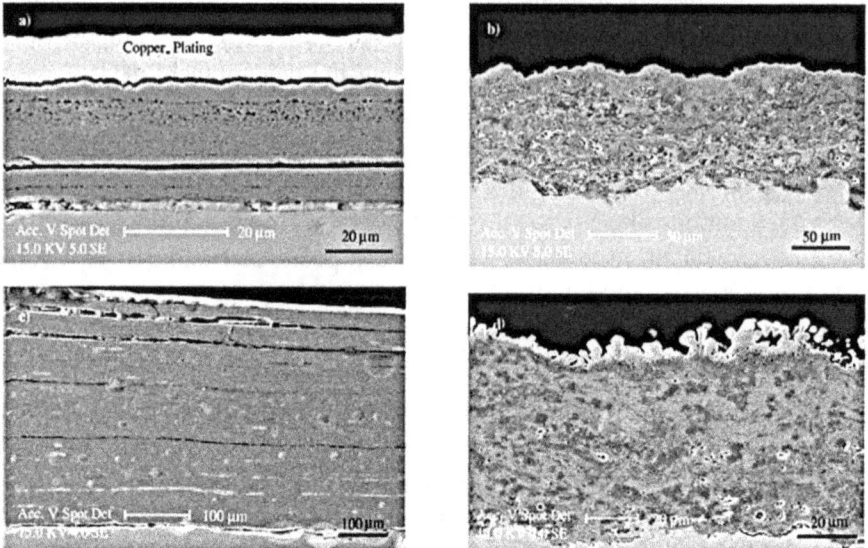

FIGURE 9.5   Cross-section of Ti with and without composite coating: (a) 700°C, 400 hr; Ti; (b) 700°C, 400 hr, coated Ti; (c) 800°C, 240 hr, Ti; (d) 800C, 240 hr, coated Ti [31].

9.2.2   STUDIES RELATED TO NI-CR COATINGS

Structural components operating at higher temperatures degrade due to high temperature oxidation. As an alternate to this issue, nickel-based alloys that contains higher content of Cr will be the candidate choice as a coating material for the protection of these components [17–22]. Reason is the continuous Cr_2O_3 protective scale that prevents the diffusion of corrosive species into the substrate. Several investigations done by different researchers confirmed that Ni-Cr nanostructured alloy coating displays multiple set of properties like higher hardness, ductility, strength, wear, and corrosion resistance as compared to conventional Ni-Cr based coatings [23–28].

Singh et al. [29] studied microstructure and oxidation demeanor of nano-structured Ni-20Cr coating deposited on SA-516 grade 70 steel by using cold spray technique. Oxidation experiments were conducted at 900°C up to 50 hr in air. Coatings were examined by various characterization techniques (TEM, SEM, EDX, SAED, and FIB) before and after oxidation experiments. Authors observed that no cracking and spalling of scale takes place in coating after exposure. High-temperature oxidation resistance was provided by thin layer of chromium oxide along with nano-sized chromium oxide present within the nanostructured matrix [29]. Furthermore, nano-sized Cr_2O_3 precipitates surround the nickel metallic grains and will serve as grain pinning objects. It was concluded that nanostructured Ni-20Cr coating exhibit enhanced oxidation resistance at 900°C. Nano Ni-20Cr (powder size 42 nm and 57 nm) coatings were deposited by using cold spray process onto T-22 and SA-516 boiler steels. Main purpose of the study is to examine the high temperature erosion-corrosion demeanor of nanostructured coatings. It was reported that after 1500 hr of exposure in actual conditions of thermal power plant, Ropar-India only marginal loss in thickness was observed. This may be attributed to the fact that with decrease in powder size micro-hardness of the coating increases as compared to conventional powders and also due to the presence of protective oxides such as NiO and Cr_2O_3 [30].

Senthilkumar et al. [23] examine the effect of thermal cycle on Ni-Cr nanostructured coatings deposited by HVOF process. Ni-Cr based nanostructured coatings and bare coatings were exposed to a temperature of 650°C. By using various characterization techniques both coated and uncoated samples were characterized for variations in microhardness, metallurgical, and phase changes. Nanostructured coatings have higher metallurgical and microhardness properties than conventional coated samples due to dense and defect-free microstructure [23].

9.2.3 OXIDATION OF NANOCOATINGS

High-temperature oxidation is a serious and complex phenomena leading to degradation of structural components like super-heaters and re-heaters in coal-based power plants. It should be prevented at early stages to minimize the losses by using suitable measures. As a remedial measure, in order to prevent engineering components from oxidation/hot corrosion a protective barrier must develop in the form of thin films such as Cr_2O_3 and Al_2O_3. The main advantage of nanocoatings is their capability to promote selective oxidation. However, unfortunately the content of Al and Cr should neither be high nor be too low. In a coating of Ni-20Cr-Al the appropriate percentage of Al should be more than 6 wt% to form a dense and complete Al_2O_3 scale [31]. If it is less than 6 wt% it should be failed to protect the substrate as complex oxides may develop in the form of Cr_2O_3, internal Al_2O_3 and $NiCr_2O_4$. As a result high rates of reaction occur leading to poor oxidation resistance.

Moreover, by using nano coatings the quantity of Al required to form a complete Al_2O_3 scale also reduces than used in conventional coating. Experimental results already shows that when the size of grains in the coating system of Ni-20Cr-Al was in range of 60 nm, then alloy containing 2 wt% Al develops a complete protective oxide scale of α-Al_2O_3 at a temperature of 1000°C in the presence of air. This concentration of Al is only 1/3 which was required in Ni-20Cr alloy having normal grain size. SEM micrographs of the Ni-20Cr-2Al samples oxidized in air displays different morphologies for different grain sizes as shown in Figure 9.6.

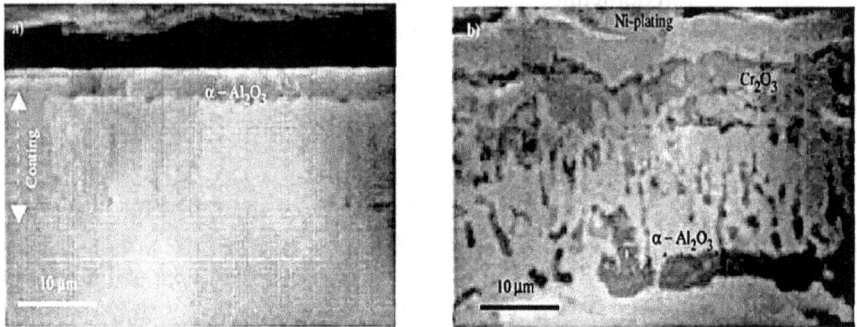

FIGURE 9.6 Cross-section of Ni-20Cr-2Al coating samples after oxidation at 1000°C for 100 hrs: (a) coating grain size 60 nm; (b) coating grain size 370 nm [31].

It is clear from Figure 9.6. That despite of same coating composition, i.e.; (Ni-20Cr-2Al), the coating system that has less grain size develops a complete, protective, and dense α-Al_2O_3 scale and proves to be better than one with conventional grain size. This indicates that oxidation resistance is clearly a function of grain size. Relationship between the grain size of the coating, oxide products, and aluminum content is shown in Figure 9.7. Indicating that nano structure clearly promotes selective oxidation of aluminum [31] by short circuit diffusion of Al through grain boundaries.

9.2.4 ENHANCED OXIDE SCALE SPALLATION PERFORMANCE

Structural components working under harsh environments rely on the formation of protective oxide film to combat degradation of the material from further oxidation. This developed oxide scale has good adhesion with substrate metal, strong, and tough. A common problem occurring in the alloys is that the protective oxide scale is not stable and as a result, cracking and spallation of the scale occurs after a certain duration under aggressive environments. Enhanced oxide scale spallation resistance is exhibited by nanostructure coatings. Scale spallation is influenced by various mechanical and chemical factors.

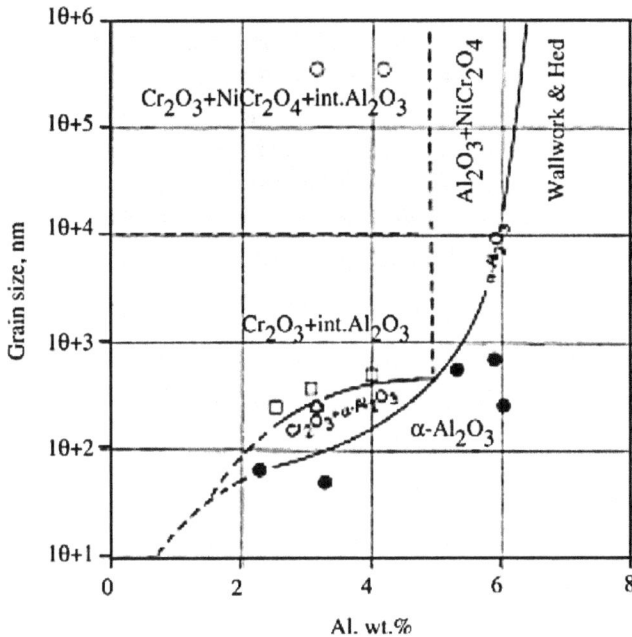

FIGURE 9.7 Relationship between the grain size of the coating, oxide products, and aluminum content for Ni-20Cr-Al at 1000°C in air [31].

The mechanism that is exhibited by nano coatings to enhance the scale spallation resistance is also complex. Different mechanisms were suggested on the basis of experimental results. First, a nano coating aids in promoting selective oxidation that results in formation of protective oxides. Secondly, at high temperatures fine grain coatings exhibited a faster creep rate resulting in release of the stresses present in the scales and therefore reduce scale spallation. Thirdly, oxides that develop on the nano coatings are pegged on to grain boundaries leading to formation of complex interface. This 'micro pegging' affects results in better adhesion of the scale to substrate.

9.3 CONCLUSION

In view of the ever-increasing energy demand, the development of sustainable methods to generate power are the most critical and urgent challenge to mankind. Growing efficiency and the latest methods through the nano-technological know-how play a major part in the innovation needed for the energy sector. Structural components made up of nanomaterials have a good service life and possess good mechanical properties than their conventional-sized counterparts. This review chapter demonstrated that nanostructured coatings possess enhanced high-temperature oxidation resistance. High-density grain boundaries occur as fast diffusion paths and promote selective oxidation of the protective oxide scale. Resistance to scale spallation and scale adhesion is improved by micro pegging along with stress release phenomena. The theories developed from this work helps in designing coatings for high-temperature oxidation and energy-related applications.

KEYWORDS

- **high-velocity oxygen fuel**
- **liquid phase epitaxy**
- **molecular beam epitaxy**
- **nanocomposite**
- **radio frequency**
- **selective layer sintering**

REFERENCES

1. Dupont, C., & Oberthür, S., (2015). Decarbonization in the EU: Setting the scene. In: *Anonymous Decarbonization in the European Union* (pp. 1–24). Springer.
2. Nielsen, H. P., Frandsen, F. J., Dam-Johansen, K., & Baxter, L. L., (2000). Implications of chlorine-associated corrosion on the operation of biomass-fired boilers. *Prog. Energy Combust. Sci., 26*(3), 283–298.
3. Prashar, G., & Vasudev, H. (2020). Hot corrosion behavior of super alloys. *Materials Today: Proceedings.* https://doi.org/10.1016/j.matpr.2020.02.226 (accessed on 12 December 2020).
4. He, J. H., Ice, M., & Lavernia, E. J., (2000). Synthesis of nanostructured Cr_3C_2-25(Ni20Cr) coatings [J]. *Metallurgical and Materials Transactions A, 31*, 555–564.
5. Tellkamp, V. L., Lau, M. L., Fabel, A., & Lavernia, E. J., (1997). Thermal spraying of nanocrystalline Inconel 718 [J]. *Nanostructured Materials, 9*, 489–492.
6. Lau, M. L., Jiang, H. G., Nuchter, W., & Lavernia, E. J., (1998). Thermal spraying nanocrystaline Ni coatings [J]. *Phys. Status Solidi A, 166*, 257–268.
7. Lau, M. L., Strock, E., Fabel, A., Lavernia, C. J., & Lavernia, E. J., (1998). Synthesis and characterization of nanocrystalline Co-Cr coatings by plasma spraying [J]. *Nanostructured Materials, 10*, 723–730.

8. He, J. H., Ice, M., Dallek, S., & Lavernia, E. J., (2000). Synthesis of nanostructured WC-12pct Co coating using mechanical milling and high velocity oxygen fuel thermal spraying [J]. *Metallurgical and Materials Transactions A, 31*, 541–553.

9. Stewart, D. A., Shipway, P. H., & Mccartney, D. G., (2000). Microstructural evolution in thermally sprayed WC-Co coatings: Comparison between nanocomposite and conventional starting powders [J]. Acta *Mater., 48*, 1593–1604.

10. Grosdidier, T., Tidu, A., & Liao, H. L., (2001). Nanocrystalline Fe40Al coating processed by thermal spraying of milled powder [J]. *Scripta Mater., 44*, 387–393.

11. Ajdelsztajn, L., Picas, J. A., Kim, G. E., Bastian, F. L., Schoenung, J., & Provenzano, V., (2002). Oxidation behavior of HVOF sprayed nanocrystalline NiCrAlY powder [J]. *Mater. Sci. Eng. A, 338*, 33–43.

12. Lau, M. L., He, J., Schweinfest, R., Ruhle, M., Levi, C. G., & Lavernia, E. J., (2003). Synthesis and characterization of nanocrystalline Cu-Al coatings [J]. *Mater. Sci. Eng. A, 347*, 231–242.

13. Wu, Y. S., Qiu, W. Q., Yu, H. Y., Zhong, X. C., Liu, Z. W., Zeng, D. C., & LI, S. Z., (2011). Cycle oxidation behavior of nanostructured Ni60-TiB$_2$ composite coating sprayed by HVOF technique [J]. *Applied Surface Science, 257*, 10224–10232.

14. Gan, J.A., and Berndt, C.C. (2014). *Nanocomposite Coatings: Thermal Spray Processing, Microstructure and Performance*. 2014 Institute of Materials, Minerals, and Mining and ASM International Published by Maney for the Institute and ASM International. doi: 10.1179/1743280414Y.000000004.

15. Wunderlich, W., (2014). The atomistic structure of metal/ceramic interfaces is the key issue for developing better properties. *Metals, 4*, 410–427.

16. Gao, W., Li, Z., & Zhang, D., (2002). *Oxidation of Metals, 57*, 99.

17. Bala, N., Singh, H., & Prakash, S., (2017). Performance of cold sprayed Ni based coatings in actual boiler environment. *Surf. Coat. Technol., 318*, 50–61.

18. Kawahara, Y., (2007). Application of high temperature corrosion-resistant materials and coatings under severe corrosive environment in waste-to-energy boilers. *J. Therm. Spray Technol., 16*, 202–213.

19. Fantozzi, D., Matikainen, V., Uusitalo, M., Koivuluoto, H., & Vuoristo, P., (2017). Chlorine- induced high temperature corrosion of Inconel 625 sprayed coatings deposited with different thermal spray techniques. *Surf. Coat. Technol., 318*, 233–243.

20. Lopez, A. J., Proy, M., Utrilla, V., Otero, E., & Rams, J., (2014). High-temperature corrosion behavior of Ni-50Cr coating deposited by high velocity oxygen-fuel technique on low alloy ferritic steel. *Mater. Des., 59*, 94–102.

21. Song, B., Pala, Z., Voisey, K. T., & Hussain, T., (2017). Gas and liquid-fueled HVOF spraying of Ni50Cr coating: Microstructure and high temperature oxidation. *Surf. Coat. Technol., 318*, 224–232.

22. Sadeghimeresht, E., Reddy, L., Hussain, T., Markocsan, N., & Joshi, S., (2018). Chlorine-induced high temperature corrosion of HVAF-sprayed Ni-based alumina and chromia forming coatings. *Corros. Sci., 132*, 170–184.

23. Senthilkumar, V., Thiyagarajan, B., Duraiselvam, M., & Karthick, K., (2015). Effect of thermal cycle on Ni-Cr based nanostructured thermal spray coating in boiler tubes. *Transactions of Nonferrous Metals Society of China, 25*, 1533–1542.

24. He, & Schoenung, J. M., (2002). Review-nanostructured coatings. *Mater. Sci. Eng. A, 336*, 274–319.

25. Kumar, M., Singh, H., Singh, N., Hong, S. M., Choi, I. S., Suh, J. Y., Chavan, N. M., Kumar, S., & Joshi, S. V., (2015). Development of nano-crystalline cold sprayed Ni-20Cr coatings for high temperature oxidation resistance. *Surf. Coat. Technol., 266,* 122–133.

26. Ivannikov, A. Y., Kalita, V. I., Komlev, D. I., Radyuk, A. A., Bagmutov, V. P., Zakharov, I. N., & Parshev, S. N., (2016). The effect of electromechanical treatment on structure and properties of plasma sprayed Ni-20Cr coating. *J. Alloys Compd., 655,* 11–20.

27. Lavernia, E. J., Lau, M. L., & Jiang, H. G., (1998). Thermal spray processing of nanostructured materials. In: Chow, G. M., & Noskova, N. I., (eds.), *Nanostructured Materials* (pp. 283–302). Springer, Dordrecht.

28. Kumar, M., Singh, H., & Singh, N., (2015). Fire side erosion-corrosion protection of boiler tubes by nanostructured coatings. *Mater. Corros., 66,* 695–709.

29. Swaminathan, S., Hong, S. M., Kumar, M., Jung, W. S., Kim, D. I., Singh, H., & Choi, I. S., (2019). Microstructural evolution and high temperature oxidation characteristics of cold sprayed Ni-20Cr nanostructured alloy coating. *Surface and Coatings Technology, 362,* 333–344. https://doi.org/10.1016/j.surfcoat.2019.01.112.

30. Kumar, M., Singh, H., Singh, N., & Joshi, (2015). Erosion-corrosion behavior of cold-spray nanostructure Ni-20Cr coatings in actual boiler environment. *Wear,* 1035–1043.

31. Wei, G., & Zhengwei, L., (2004). Nanostructured alloy and composite coatings for high temperature applications. *Material Research, 7*(1), 175–182.

CHAPTER 10

Molecular Dynamics Simulation for Finding Variation in Young's Modulus of Defective Single-Walled Carbon Nanotubes

MANISH DHAWAN[1] and SUMIT SHARMA[2]

[1]*Associate Professor, Department of Mechanical Engineering, Lovely Professional, University, Phagwara – 144411, Punjab, India, E-mail: mds_78@rediffmail.com*

[2]*Assistant Professor, Department of Mechanical Engineering, Dr. B. R. Ambedkar National Institute of Technology, Jalandhar, Punjab, India*

ABSTRACT

A perfect carbon nanotube (CNT) exhibits extraordinary mechanical and thermal properties. However, the presence of vacancy and Stone-Wales (SW) defects in CNTs has been determined experimentally. In this study, the effect produced by multiple SW and vacancy defects on Young's modulus of single-walled carbon nanotubes (SWCNTs) has been studied using molecular dynamics (MD) simulation. A (5,5) armchair SWCNT of diameter 0.67 nm has been used in this work. To study the positional effects of defects, a center and a corner position have been selected and multiple defects are placed on these positions. It has been found that the reduction in Young's modulus is more in the case of vacancy defects. Simulated results indicate that the defects placed at the center significantly decrease the modulus in comparison to SW defects.

10.1   INTRODUCTION

The first report on carbon nanotubes (CNTs) was published by Iijima in 1991 [1]. Since their discovery, these small nano-structures have attracted much attention due to their unique properties. The carbon arc-discharge methods [1, 2], laser vaporization of graphite electrode, and the chemical vapor-deposition methods from various carbon precursors [39] are the three main synthesis methods of single-walled carbon nanotubes (SWCNTs). High elastic modulus of CNTs has been found in both experimental [3, 4] and theoretical [5, 6] studies. Treacy et al. were among the first to study Young's modulus of CNTs experimentally by using a vibrating beam model. Their results show the average Young's modulus of 1.8 TPa for 11 MWCNTs. Krishnan [7] used the same method as Treacy et al. for studying Young's modulus of SWCNTs and reported the average Young's modulus of 1.25 TPa for 27 CNTs. By measuring bending force of MWCNTs using atomic force microscopy (AFM), Wong [4] reported Young's modulus of 1.28 ± 0.59 TPa.

In addition to experimental methods, there are a significant number of theoretical methods to study the mechanical behavior of CNTs. These methods are divided into three main categories as molecular dynamics (MD), Monte Carlo (MC), and Ab initio. An empirical force-constant model was used by Lu [5] to study the elastic properties of CNTs and carbon nano-ropes. Their studies show Young's modulus of about 1 TPa and shear modulus of 0.5 TPa for radii ranging from 0.34 to 13.5 6nm. A quenched MD simulation technique was used by Cornwell [8] to study the elastic behavior of SWCNTs in compression. A Tersoff-Brenner potential was used by them to provide interatomic interactions. Study of the above experimental and theoretical reports indicates that perfect CNT structures possess high elastic modulus than any other material. However, studies of Ebbesen and Takada [9] show that CNTs are not as perfect as they were once thought to be. Mainly, the defects in the CNT structures occur at the time of chemical synthesis [10], the purification process in chemical treatment [11], or when the CNTs are subject to irradiation [40]. The existence of vacancies and stone Wales (SW) defects are experimentally observed by researchers [12, 13].

In recent years, by applying theoretical methods a significant work has been done to study the mechanical behavior of the CNTs containing SW and vacancy defects. By employing MD simulation, Sammalkorpi [14] studied the effect produced by vacancy-related defects on the mechanical behavior of SWCNTs. Their studies predicted 3% reduction in Young's modulus when a relatively high defect density of one vacancy per 50 Å was introduced. A MD simulation technique was used by Xin [15] to study the buckling behavior of zigzag and armchair SWCNTs. Their results show that the buckling mode was significantly influenced by each small defect. The role of vacancy defects in the fracture of CNTs was studied by Mielke et al. [16] using density functional theory and semiempirical methods, and molecular mechanics (MM) calculations with a Tersoff-Brenner potential. By adapting three different stress measures at atomic scales and introducing strain measures as energetically conjugate quantities, Chandra et al. [17] studied the local properties of SW defects in SWCNTs under axial tension. The stiffness of defects in their results was reduced by about 30–50% and was dependent on a number of factors such as the chirality, the diameter, and the loading conditions. An atomistic simulation technique was used by Lu and Bhattacharya [18] to study the effect of randomly occurring SW defects on the mechanical properties of (6,6) armchair and (10,10) zigzag SWCNTs. Belytschko et al. [19] used a MM approach along with the Morse and Brenner potentials to study the fracture of a CNT. Their observations showed that even a single missing atom can reduce the strength of a CNT by about 25%. Rafiee and Mahdavi [20] used MD simulation to study the influence of vacancy defect on Young's modulus of SWCNTs. A 10% reduction was found by them in Young's modulus in the presence of six vacancy defects.

In previous years, the main focus was to study the mechanical properties of CNTs in the presence of defects. To the best of the author's knowledge, very little literature has discussed the positional effect of defects on Young's modulus of SWCNTs. In this study, a MD technique has been used with COMPASS (condensed phase optimized molecular potential for atomistic simulation studies) force field to predict Young's modulus of CNTs having SW and vacancy defects. The objective of this article is to reveal the positional effect of SW and vacancy defects on Young's modulus of SWCNTs and provide the result for a better understanding of behavior of CNTs having defects.

10.2   DEFECTS FORMATION

In this work, Young's modulus of CNTs with two types of structural defects, SW, and vacancy have been studied by applying MD simulations. The formation of defects has been explained in subsections.

10.2.1 STONE-WALES (SW) DEFECT

To create a SW defect, four neighboring hexagons of the CNT were selected and by a 90° rotation, were converted into 2 pentagons and two heptagons (5-7-7-5). The formation of SW defect has been shown in Figure 10.1. One SW defect was placed at the middle and rest was distributed along the length of the CNT.

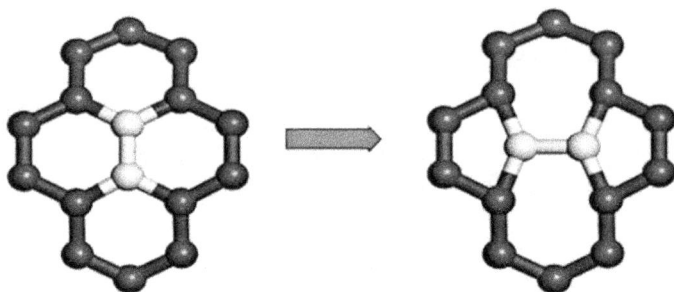

FIGURE 10.1   Formation of Stone-Wales defect in the hexagonal lattice of CNT structures.

10.2.2 VACANCY DEFECT

To create a vacancy defect, C-C bonds were removed from the pristine hexagonal CNT structure. The formation of vacancy defect has been shown in Figure 10.2. One atom was removed at the middle of the CNT structure for studying the effect of one vacancy. For creating another 2–4 vacancy defects, atoms were removed in the same manner from the CNTs.

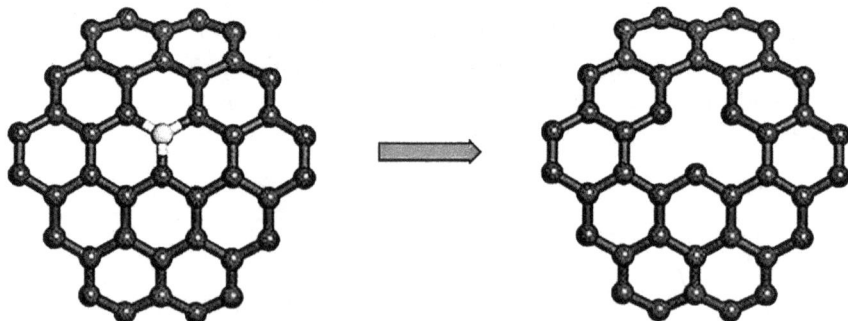

FIGURE 10.2   Formation of vacancy defect in CNT structures by removing one atom.

10.3   THE FORCE FIELD

In MM and MD approaches, the average description of the existing interactions among various atoms in a molecule or a group of molecules in terms of functions and parameters sets are considered as a force field. The force field parameters are typically obtained either from ab initio or semi-empirical quantum mechanical calculations or by fitting to experimental data such as neutron, X-ray, and electron diffraction, NMR, infrared, Raman, and neutron spectroscopy, etc.

Generally, the terms in any force field that describes the total potential energy are following:

$$E_{total} = E_{valence} + E_{cross-term} + E_{non-bond} \tag{1}$$

The sum of the valence, cross-term, and non-bond interaction energies of a system of interacting particles can be expressed as the total potential energy of a system.

$$E_{valence} = E_{bond} + E_{angle} + E_{torsion} + E_{oop} + E_{UB} \tag{2}$$

The valence energy consists of a bond stretching term, E_{bond}, a two-bond angle term, E_{angle}, a dihedral bond-torsion term, $E_{torsion}$, an inversion term, E_{oop}, and a Urey-Bradlay term, E_{UB}.

$$E_{cross-term} = E_{bond-bond} + E_{angle-angle} + E_{bond-angle} + E_{end-bond-torsion}$$
$$+ E_{middle-bond-torsion} + E_{angle-angle-torsion} \tag{3}$$

$E_{cross-term}$, the cross-term interacting energy generally includes: $E_{bond-bond}$ which describes the stretch-stretch interactions between two adjacent bonds, the term $E_{angle-angle}$ describes the bend-bend interactions between two valence angles which are associated with a common vertex atom, $E_{bond-angle}$ indicates the stretch-bend interactions between a two-bond angle and one of its bonds, stretch-torsion interactions between a dihedral angle and one of its end bonds given by term, $E_{end-bond-torsion}$, $E_{middle-bond-torsion}$, which is stretch-torsion interactions between a dihedral angle and its middle bond, $E_{angle-torsion}$ defines bend-torsion interactions between a dihedral angle and one of its valence angles and the term $E_{angle-angle-torsion}$ represents the end-bend-torsion interactions between a dihedral angle and its two valence angles.

$$E_{non-bond} = E_{vdW} + E_{Coulomb} + E_{H-bond} \tag{4}$$

The non-bonded interaction term, $E_{\text{non-bond}}$, describes the interactions between non-bonded atoms and includes the van der Waals energy, E_{vdW}, the Coulomb electrostatic energy, E_{Coulomb}, and the hydrogen bond energy which is $E_{\text{H-bond}}$.

The mechanical and structural properties of a system can be studied by MD simulations which provide the atomic motions of individual atoms as a function of time. The inter-atomic forces are used to govern the kinematics of atoms in MD simulations. In this study, the COMPASS force field has been used for providing inter-atomic interactions during MD simulation. This force field has been verified by using condensed-phase properties and it can be applied to study the mechanical properties of CNTs [21].

The COMPASS force field includes the following terms for the total potential energy [22, 23]:

$$
\begin{aligned}
E_{\text{valence}} = &\sum_{b}[K_2(b-b_0)^2 + K_3(b-b_0)^3 + K_4(b-b_0)^4] \\
&+\sum_{\theta}\left[H_2(\theta-\theta_0)^2 + H_3(\theta-\theta_0)^3 + H_4(\theta-\theta_0)^4\right] \\
&+\sum_{\varnothing}[V_1\left[1-\cos(\varnothing-\varnothing_1^0)\right] + V_2[1-\cos(2\varnothing-\varnothing_2^0)] \\
&+V_3[1-\cos(3\varnothing-\varnothing_3^0)]] + \sum_{\chi}K_x\chi^2 + E_{UB}
\end{aligned}
\tag{5}
$$

$$
\begin{aligned}
E_{\text{cross-term}} = &\sum_{b}\sum_{b'}F_{bb'}(b-b_0)(b'-b_0') + \sum_{\theta}\sum_{\theta'}F_{\theta\theta'}(\theta-\theta_0)(\theta'-\theta_0') \\
&+\sum_{b}\sum_{\theta}F_{b\theta}(b-b_0)(\theta-\theta_0) + \sum_{b}\sum_{\theta}F_{b\theta}(b-b_0) \\
&\times[V_1\cos\varnothing + V_2\cos2\varnothing + V_3\cos3\varnothing] + \sum_{b'}\sum_{\theta}F_{b'\theta}(b'-b_0')(b'-b_0') \\
&\times[F_1\cos\varnothing + F_2\cos2\varnothing + F_3\cos3\varnothing] + \sum_{\theta}\times\sum_{\varnothing}F_{\theta\varnothing}(\theta-\theta_0) \\
&\times[V_1\cos\varnothing + V_2\cos2\varnothing + V_3\cos3\varnothing] + \sum_{\varnothing}\sum_{\theta}\sum_{\theta'}K_{\varnothing\theta\theta'}\cos\varnothing \\
&\times(\theta-\theta_0)\times(\theta'-\theta_0')
\end{aligned}
\tag{6}
$$

$$
E_{\text{non-bond}} = \sum_{i>j}\left[\frac{A_{ij}}{r_{ij}^9} - \frac{B_{ij}}{r_{ij}^6}\right] + \sum_{i>j}\frac{q_iq_j}{\varepsilon r_{ij}} + E_{H-bond}
\tag{7}
$$

where; q defines the atomic charge, ε is the dielectric constant, and r_{ij} is the i-j atomic separation distance. The lengths of two adjacent bonds are given by b and b', θ is the two bond-angles, $\varnothing$ is the dihedral torsion angle, χ and is the out-of-plane angle. $b_{0,ki(i=2-4)}$, θ_0, H_i ($I = 2 - 4$), $\varnothing_i^0$ ($i = 1 - 3$), V_i($i = 1 - 3$), F_{bb}'b_0', $F_{\theta\theta}$', θ_0', $F_{b\theta}$, $F_{b\theta}$ $F_{b'\theta}$, F_i($i = 1 - 3$), $F_{\theta\theta}$, $K_{\varnothing\theta\theta}$, A_{ij} and B_{ij} are the parameters which depend on the system.

10.4 MODELS AND METHOD

Firstly, a (5,5) armchair SWCNT of diameter 0.67 nm and length of about 6.1 nm was generated using Materials Studio 7.0. The pristine structure of a (5,5) armchair SWCNT is shown in Figure 10.3. The first step of any simulation study starts with optimization which minimizes the energy of a structure. A conjugate gradient algorithm [24] with 5000 iterations has been used for this study. Geometry optimization plays an important role to obtain the accurate results of the simulation. If provided convergence criteria are not satisfied in this step it leads to the error, in the final result. Figures 10.4 and 10.5 shows the structure of 4 SWCNTs with 1–4 SW and vacancy defects, respectively. In both cases, the single defect is created at the middle and other defects are modeled by taking equal distance to the middle defect.

FIGURE 10.3   A perfect (5,5) CNT with a diameter of 0.67 nm and a length of 6.14 nm.

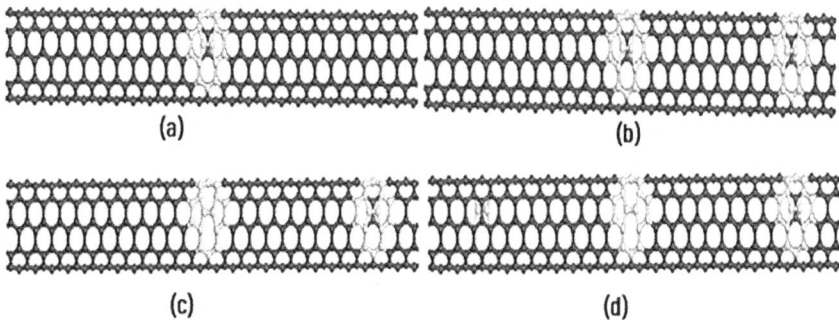

FIGURE 10.4   SWCNT (5,5) with different number of Stone-Wales defects: (a) 1 SW defect; (b) 2 SW defects; (c) SW defects; (d) 4 SW defects.

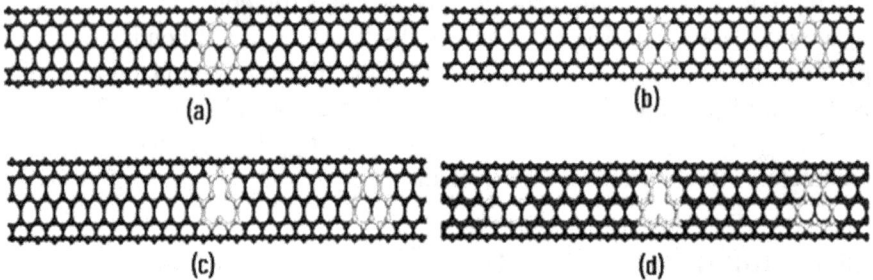

FIGURE 10.5 SWCNT (5,5) with different number of vacancy defects: (a) 1 vacancy defect; (b) 2 vacancy defects; (c) 3 vacancy defects; (d) 4 vacancy defects.

Figure 10.6 defines the places on the CNT structure which has been used to study the positional defects. Once a structure is optimized it can be subjected to MD simulation. The system is equilibrated over constant volume (V) and constant temperature (T) canonical ensemble (NVT) at the room temperature of 298 K for 50 ps. The time step for the simulation was set as 1 fs and the temperature during the simulation was controlled by Andersen's thermostat [25]. Periodic boundary conditions were provided in each MD simulation to increase the accuracy. The atomic configuration, atomic velocities, and other information of structure get recorded in the trajectory file when the structure is subjected to finite simulation time. Trajectory file is considered as the main product of the MD simulation. Once the trajectory file of simulated structure has been obtained it can be used to calculate the mechanical and structural properties of CNTs, i.e., Young's modulus, shear modulus, Poisson's ratio, and radial distribution function (RDF), etc. In this study, a small strain of 0.001 has been applied to calculate Young's modulus of CNT structures. Other parameters of this step have been listed in Table 10.1.

TABLE 10.1 Mechanical Properties Parameters

S. No.	Parameters	Value
1.	Number of strains	6
2.	Maximum strain	0.001
3.	Maximum number of iterations	10
4.	Force field	COMPASS
5.	Repulsive cut-off	6 Å

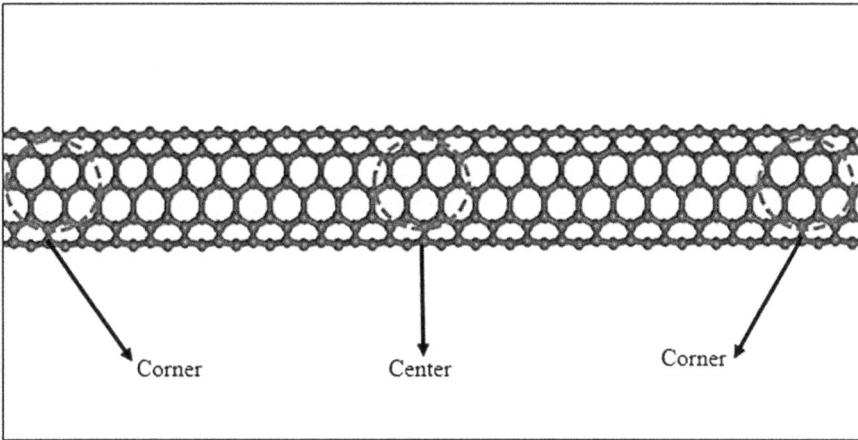

FIGURE 10.6 Schematic of a (5,5) CNT showing the center and corner positions.

10.5   RESULTS AND DISCUSSIONS

This section has been divided into two parts. In the first part, the results of SWCNTs with 1–4 SW and vacancy defects have been given with comparative discussion. The second part discusses the results of CNT with positional effect.

10.5.1   EFFECT OF INCREASING SW AND VACANCY DEFECTS ON THE YOUNG'S MODULUS

Figure 10.7 illustrates the energy variation in CNT structures in the presence of defects. The introduction of 1 SW defect increases the energy of the CNT about 1.51%. On the contrary, 1.57% reduction has been observed when one atom is removed from the perfect CNT structure. The presence of 4 SW defects increases the potential energy about 5.94% when compared to the pristine structures. It has been found that as the number of SW defects increases, more time is required to stabilize the potential energy in the CNT structures. From Figure 10.7 it can be concluded that the formation of SW defects in CNT structures lead to an increase in energy. Sharma et al. [26] used MD simulation and observed that with the increase of SW defects potential energy increases. In the case of vacancy defects, it has been found that CNT with 1 vacancy defect takes less time to the energy stabilization in comparison to 4 defects. Table 10.2 shows the comparative results of Young's modulus of (5,5) armchair value obtained from the present study and those reported in previous works. The Young's modulus of the perfect CNT is obtained as of 911.20 GPa, which makes a good agreement with 948 GPa obtained by Yuan and Liew [27] using MD simulation. Using an empirical force constant model Lu [5] reported Young's modulus of (5,5) armchair CNT as 971 GPa. Results obtained from DFT [16] of 960 GPa and FEM [28] of 970 GPa are found to be close to the results of the present study.

Figure 10.8 shows the variation in Young's modulus in the presence of SW and vacancy defects. It can be concluded that vacancy defects have more influence on Young's modulus in comparison to SW defects. Table 10.3 shows the percentage reduction in Young's modulus of SWCNTs in the presence of 1–4 SW and vacancy defects. It has been observed that 1 and 4 SW defects decrease Young's modulus of (5,5) armchair SWCNT about 0.95% and 2.48%, respectively. In the case of vacancy defects, 1.78% and 6.40% reduction has been found in the presence of 1 and 4 defects, respectively. Rafiee and Pourazizi [29] used FEM modeling and reported the 0.2% reduction with 1 SW defect in (20,20) armchair SWCNT, which makes a good agreement with the present calculations. The less reduction obtained by [32] was due to the size of the SWCNT. Talukdar and Mitra [30] used MD simulation with Brenner's bond-order potential and reported 5% reduction with 1 SW defect in (8,8) armchair SWCNT. The higher results may be due to the use of Tersoff-Brenner potential. Islam et al. [34] used MD simulation to study the mechanical properties of (6,6) armchair SWCNT in the presence of 1–5 SW and vacancy defects. In the presence of 5 SW and vacancy defects, they observed 3.2% and 4.5% reduction, respectively in longitudinal Young's modulus of SWCNTs. The results obtained have good agreement with the present calculations of 2.8% reduction in the presence of 4 SW defects.

A comparison of Young's modulus values of vacancy defective CNTs with other researchers has been given in Table 10.4. Rafiee and Pourazizi [29] used FEM for different CNTs and found Young's modulus in the range of 0.27–0.67 and 0.62 for 1 and 3 defects, respectively. Using an FEM approach Ghavamian et al. [31] reported 1.8% reduction in modulus for (10,10) armchair SWCNT. Using MD simulation, 0.9% reduction has been found by Saxena [32] for (10,10) armchair SWCNT. Sharma et al. [33] used the same force field as used in this study and reported the 15.6% reduction in modulus for a (9,9) armchair SWCNT. Chen et al. [34] used MD simulation and reported 4.8% reduction in the presence of 6 vacancy

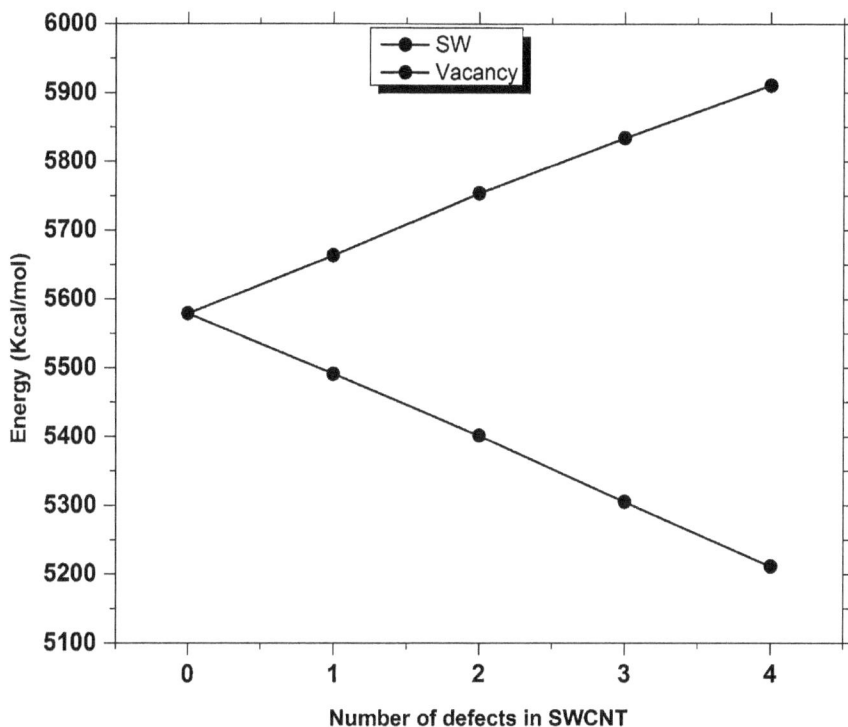

FIGURE 10.7 Energy variation in the presence of varying SW and vacancy defects w.r.t Pristine CNT.

TABLE 10.2 Comparison of Young's Modulus of a (5,5) Armchair SWCNT with Previous Studies

Researcher (s)	Young's Modulus (GPa)	Method
Lu [5]	971	Empirical force constant model
Chwal [28]	970	FEM
Yuan [27]	948	MD
Mielke [41]	960	DFT
This Study	911.20	MD

defects for an (8,8) armchair SWCNT. Using MD simulation Yuan and Liew [42] reported the 7.9 and 15.8% reduction in modulus by 1 and 2% vacancy defects, respectively. The above studies using FEM and MD simulation methods support the calculations of the present study. From the above discussion, it can be concluded that there is a gradual degradation

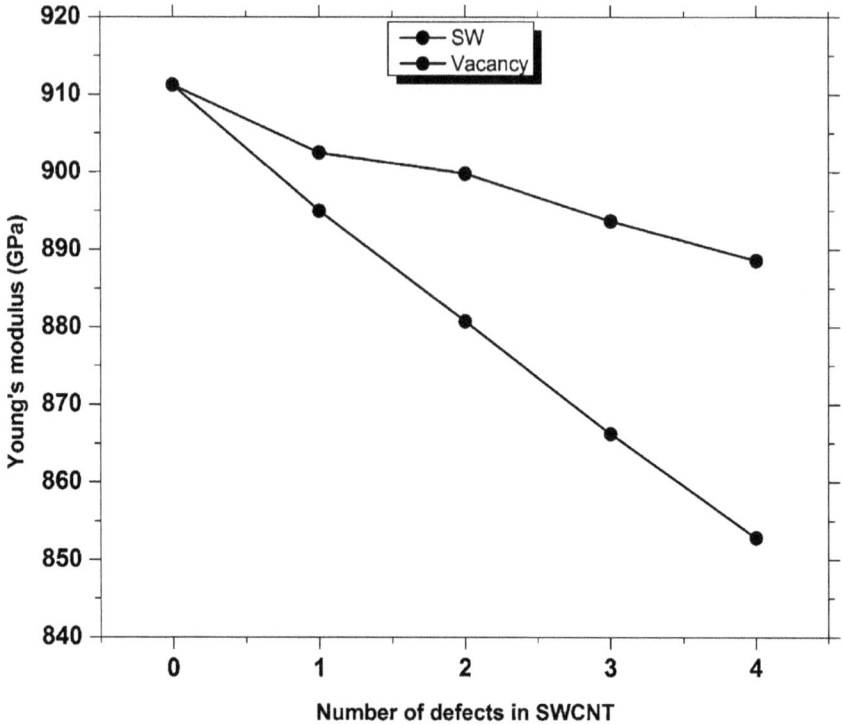

FIGURE 10.8   Variation in Young's modulus in the presence of SW and vacancy defects.

TABLE 10.3   Percentage Reduction in Young's Modulus in the Presence of 1–4 SW and Vacancy Defects

Number of Defects	Percentage Reduction in Modulus	
	SW	Vacancy
1	0.95	1.78
2	1.25	3.34
3	1.92	4.93
4	2.48	6.40

in Young's modulus in the presence of defects. Simulation results show that the effect of vacancy defect in degrading Young's modulus is more pronounced than the SW defects.

TABLE 10.4 Comparison of Calculated Young's Modulus of Vacancy Defected CNTs with Other Studies

Researcher(s)	Method	Tube Type	Number of Vacancy Defect	Percentage Reduction in Modulus
Chen [34]	MD	Armchair (8,8)	6	4.8
Sharma [33]	MD	Armchair (9,9)	1	2.9
	COMPASS		4	15.6
Saxena [32]	MD	Armchair (10,10)	1	0.9
Ghavamian [31]	FEM	Armchair (10,10)	2	1.8
Rafiee [29]	FEM	Armchair	1	0.27–0.67
			3	0.6–2
This Study	MD	Armchair (5,5)	1	1.78
			2	3.34
			3	4.93
			4	6.40

10.5.2 EFFECT OF POSITIONS WITH VARYING SW AND VACANCY DEFECTS

In this section, the positional effect of SW and vacancy defects on Young's modulus of a (5,5) armchair CNT has been discussed. The two positions considered have been shown in Figure 10.6. Multiple defects have been created at the center and corner with random order. In Tables 10.5 and 10.6, the percentage reduction in Young's modulus has been given w.r.t to pristine CNT. A center SW and vacancy defect reduce Young's modulus about 0.95 and 1.78%, respectively. However, defects created at corner in both the cases show less reduction in Young's modulus in comparison to center defects. The formation of two SW defects follow the same trend as 1 SW defect, but results obtained for vacancy defects are contrary to this. In both the cases, defects which have been placed in each corner of the CNTs show less reduction in comparison to 2 defects at the same corner. The reduction in modulus has been increased by 0.08% for SW and 0.66% for vacancy defects in comparison to 1 defect at each corner when the one defect is placed at one corner and another at the center. Further, the reduction has been increased by 0.67% for SW and 1.59% for vacancy defects when two defects are placed at the center and one in a corner. From the above discussion and results of Tables 10.5 and 10.6, it can be concluded that the defects placed at the center have more influence than the corner defects. The Young's modulus decrease more rapidly in the case of vacancy defects and the defects placed at the center.

TABLE 10.5 Percentage Reduction in the Young's Modulus of SWCNT in the Case of Positional Effect of SW Defects

SW Defects Position	Percentage Reduction in Modulus
1 Defect at Center	0.95
1 Defects at Corner	0.66
2 Defects at Center	1.99
2 Defects at Corner	1.72
1 Defect at each Corner	1.17
1 Defect at Center and one in Corner	1.25
2 Defects at Center and one in Corner	1.92

TABLE 10.6 Percentage Reduction in the Young's Modulus of SWCNT in the Case of Positional Effect of Vacancy Defects

Vacancy Defects Position	Percentage Reduction in Modulus
1 atom missing at Center	1.78
1 atom missing at Corner	1.39
2 atom missing at Center	3.13
2 atom missing at Corner	3.46
1 atom missing at each Corner	2.68
1 atom missing at Center and one in Corner	3.34
2 atom missing at Center and one in Corner	4.93

10.6 CONCLUSIONS

In summary, a MD simulation with the COMPASS force field has been used to study the variation in Young's modulus of a (5,5) armchair SWCNT in the presence of defects. Two types of structural defects, vacancy, and SW have been modeled. Simulated results show that the influence of vacancy defects is considerably higher than that of SW. A gradual reduction in Young's modulus has been observed with increasing number of defects. Two positions, center, and corners of CNTs have been selected and multiple defects were placed in random order to study the effects of position. Further, it has been found that the defects placed at the center position decrease the modulus more in comparison to corner defects. This study will help the researchers to understand the positional effect produced by SW and vacancy defects as well as provide the data for experimental studies.

KEYWORDS

- **atomic force microscopy**
- **carbon nanotubes**
- **molecular dynamics**
- **molecular mechanics**
- **radial distribution function**
- **single-walled carbon nanotubes**

REFERENCES

1. Iijima, S., (1991). Helical microtubules of graphitic carbon. *Nature, 354*(6348), 56–58.
2. Oberlin, A., Endo, M., & Koyama, T., (1976). Filamentous growth of carbon through benzene decomposition. *Journal of Crystal Growth, 32*(3), 335–349.
3. Treacy, M. M. J., Ebbesen, T. W., & Gibson, J. M., (1996). Exceptionally high Young's modulus observed for individual carbon nanotubes. *Nature, 381*(6584), 678–680.
4. Wong, E. W., Sheehan, P. E., & Lieber, C. M., (1997). Nanobeam mechanics: Elasticity, strength and toughness of nanorods and nanotubes. *Science, 277*(5334), 1971–1975.
5. Lu, J. P., (1997). Elastic properties of carbon nanotubes and nanoropes. *Physical Review Letters, 79*(12), 1297–1300.
6. Ruoff, R. S., Qian, D., & Liu, W. K., (2003). Mechanical properties of carbon nanotubes: Theoretical predictions and experimental measurements. *Comptes. Rendus. Physique, 4*(9), 993–1008.
7. Krishnan, A., Dujardin, E., Ebbesen, T., Yianilos, P., & Treacy, M., (1998). Young's modulus of single-walled nanotubes. *Physical Review B, 58*(20), 14013–14019.
8. Cornwell, C. F., & Wille, L. T., (1997). Elastic properties of single-walled carbon nanotubes in compression. *Solid State Communication, 101*(8), 555–558.
9. Ebbesen, T. W., & Takada, T., (1995). Topological and SP^3 defect structures in nanotubes. *Carbon, 33*(7), 973–978.
10. Gao, R., Wang, Z. L., Bai, Z., De Heer, W. A., Dai, L., & Gao, M., (2000). Nanomechanics of individual carbon nanotubes from pyrolytically grown arrays. *Physical Review Letters, 85*(3), 622–625.
11. Andrews, R., Jacques, D., Qian, D., & Dickey, E. C., (2001). Purification and structural annealing of multiwalled carbon nanotubes at graphitization temperatures. *Carbon, 39*(11), 1681–1687.
12. Hashimoto, A., Suenaga, K., Gloter, A., Urita, K., & Iijima, S., (2004). Direct evidence for atomic defects in graphene layers. *Nature, 430*(7002), 870–873.

13. Suenaga, K., Wakabayashi, H., Koshino, M., Sato, Y., Urita, K., & Iijima, S., (2007). Imaging active topological defects in carbon nanotubes. *Nature Nanotechnology, 2*(6), 358–360.

14. Sammalkorpi, M., Krasheninnikov, A., Kuronen, A., Nordlund, K., & Kaski, K., (2004). Mechanical properties of carbon nanotubes with vacancies and related defects. *Physical Review B, 70*(24), 245416–245423.

15. Xin, H., & Han, Q., (2012). Mechanical properties of perfect and defective zigzag single-walled carbon nanotubes under axial compression. *Journal of Shanghai Jiaotong University (Science), 17*(5), 545–551.

16. Mielke, S. L., Troya, D., Zhang, S., Li, J. L., Xiao, S., Ruoff, R. S., Schatz, G. C., & Belytschko, T., (2004). The role of vacancy defects and holes in the fracture of carbon nanotubes. *Chemical Physics Letters, 390*(4–6), 413–420.

17. Chandra, N., Namilae, S., & Shet, C., (2004). Local elastic properties of carbon nanotubes in the presence of stone-wales defects. *Physical Review B, 69*(9), 1–12.

18. Lu, Q., & Bhattacharya, B., (2005). Effect of randomly occurring stone-wales defects on mechanical properties of carbon nanotubes using atomistic simulation. *Nanotechnology, 16*(4), 555, 566.

19. Belytschko, T., Xiao, S. P., Schatz, G. C., & Ruoff, R., (2002). Atomistic simulations of nanotube fracture. *Physical Review B, 65*(23), 235430–235437.

20. Rafiee, R., & Mahdavi, M., (2015). Molecular dynamics simulation of defected carbon nanotubes. *Proceedings of the Institution of Mechanical Engineers, Part L: Journal of Materials: Design and Applications, 230*(2), 654–662.

21. Loup, V., (1967). Computer experiments on classical fluids. I. Thermodynamical properties of Lennard-Jones molecules. *Physical Review, 159*(1), 98–103.

22. Sun, H., (1998). COMPASS: An ab initio force-field optimized for condensed-phase applications-overview with details on alkane and benzene compounds. *Journal of Physical Chemistry B, 102*(38), 7338–7364.

23. Sun, H., Ren, P., & Fried, J. R., (1998). The COMPASS and validation. *Computational and Theoretical Polymer Science, 8*(1), 229–246.

24. Polyak, B. T., (1969). The conjugate gradient method in extreme problems. *Computational Mathematics and Mathematical Physics, 9*(4), 94–112.

25. Andersen, H. C., (1980). Molecular dynamics simulations at constant pressure and/or temperature. *Journal of Chemical Physics, 72*(4), 2384–2393.

26. Sharma, K., Tomar, A., Saxena, K. K., & Shukla, M., (2011). Molecular dynamics evaluation of mechanical properties of carbon nanotubes with number of Stone-Wales defects. In: *International Conference on Nanoscience, Engineering and Technology (ICONSET 2011)* (pp. 247–251).

27. Yuan, J., & Liew, K. M., (2009). Effects of vacancy defect reconstruction on the elastic properties of carbon nanotubes. *Carbon, 47*(6), 1526–1533.

28. Chwal, M., (2011). Influence of vacancy defects on the mechanical behavior and properties of carbon nanotubes. *Procedia Engineering, 10,* 1579–1584.

29. Rafiee, R., & Pourazizi, R., (2014). Evaluating the influence of defects on Young's modulus of carbon nanotubes using stochastic modeling. *Materials Research, 17*(3), 758–766.

30. Talukdar, K., & Mitra, A. K., (2010). Influence of odd and even number of Stone-Wales defects on the fracture behavior of an armchair single-walled carbon nanotube under axial and torsional strain. *Molecular Simulation, 36*(6), 409–417.

31. Ghavamian, A., Rahmandoust, M., & Ochsner, A., (2012). A numerical evaluation of the influence of defects on the elastic modulus of single and multi-walled carbon nanotubes. *Computational Materials Science, 62*(1), 110–116.
32. Saxena, K. K., & Lal, A., (2012). Comparative molecular dynamics simulation study of mechanical properties of carbon nanotubes with number of stone-wales and vacancy defects. *Procedia Engineering, 38*(1), 2347–2355.
33. Sharma, K., Saxena, K. K., & Shukla, M., (2012). Effect of multiple stone-wales and vacancy defects on the mechanical behavior of carbon nanotubes using molecular dynamics. In: *International Conference on Modelling Optimization and Computing,* (Vol. 38, pp. 3373–3380).
34. Chen, L., Zhao, Q., Gong, Z., & Zhang, H., (2010). The effects of different defects on the elastic constants of single-walled carbon nanotubes. *Nano/Micro Engineered and Molecular Systems (NEMS), 5th IEEE International Conference* (pp. 777–780).
35. Colomer, J. F., Piedigrosso, P., Willems, I., Journet, C., Bernier, P., Van, T. G., Fonseca, A., & Nagy, J. B., (1998). Purification of catalytically produced multi-wall nanotubes. *Journal of the Chemical Society, Faraday Transactions, 94*(24), 3753–3758.
36. Hernández, E., Goze, C., Bernier, P., & Rubio, A., (1998). Elastic properties of C and $B_xC_yN_z$ composite nanotubes. *Physical Review Letters, 80*(20), 4502–4505.
37. Islam, M. Z., Mahboob, M., & Lowe, R. L., (2016). Mechanical properties of defective carbon nanotube/polyethylene nanocomposites: A molecular dynamics simulation study. *Polymer Composites, 37*(1), 305–314.
38. Terrones, M., Banhart, F., Grobert, N., Charlier, J. C., Terrones, H., & Ajayan, P. M., (2002). Molecular junctions by joining single-walled carbon nanotubes. *Physical Review Letters, 89*(7), 075505–075508.
39. Tendeloo, G., & Nagy, J., (1998). Purification of catalytically produced multi-wall nanotubes. Journal of the Chemical Society, *Faraday Transactions, 94*(24), 3753–3758.
40. Terrones, M., Banhart, F., Grobert, N., Charlier, J.-C., Terrones, H., & Ajayan, P., (2002). Molecular junctions by joining single-walled carbon nanotubes. *Physical Review Letters, 89*(7), 075505.
41. Mielke, S. L., Troya, D., Zhang, S., Li, J.-L., Xiao, S., Car, R., Ruoff, R. S., Schatz, G. C., & Belytschko, T., (2004). The role of vacancy defects and holes in the fracture of carbon nanotubes. *Chem Phys Lett, 390*(4–6), 413–420.
42. Yuan, J., & Liew, K. M., (2009). Effects of vacancy defect reconstruction on the elastic properties of carbon nanotubes. *Carbon, 47*(6), 1526–1533.

CHAPTER 11

Design and Development of Efficient Semiconductor-Based Surface Acoustic Wave Gas Sensor: A Systematic Review

PAYAL PATIAL

Department of Electronics and Communication Engineering, Chandigarh University, Gharuan, Kharar, Mohali – 140413, Punjab, India, E-mail: payal.patial@gmail.com

ABSTRACT

Leakage of toxic gases causing fire, explosions, and even affects surviving conditions of living beings. Highly sensitive nature of SAW-based sensor is able to detect small amount of change in chemical quantity. So, SAW-based sensors are helpful and reduce the chances of disaster. The design, analysis, optimization, and simulation of SAW-based gas sensors are exhibited. The thin film used in SAW-based sensors adsorbs the chemical particles from the environment, the density of the thin film increases, which results in the frequency shift in the applied signal. SAW-based sensors can detect small change in frequency even at small loading effect, this effect occurs when small number of particles is absorbed by thin film. Sensitivity of SAW-based sensor can further be enhanced by varying the various parameters of implemented sensing module, i.e., substrate, and thin film thickness, IDT legs variation. By considering different material for substrate (Quartz, Lithium Neonates, and Lithium Tantalite's) and varying their thickness, enhanced sensitivity can be achieved. Sensing module can be implemented on COMSOL metaphysics 5.3a.

11.1 INTRODUCTION

Lord Rayleigh invented the concept and theory of surface acoustic wave (SAW) gas sensor, in 1885. He explained the propagation of wave along the surface of sensor [25]. A highly sensitive sensor, working on sensing principle which is dependent on the mechanical wave when it is transmitted on surface, the frequency is changed, here the substance if piezoelectric. The members of SAW gas sensor based sensors are piezoelectric substrate, thin film, and inter digitized transducers (IDT). IDTs are fabricated over piezoelectric and porous thin films are placed over IDTs, thickness, and materials. Thin film is the critical component for sensitivity measurement, when thin film of the sensor is exposed to absorb the chemical particles from the environment. Chemical particle adsorbed by thin film leads to, this result in the frequency change of applied signal. SAW gas sensors are used to sense gas phase and liquid phase. Parameters which are responsible to increase the sensitivity, mass, and elasticity, etc., are described [7]. The structure of SAW gas sensor consists of a thin film, intermediate layer, and piezoelectric substrate [2, 8, 10, 16, 18, 28, 40, 41].

11.2   THE INTER DIGITATED TRANSDUCER (IDTS)

Inter digitized transducers (IDTs) consist of input and output transducers, input section converts the electrical signal into mechanical signal which is further propagated over the piezoelectric substrate and converted back to the electrical signal from mechanical signal by using output transducer. The shift in frequency is used for analyzing sensitivity of the gas sensor.

Frequency Shift = Input signal frequency – Output signal frequency   (1)

The toxic gas is absorbed by thin film of SXFA (flour alcohol polysiloxane) and sensitivity comes out to be 0.65 kHz/mg/m^3; when thin-film SXFA is placed in between IDTs then sensitivity comes out to be 2.1 kHz/mg/m^3 (Figures 11.1 and 11.2) [21].

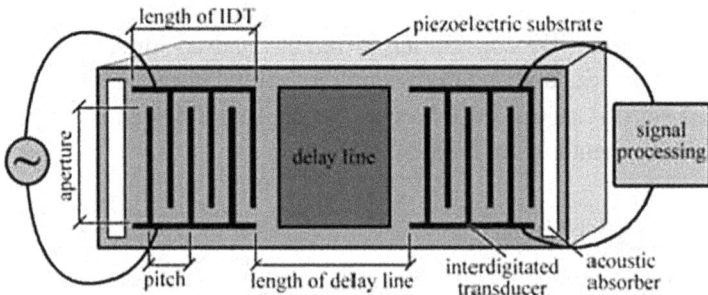

FIGURE 11.1   Schematic diagram of SAW-based gas sensor.

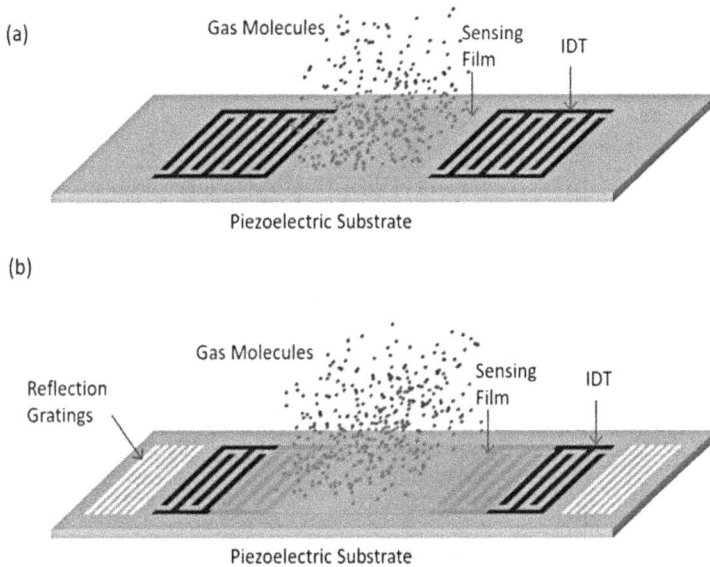

FIGURE 11.2 (a) Thin-film place between IDTs; (b) Thin-film placed over IDT.

The sensitivity of sensor depends on the thin film, as thin film adsorbs more toxic particles then there is change in density of thin film which further contributes to sensitivity of sensor. Surface to volume ratio should be more to achieve high sensitivity. That's why when we use thin film consist of Niño rods as compared to layered thin film, we get more sensitivity.

11.3 RESULTS AND DISCUSSION

The process of identify the valid publications is depicted in Figure 11.3. After removing 53 duplicate publications, 315 publications were the valid publications. While reading the abstracts and conclusion, 93 publications retained out of 222 publications, by our two-team members. After analyzes the methods and materials, 73 publications are retained.

After the full article review, 36 additional publications were excluded because either (1) they failed to perform content analysis; or (2) gas sensing was not their primary focus. For systematic review, 25 manuscripts were added, finally (Table 11.1).

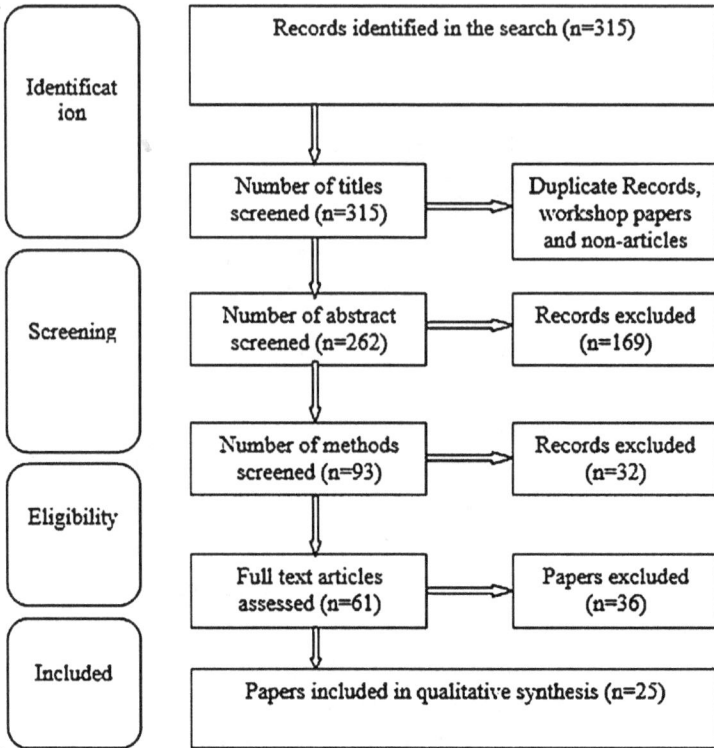

FIGURE 11.3   Illustration of the steps used in the literature search.

In Ref. [25], using Langmuir-Blodgett film, an acoustic wave-based chemical sensor is designed. Resonator is fabricated over ST-Quartz, whose thickness is 0.5 mm. resonator is placed between reflectors stripes, which are made up of aluminum. Outcome showed us that change in mass loading effect will change the resonance frequency.

In Ref. [7], after analyzing and reviewing various acoustic wave sensors, it is found that operating frequency for SAW-based sensor lies between 40–100 (MHz). As the research in this field progress, operating frequency for SAW-based sensors has increased.

In Ref. [25], IDTs are made up of aluminum, when polyvinyl alcohol film is placed over it and for detecting relative humidity, sensor is designed. It is found that efficiency of PVA thin layer reduces with time due to low aging of PVA layer, which further alters the performance of sensor.

In Ref. [2], SAW gas sensor using $LiNbO_3$ piezoelectric substrate and zinc oxide thin film is designed. Comparison of simulated and experimental results

TABLE 11.1 Summary of Publications with Respect to Their Technology Used, Size, and Material Used

SL. No.	Technology Used		Frequency		Size			Material Used	
	Design	Simulation	Operating Frequency	Frequency Shift	Thickness	Finger Width	Substrate	IDT	Thin Film
1.	–	–	–	–	3000 A	–	–	–	–
2.	–	–	110 MHz	–	1 mm	–	–	–	–
3.	–	–	–	11.5 MHz	–	–	–	Al	–
4.	–	–	220.75–436 MHz	–	IDT = 200 nm	–	–	Al	ZnO
5.	–	SEM	–	–	–	–	LiNbO$_3$	Ni	–
6.	–	–	–	–	–	–	Quartz ST LiNbO$_3$ LiTAO$_3$	–	–
7.	–	FEM	–	–	Film = 137 μm	–	LiNbO$_3$	–	–
8.	–	–	–	–	–	–	LibNO$_3$	Ni	–
9.	–	FEM	158 MHz	–	–	–	ST Quartz	–	–
10.	–	–	40 MHz	–	Pd = 20 nm	–	–	–	Pd
11.	–	–	–	–	–	–	LiNbO$_3$	–	–
12.	–	FEM	130.2 MHz	–	ZnO = 1 μm hGan = 1.5 μm hSiC = 24 μm	–	LiNbO$_3$	–	ZnO
13.	–	–	–	–		–	SiC	Au	–
14.	–	–	100 MHz	–	IDT = 1 μm	0.5–2 μm	–	–	–
15.	–	–	–	–	–	–	Quartz ST LiNbO$_3$ LiTAO$_3$	–	–
16.	–	FEM	427.88–469.19 MHz	–	Si = 20 μm	–	Si	Al	–
17.	–	–	–	–	–	–	–	–	–
18.	–	XRD FESEM	–	115.9 KHz	–	–	–	–	–
19.	–	–	50 MHz	–	–	20 μm	LiNbO$_3$	–	–
20.	–	FEM	303 MHz	–	SXFA Film = 65 nm	–	LiNbO$_3$	Ni	–
21.	–	–	–	–	–	–	–	–	–
22.	–	–	–	–	–	–	–	–	–
23.	–	–	–	–	236 nm	–	–	–	–
24.	–	–	35000–12,500 HZ	–	–	–	–	–	–
25.	–	–	–	–	–	–	–	–	–

for wave velocity is carried out. Simulation results showed that transmitted wave penetrates into Si, when thickness of ZnO is kept small. Therefore, at small thickness of ZnO, Si parameters play a crucial role in deciding velocity.

In Ref. [18], high aspect ratio electrodes are fabricated by using nickel electroplating. It is found that highly compact IDTs are useful to reduce phase velocity and to attain good sensitivity. Dimensions of electrodes are 6 μm wide and 30 μm high and have a 0.5 metallization aspect ratio.

In Ref. [40], a change in the velocity of a SAW-based sensor is investigated by using the perturbation method. The wave velocity changes are due to a change in the surface pressure and temperature of the piezoelectric substrate when mechanical stress is applied to it. In piezoelectric substrate, by including temperature or pressure coefficients, the relationship between resonant frequency and temperature/pressure.

In Ref. [10], FEM simulation is carried out with mass loading effect and without mass loading effect, their comparison is done. Mass loading effect came to existence when porous thin film adsorb small amount of gas molecules, due to this density of thin-film increases and frequency shift is obtained. In this chapter, an impedance loader SAW sensor is designed.

In Ref. [8], SAW-based sensor with LINbO$_3$ substrate is fabricated, the material used for the fabrication of sensor is Nickel. This chapter gives idea to fabricate the SAW sensor with high aspect ratio of Inter Digitated Transducers. The concept of HAR electrode is also used to compute the results.

In Ref. [41], FEM technique for organic vapor sensor is considered and simulation is performed. Results of simulations are compared in the presence and absence of vapor. Sensitivities for diff-diff concentrations of vapors are obtained. To estimate the polymer coated sensor response, LSER method is used with FEM simulation.

In Ref. [16], the sensor using LiNbO$_3$ piezoelectric substrate and palladium thin film of 20 nm thickness is fabricated. The layer made up of 720 nm and 20 nm Pd substrate is fabricated which shows high sensitivity.

In Ref. [20], a sensor with 41°YX LiNbO$_3$ substrate is fabricated, having two thin films to sense the gases experimentally. Teflon AF 2400 thin film used to sense the presence of CO$_2$ and Indium Tin Oxide to sense NO$_2$. It is found that ITO film has high change in its electrical properties.

In Ref. [30], a humidity sensor is fabricated having ZnO, AlN, and Si, respectively. Si served as substrate layer. ZnO thin film is porous and shows a good response to absorb the water vapors in a humid environment.

In Ref. [17], Piezoelectric GaN layer and SiC as the substrate layer is used to design SAW-based sensor. Modeling and simulation are carried out using FEM. In this chapter, model and transient analysis are carried out, and it is found that when we have to determine Eigen frequencies of the system, model analysis is used. When we have to determine the frequency of excitation, then the transient analysis is used.

In Ref. [14], a multilayer structure using $LiNbO_3$ and ZnO is used to enhance sensitivity. Sensitivity can be enhanced by varying aspect ratio of nanorods, and other approachable ways are gap between consecutive electrodes and intermediate layer thickness. In this chapter, layered thin film and the nanostructured thin film is used to design sensor, and their comparison is carried out. It is found that nanostructured thin film adsorbs more gas particles. Therefore, the change in density is more, so as the sensitivity.

In Ref. [24], this chapter deals with the selection of piezoelectric substrates. Various materials are compared with each other by considering the properties like, dielectric constant, maximum working temperature, velocity, and TCF. The importance of piezoelectric material with their cutting angles and temperature dependencies are discussed. Results of sensitivities for sensor operating at 433.92 MHz frequency is used to detect gases like ethanol, ethyl acetate, and toluene are presented.

In Ref. [19], this chapter deals with the physics of propagation of acoustic wave over the surface. By varying the temperature range from 0° to 150° FEM simulation is carried out. Simulation results showed that Eigen frequencies decrease with increase in temperature linearly. Thin film used in this SAW-based sensor is Aluminum Nitride.

In Ref. [13], FEM simulation is carried out by simulation software by using $LiNbO_3$ substrate having 20 μm thicknesses. SPUDT structure of IDT is used using three reflectors. Effect of IDT structure and input signal on sensor output is also discussed. It is found that when we use 50 IDTs then we got highest amplitude of reflected wave.

In Ref. [39], SnO_2 thin film is deposited over $LiNbO_3$ which has 128 YX cut by using sol-gel method and magnetron sputtering. To improve the sensitivity of SAW sensor bi-layer, thin film is used. Bi-layer is made up of two layers in which one layer is pure SnO_2, and another is a palladium layer, which is highly dispersed. Palladium nanoparticles size increased lead to an increase in thickness, which reduces the sensing capabilities of SnO_2 thin film.

In Ref. [32], a SAW-based sensor is designed to measure longitudinal strain. $LiNbO_3$ diaphragm is used as sensing element. Operating frequency of the sensor is around 50 Mhz. the strain sensitivity of the sensor increased by using the novel design.

In Ref. [21], Sensitivity increases by 3.2 times when thin film is placed in between inter digital transducers instead of placing it over the module. The absorption of DMMP (*dimethyl methylphosphonate*) gas changes density and thickness of SXFA film and affect the SAW propagation velocity, hence the resonance frequency of the device.

In Ref. [23], an earlier SAW-based sensor is operated at a temperature ranging from 100°C to 200°C. Whenever there is an increase in operating temperature, then material properties started behaving in a different way. In this chapter, the experiment is carried out by varying the metallization and secondly by using diffusion barriers. FEM simulation is used to study the high-temperature characteristics of such devices. To operate the device at a very high-temperature experimental deformation analysis is carried out to analyze high-temperature characteristics.

In Ref. [37], this chapter's theoretical study is carried out of a micro rate sensor, which is based on the shear horizontal acoustic wave, which is propagating on the surface. This is known as a gyroscopic effect. It is discussed in this chapter that if we vary the metal thickness, change in electrochemical coupling factor, and change in velocity is investigated.

In Ref. [25], the sensor is designed to operate in a high temperature and pressure environment. The thin film used in SAW gas sensor used to absorb the toxic particle made up of carbon nanotubes. The second model consists of ZnO and CuO thin film to adsorb. The frequency shift is calculated in the presence and absence of toxic particles, and results are presented. It is observed that the sensitivity of the second model is more as compared to model having thin film of single material.

In Ref. [33], sensor is developed to sense ammonia gas by using CO_3O_4/SiO_2 sensing film. Sol gel method is used to deposit CO_3O_4/SiO_2 on substrate. Frequency shift of 3500 Hz is obtained while sensing 1 ppm of ammonia gas. When the concentration of ammonia gas is increased to 12500 Hz were observed. Therefore, it can be concluded that when the concentration of toxic gas increases, then the frequency shift is observed more.

In Ref. [1], electro-mechanical waves used in SAW gas sensors are considered. In this chapter, a review of the SAW gas sensor to sense a chemical response. Discussion is carried out for SAW gas sensor for sensing gaseous phase as well liquid phase.

11.4   CONCLUSION

We would acknowledge various limitations in this systematic review. First, to include eligible and related publications, we use various SAWL-related keywords. However, using such a process, many publications may be missed. Second, we did not include 53 workshop articles and 2 after full article review, which could be considered in a future review. Third, we ignored many studies that focused on gas sensors but neglected to investigate or discuss SAW gas sensors. In addition, some studies that investigated CNG and PNG Gas sensors were excluded because there was no further investigation on their impact. Finally, the Zinc oxide based gas sensor methods is an important consideration, 131 but was beyond the focus of this systematic review.

KEYWORDS

- *dimethyl methyl phosphonate*
- finite element method
- inter digitized transducers
- piezoelectric
- *polyvinyl acetate*
- surface acoustic wave

REFERENCES

1. Abraham, N., et al. (2019). "Simulation studies on the responses of ZnO-CuO/CNT nanocomposite based SAW sensor to various volatile organic chemicals." *Journal of Science: Advanced Materials and Devices 4*(1), 125–131.
2. Ahmadi, S., et al. (2004). Characterization of multi-and single-layer structure SAW sensor [gas sensor]. SENSORS, 2004 IEEE, *IEEE*.
3. Aslam, M., et al. (2018). "FEM analysis of sezawa mode SAW sensor for VOC based on CMOS compatible AlN/SiO$_2$/Si multilayer structure." *Sensors 18*(6), 1687.
4. Bao, Z., et al. (2015). Highly sensitive strain sensors using surface acoustic wave on aluminum nitride thin film for wireless sensor networks. 2015 Transducers-2015 18th International Conference on Solid-State Sensors, Actuators and Microsystems (TRANSDUCERS), *IEEE*.
5. Bo, L., et al. (2016). "Surface acoustic wave devices for sensor applications." *Journal of Semiconductors 37*(2), 021001.
6. Cittadini, M., et al. (2015). "ZnO nanorods grown on ZnO sol–gel seed films: characteristics and optical gas-sensing properties." *Sensors and Actuators B: Chemical 213*, 493–500.
7. Dorozhkin, L., & Rozanov, I. (2001). "Acoustic wave chemical sensors for gases." *Journal of Analytical Chemistry, 56*(5), 399–416.

8. Dühring, M. B., et al. (2009). "Energy storage and dispersion of surface acoustic waves trapped in a periodic array of mechanical resonators." *Journal of Applied Physics 105*(9), 093504.

9. Elhosni, M., et al. (2016). "Magnetic field SAW sensors based on magnetostrictive-piezoelectric layered structures: FEM modeling and experimental validation." *Sensors and Actuators A: Physical 240*, 41–49.

10. Dühring, M. B., et al. (2009). "Energy storage and dispersion of surface acoustic waves trapped in a periodic array of mechanical resonators." *Journal of Applied Physics 105*(9), 093504.

11. García-Gancedo, L., et al. (2013). "Direct comparison of the gravimetric responsivities of ZnO-based FBARs and SMRs." *Sensors and Actuators B: Chemical 183*, 136–143.

12. Go, D. B., et al. (2017). "Surface acoustic wave devices for chemical sensing and microfluidics: a review and perspective." *Analytical Methods 9*(28), 4112–4134.

13. Ha, N. T., et al. (2017). A study of the effect of IDTs and input signals on the amplitude of propagation waves of the passive SAW structure. 2017 International Conference on Information and Communication Technology Convergence (ICTC), IEEE.

14. Hasan, M. N., et al. (2017). "Simulation and fabrication of SAW-based gas sensor with modified surface state of active layer and electrode orientation for enhanced H 2 gas sensing." *Journal of Electronic Materials 46*(2), 679–686.

15. Hu, M., & F. L. Duan (2018). "Design, fabrication and characterization of SAW devices on LiNbO$_3$ bulk and ZnO thin film substrates." *Solid-State Electronics 150*, 28–34.

16. Jakubik, W. P. (2011). "Surface acoustic wave-based gas sensors." *Thin Solid Films 520*(3), 986–993.

17. Kutiš, V., et al. (2012). "Modelling and simulation of SAW sensor using FEM." *Procedia Engineering 48*, 332–337.

18. Laude, V., et al. (2006). "Surface acoustic wave trapping in a periodic array of mechanical resonators." *Applied Physics Letters 89*(8), 083515.

19. Li, Y., et al. (2017). FEM simulation of SAW temperature sensor based on Al/AlN/Si structure. 2017 Chinese Automation Congress (CAC), *IEEE*.

20. Lim, C., et al. (2011). "Development of SAW-based multi-gas sensor for simultaneous detection of CO$_2$ and NO$_2$." *Sensors and Actuators B: Chemical 154*(1), 9–16.

21. Lukose, V., & H. B. Nemade (2019). "Finite element simulation of one-port surface acoustic wave resonator with thick interdigital transducer for gas sensing." *Microsystem Technologies 25*(2), 441–446.

22. Meulendyk, B. J., et al. (2010). "Hydrogen fluoride gas detection mechanism on quartz using SAW sensors." *IEEE Sensors Journal 11*(9), 1768–1775..

23. Mrosk, J. W., et al. (2001). "Materials issues of SAW sensors for high-temperature applications." *IEEE Transactions on Industrial Electronics 48*(2), 258–264.

24. Mujahid, A., & Dickert, F. L. (2017). "Surface acoustic wave (SAW) for chemical sensing applications of recognition layers." *Sensors 17*(12), 2716.

25. Nomura, T., et al. (1998). "Chemical sensor based on surface acoustic wave resonator using Langmuir-Blodgett film." *IEEE Transactions on Ultrasonics, Ferroelectrics, and Frequency Control 45*(5), 1261–1265.

26. Pang, H.-F., et al. (2013). "Love mode surface acoustic wave ultraviolet sensor using ZnO films deposited on 36 Y-cut LiTaO3." *Sensors and Actuators A: Physical 193*, 87–94.

27. Park, J.-K., et al. (2018). "Real-time humidity sensor based on microwave resonator coupled with PEDOT: PSS conducting polymer film." *Scientific Reports 8*(1), 439.
28. Penza, M., & G. Cassano (2000). "Relative humidity sensing by PVA-coated dual resonator SAW oscillator." *Sensors and Actuators B: Chemical 68*(1–3), 300–306.
29. M. Penza *et al.*, "Surface acoustic wave 915 MHz resonator oscillator gas sensors using SnO 2 nanowires-based nanocomposite layer," in *SENSORS, 2008 IEEE*, 2008: IEEE, pp. 204–207.
30. Phan, D., & G. Chung (2011). FEM modeling SAW humidity sensor based on ZnO/IDTs/AlN/Si structures. 2011 16th International Solid-State Sensors, Actuators and Microsystems Conference, IEEE.
31. Qin, L., et al. (2010). "Analytical study of dual-mode thin film bulk acoustic resonators (FBARs) based on ZnO and AlN films with tilted c-axis orientation." *IEEE Transactions on Ultrasonics, Ferroelectrics, and Frequency Control 57*(8), 1840–1853.
32. Ruppel, C. C. (2017). "Acoustic wave filter technology–A review." *IEEE Transactions on Ultrasonics, Ferroelectrics, and Frequency Control 64*(9), 1390–1400.
33. Tang, Y.-L., et al. (2014). "Highly sensitive room-temperature surface acoustic wave (SAW) ammonia sensors based on Co_3O_4/SiO_2 composite films." *Journal of Hazardous Materials 280*, 127–133.
34. Tang, Y., Li, Z., Zu, X., Ma, J., Wang, L., Yang, J., & Yu, Q. (2015). Room-temperature NH_3 gas sensors based on Ag-doped γ-Fe_2O_3/SiO_2 composite films with sub-ppm detection ability. *Journal of Hazardous Materials, 298*, 154–161.
35. Tigli, O., & Zaghloul, M. E. (2005). Design and fabrication of a novel_SAW bio/chemical sensor in CMOS. SENSORS, 2005 IEEE, *IEEE*.
36. Wang, S.-Y., et al. (2015). "Surface acoustic wave ammonia sensor based on ZnO/SiO2 composite film." *Journal of Hazardous Materials 285*, 368–374.
37. Wang, W., et al. (2009). Theoretical sensitivity evaluation of a shear-horizontal SAW based micro rate sensor. 2009 IEEE International Ultrasonics Symposium, *IEEE*.
38. Wang, X., et al. (2012). "Development of a SnO2/CuO-coated surface acoustic wave-based H2S sensor with switch-like response and recovery." *Sensors and Actuators B: Chemical 169*, 10–16.
39. Yang, L., et al. (2017). "The investigation of hydrogen gas sensing properties of SAW gas sensor based on palladium surface modified SnO_2 thin film." *Materials Science in Semiconductor Processing 60*, 16–28.
40. Zhang, X., et al. (2007). "Optimal selection of piezoelectric substrates and crystal cuts for SAW-based pressure and temperature sensors." *IEEE Transactions on Ultrasonics, Ferroelectrics, and Frequency Control 54*(6), 1207–1216.
41. Zhao, Y.-G., et al. (2009). "FEM modeling of SAW organic vapor sensors." *Sensors and Actuators A: Physical 154*(1), 30–34.
42. Zheng, P., et al. (2009). Multiphysics simulation of the effect of sensing and spacer layers on SAW velocity. COMSOL Conference, Boston.
43. Zhu, Y., et al. (2017). "Synthesis of functionalized mesoporous TiO2-SiO2 with organic fluoroalcohol as high performance DMMP gas sensor." *Sensors and Actuators B: Chemical 248*, 785–792.

Index

α

α-eleostearic acid, 78
α-linolenic acid, 37
α-lipoic acid (ALA), 70

β

β-carotene, 70
β-tubulin, 37

A

Abraxane, 16, 55
Acetyl-L-carnitine (ALC), 70, 88
Acoustic
  carrier signal, 9
  signal, 9, 12, 18
  transmitters, 13
  wave, 248, 251, 252
Activated immune cells, 82
Active
  pharmaceutical ingredients, 32
  scavenger system, 13
Acute cardiovascular disease, 66
Adaptive immune systems, 87
Adenoviruses, 82
Adhesion, 222–224
Adsorption, 33, 104, 106, 118, 141, 148, 172
Affluent epithelial diffusion, 33
Agglomeration, 166, 175
Albumin, 16, 101, 108, 109, 118, 124, 126, 127
Alcoholism, 181
Aldose reductase, 67, 71
Alginate gel encapsulated chitosan-coated nanocore (ACNC), 110, 121, 125, 127
Allergen, 106, 124
Allodynia, 66, 80
Alpha
  helices, 100
  tocopherol, 78, 182
Aluminum nitride, 251
Alzheimer's disease, 8, 179, 180

American society for testing materials (ASTM), 87, 88
Amino acids, 18, 70, 125
Ammonium, 56, 113, 115
  bicarbonate, 56
  carbonate, 56
Amorphization, 115
Amphiphilic
  block copolymers, 36
  moieties, 73
  surfactants, 36
Amphotericin B, 33
Amputations, 66
Amyloid β-protein, 8
Anatomical membranes, 33, 36, 39, 43
Andersen's thermostat, 234
Angiotensin, 150, 151
Annihilation, 40, 41, 43, 56
Anthrone, 105, 116, 117
Anti-angiogenic activities, 36
Antibody conjugates, 7
Anticancer
  agents, 1, 8, 33
  drugs, 33, 35, 37
  medicament, 153
  properties, 37
Antidepressants, 70
Antidiabetic
  activity, 79
  drug, 75, 136
  lipophilic substances, 78
Antigen
  conveyance frameworks, 56
  delivery, 123, 125, 126
  presenting cell (APCs), 124, 127
Antineoplastic agents, 33
Antioxidant, 69, 70, 78, 83, 84
Anti-psychotic drug, 122
Antitumor activity, 148, 180, 181
Aphaeresis, 13
Apoptosis, 83, 179

Aquasome, 97–99, 101, 102, 104–115,
    117–127
  composition
    bioactive molecule, 101
    coating material, 101
    core material, 101
  preparation method, 102
    core coating, 104
    core preparation, 102
    drugs immobilization, 104
  unique properties, 99
Aqueous
  dispersion, 75, 183
  fluorescent solution, 173
  media during dilution, 131
  organic solution, 173
  suspension, 18, 183
Artificial
  oxygen carrier nanobot, 14
  phagocytes, 14
  respirocyte, 18, 20
Astrocytes, 80
Atherosclerosis, 14, 19, 181
Atomic
  force microscopy (AFM), 175, 187, 228,
    241
  nanotechnology, 55
  scale, 6, 229
  velocities, 234
Autoclaving, 173, 186
Autonomous
  nervous system, 67
  neuropathy, 66
Axial tension, 229
Azobenzene, 35

B

Baicalin
  SLNs (BA-SLN), 176
  solution (BA-sol), 176, 187
Bend-torsion interactions, 231
Berberine nanosuspensions, 79
Bile salts, 136, 142, 149
Bioactive molecule, 98–101, 126
Biocompatibility, 34, 37, 39, 41, 74
Biocompatible liposomal surfaces, 33
Bioconjugation, 38, 42
Biodegradability, 35, 37, 41, 74

Biodegradable
  drug delivery vectors, 32
  polymers, 35, 86
Biological
  membranes, 32, 73
  probing, 38
  tolerance, 37
Biomacromolecules, 38, 41, 42, 146
Biomarkers, 38, 72
Biomass, 215
Biomolecules, 38, 75, 84, 170
Biopharmaceuticals, 81, 146
Biosensing, 31, 32
Biphasic dose-response interaction, 84
Bipolar affinity, 140
Blockbuster drugs, 32
Blood
  borne
    devices, 8
    medical nanobots, 7
  brain barrier (BBB), 15, 20, 31, 32, 36,
    43, 179
  capillaries, 7, 9, 11, 12, 17, 19
  cells, 17, 19
  glucose level, 16
  stream, 10, 18
Bottom-up technology, 79
Bovine serum albumin (BSA), 108, 114,
    118, 124, 126, 127
Brain capillaries, 36
Broad-spectrum antibiotic activity, 172
Brownian motion, 12
Buckling, 229
Bupivacaine, 82
Bypass hepatic metabolism, 132

C

Calcium
  channels, 69, 70, 81
  hydrogen phosphate (CaHPO4), 114, 115
  phosphate, 98, 101, 102, 110, 111, 113
Cancer, 1, 7, 8, 14–16, 20, 33–35, 37, 40,
    41, 55, 56, 124, 145, 150, 181, 182
Carbohydrate, 66, 98, 99, 101, 102,
    104–108, 111, 112, 114, 116–119, 126,
    127
Carbon
  allotrope, 7

atoms, 17
dioxide (CO_2), 5, 18, 103, 199
nano-ropes, 228
nanotube (CNT), 20, 40, 57, 61, 227–230,
    232–237, 239–241, 252
Carboxylate groups, 38
Cardiovascular disease, 7, 66, 181, 182
Cargo
    drug, 33, 35, 39
    molecules, 41, 43
    pharmaceuticals, 32
Carotenoids, 182
Carrier
    biomolecule, 42
    drug, 34
    liposome, 34
    vector, 39
Cell
    cycle, 37
    entry mechanism, 87
    mediated immunity (CMI), 123
    membranes, 87
    penetrating peptide, 34
    repair, 14, 20
    surface area, 87
Cellobiose, 101, 104, 108, 109, 114, 115,
    117, 118
Cellular
    biological process, 72
    immune response, 123
    infiltration, 83
    labeling, 42
    migration, 42
    redox process, 84
    transfer, 36
    uptake, 34, 36, 98, 179, 181
Cellulose, 11, 145
Central nervous system, 36, 66
Centrifugation, 13, 102, 104
Ceramics, 98, 101, 102, 104, 107, 108,
    111–113, 115, 126, 127
Cerebral
    aneurysm, 8
    treatment, 8
Cerebrospinal fluid, 32
Cerium oxide (CeO_2), 84, 88
Chemotactic sensor, 11
Chemotherapeutics, 34, 37

Chemotherapy, 8, 15, 41, 54, 151
Chitosan (CS), 35, 78, 86, 88, 110, 113, 121,
    125, 127
    coated nanocores (CNC), 110, 113, 121
Chromatin, 19
Chromatography, 178
Chronic
    degenerative diseases, 182
    diabetic hyperglycemia, 65
    disease, 63
    hyperglycemia, 66
    metabolic disorder, 16
Coalescence, 138, 139, 144, 169, 175
Colloidal
    carriers, 131, 165
    drug delivery systems, 34
Complementary metal oxide semiconductor
    (CMOS), 16
Concanavalin A, 105, 116, 123
Condensed phase optimized molecular
    potential for atomistic simulation studies
    (COMPASS), 229, 232, 234, 239, 240
Conjugate gradient algorithm, 233
Connective tissues, 34
Contemporary drug delivery paradigm, 32
Conventional
    coating, 216, 221
    emulsion, 129
    intra-venous formulation, 99
Corneal
    keratitis, 173
    surface, 173, 176
Corrosion resistance, 216, 219, 220
Co-surfactant, 36, 77, 131, 132, 136, 138,
    140, 171, 186
Covalent bonding, 76
Cremophor, 16, 136
Critical
    biological events, 42
    micellar temperature, 77
Crystalline, 101, 105, 114, 115, 175
Crystallization, 169, 201
Curcumin, 33, 35, 70, 147, 180
Cyclodextrin, 82, 130, 152
Cytoplasm, 33
Cytoplasmic decoupling, 40
Cytotoxicity, 37, 42, 77, 150

D

Dehydration, 98–101, 122, 125, 126
Delivery vehicles, 34, 76
Denaturation, 99, 122, 126
Dendrimer, 1, 15, 20, 39, 40, 53, 54, 75, 76, 86, 124
Diabetes, 14, 16, 63–67, 69, 74, 83–85, 88
Diabetic neuropathy (DPN), 63–71, 73, 79–84, 88
  conventional treatment, 69, 70, 71
    hyperglycaemic control, 69
    oxidative stress reduction, 70
    pain management therapy, 69
Diagnosis, 2, 6, 14, 15, 17, 20, 41, 42, 65, 70, 72, 81, 84
Diamondoid, 7, 9, 10
Differential scanning calorimetry (DSC), 175, 187
Diffusion, 35, 125, 167, 173, 176, 180, 217, 219, 220, 222, 224, 252
Dimethylbenz[a]anthracene (DMBA), 37
Donepezil solution, 173
Dopamine, 8, 70, 81, 180
  and g-amino butyrate (GABA), 8
Dormant groups, 75
Downregulation, 37
Doxorubicin, 34–37
Drug
  carrier molecules, 20
  classes, 69
  deliberation, 32
  delivery, 1, 6, 7, 14, 15, 31–38, 42, 73–75, 77, 82, 86, 97, 98, 126, 129–132, 135, 150–153, 155, 186
    prototypes, 32
  delivery nanotechnology, 32
    liposomes, 33
    nanoemulsions, 36
    polymeric nanoparticles, 34
  design, 42
  doxorubicin, 34
  elution, 35
  free lipid core, 167
  loading, 8, 31, 98, 104, 105, 117, 118, 136, 141, 166, 167, 174
    ability, 41
    capacity, 8, 31, 136
    efficiency, 105

market, 73
metabolism, 31, 32
molecules, 20, 35, 36, 131, 139
non-compliance, 65
payload, 105
pharmacokinetics, 31, 38
precipitation, 144
release, 15, 35, 37, 81, 118, 119, 121, 123, 142, 143, 146–148, 150, 165, 166, 173
targeting, 181, 182
therapy, 65, 130
Dry emulsions, 147
Dye-sensitized solar cell, 199, 200
Dynamic light scattering (DLS), 174, 187

E

Edible oils, 137, 139
Elasticity, 246
Electrochemical coupling factor, 252
Electrodeposition, 209
Electrodes, 201, 202, 209, 250, 251
Electrolyte, 199–202, 206–209
Electron
  microphotograph, 109
  microscopy, 107, 108, 175
Electronegative atoms, 114
Electrophoretic mobility, 175
Electroporation, 42
Electro-static interaction, 77
Ellipsoids, 53
Emulsification, 75, 77, 79, 129, 132, 136–140, 143, 144, 150, 168, 173
Encapsulation, 15, 36, 37, 77, 80, 180
End-bend-torsion interactions, 231
Endocrinopathies, 66
Endocytosis, 33, 36, 39, 42, 81
Endosomal escape, 39, 40
Endosomes, 83
Endothelial
  cells, 36, 83
  hyperplasia, 66
  penetration, 41
Enhanced-permeability-and-retention (EPR), 40, 43
Entrapment efficiency (EE), 37, 165, 174, 187
Enzymatic degradation, 36, 145, 166

Enzyme
  degradation, 86
  delivery, 125
  remodeling, 38
Epidermal carcinoma cell lines, 37
Epithelial cells, 34, 86
Erosion, 35, 143, 221
Erythropoietin, 172
Exocytosis, 10

F

Fabrication, 4, 7, 10, 106, 201, 209, 250
False color technique, 111
Fatty acid, 86, 140, 143, 148, 171, 182
Faulty gene expression, 43
Femtosecond laser surgery, 17
Fickian diffusion, 35
Fill factor (FF), 200, 202, 206, 207, 209, 210
Finite element method, 253
Fluctuation, 8, 125, 136, 146
Fluorescent SLNs (F-SLN), 173
Fluorine, 10, 200
Food and Drug Administration (FDA), 55,
     56, 61, 86, 87
Fossil fuels, 199, 215, 216
Fourier transform infrared (FTIR), 104,
     112–114
Free
  drugs, 74
  fatty acids, 172
  quercetin, 180
  radicals, 67, 71, 83, 84
Frequency, 8, 16, 17, 41, 65, 66, 74, 154,
     166, 180, 217, 224, 245, 246, 248,
     250–252
Fructose intracellular accumulation, 71
Fullerene nanocomponents, 10
Functional
  groups, 16, 75
  head groups, 38

G

Gabapentin, 70
G-amino butyrate (GABA), 8
Gas
  molecules, 18, 250
  sensor methods, 253

Gastric
  fluid, 144, 146
  motility, 135, 142
Gastrointestinal
  absorption, 132
  cholinergic nerves, 16
  lipase inhibitor, 150
  tract, 36, 130, 132
Gelator, 201
Gene
  delivery, 39, 41, 76, 146
  therapy, 39, 43, 124, 184
  transfection, 31
Geometric scattering, 175
Gestational diabetes, 66
Glial cells, 80, 82
Glimepiride, 79, 150
Glioblastoma, 179
Glioma cell lines, 37
Global
  energy transition, 199
  health care industry, 2
  market, 4
Glucose, 16, 17, 66, 67, 71, 74, 76, 79, 85,
     105, 116, 122, 125
  homeostasis, 85
Glutamic acid, 33
Glutathione, 67
Glycerol, 18, 148, 172
Gold
  nanoparticles, 84
  nanoshells, 35
Graphene, 40, 58
Greenhouse effects, 199

H

Health care system, 2–4, 7–9, 16, 17, 19, 20
Hematoxylin, 178
Hemocompatibility, 34
Hemoglobin, 18, 108, 109, 124–126
Hepatocellular carcinoma (HCC), 36, 43, 181
Heptagons, 230
High
  pressure homogenization (HPH), 167,
       168, 170, 171, 187
    cold homogenization, 169
    hot homogenization, 169
  velocity oxygen fuel (HVOF), 216, 221, 224

Homogeneous matrix, 167
Homogenization, 77, 167–170, 172
Human excretory system, 13
Huntington's disease, 180
Hyaluronic acid (HA), 34, 43
Hydrogen (H), 10, 41, 100, 102, 103, 114, 115, 179, 232
Hydrolysis, 34, 56, 105, 116, 131, 135, 148
Hydrophilic
    anticancer drug caffeine, 37
    drugs, 36, 86, 131, 152
    layer, 77
    macromolecular absorption, 86
    molecules, 33, 76
Hydrophobic
    anticancer drug, 37
    compounds, 76, 100, 101
    drugs, 33, 130, 136, 165
    interactions, 77
    medicines, 74, 77
Hydrophobicity, 34
Hydroxyapatite core, 97, 106, 107, 109, 115, 116
Hyperalgesia, 66, 80
Hyperglycemia, 65, 66, 71, 75, 76, 82, 85
Hypertension, 182
Hypoglycemia, 66

I

Immune system, 7–9, 16, 39, 87
Immunization, 54, 56, 124
Immunoassays, 38
Immunodiagnostic applications, 38
Immunogenicity, 82
Immunoliposomes, 33
Immunomodulator, 182
Immunotherapy, 8, 39
Immunotoxicity, 87
In vitro
    drug release, 105
    in vivo correlation (IVIVC), 148, 149, 155
    leakage test, 78
    permeation efficiency, 174
    release studies, 118, 119
    research, 87
In vivo, 32, 34, 36, 38, 40, 42, 74, 84, 87, 88, 123, 129, 137, 139, 142, 148–151, 155, 172, 174
    activity, 36

bioavailability, 149
bioimaging, 32
detection, 38
imaging, 42, 88
study, 125
targeting, 34
Indium doped tin oxide (ITO), 200
Indomethacin, 111, 123
Inflammatory
    cells, 19, 37, 83
    degradation, 65
    reaction, 83
Infrared (IR), 35, 40, 55, 61, 114, 231
Innate immune response, 33, 39, 40, 43
Insulin, 16, 65, 66, 70, 74, 75, 78, 85, 86, 99, 122, 125
    deficiency, 65
    delivery, 85, 122
    resistance, 65, 85
Inter digitated transducer (IDT), 245–247, 251, 253
Interfacial
    energy, 138
    surface area, 132, 138
    tension, 137, 139
Inter-subject variability, 147
Intracellular
    delivery, 34
    sites, 34
    uptake, 33
Ionic
    bonds, 98, 99
    conductivity, 202, 207, 209
    interactions, 101
    surfactants, 136
Isotonic solution, 178

J

Jobless production trends, 59

K

Kinematics, 232
Kinetics, 35, 75, 84, 120, 121, 146, 174, 219

L

Lactic acidosis, 74
Lactose, 101, 105, 106, 108, 111, 112, 118, 121, 147

Lamotrigine, 70, 81
Langmuir-Blodgett film, 248
Lattice system, 18
Leukocytes, 9
Ligands, 81, 145
Light microscopy, 143
Lipid
  based drug delivery system (LBDDS),
    131, 155
  bilayers, 55
  carriers, 131, 142
  formulation classification system (LFCS),
    137, 155
  medium-chain lipids, 136
  peroxidation, 84, 151
Lipophilic
  cargo drug molecules, 33
  drug, 36, 131–133, 136, 137, 140, 167,
    169, 170, 173
  phases, 140
Lipophilicity, 32, 130
Liposome, 7, 8, 20, 33, 34, 39, 42, 55,
  74–76, 80, 82, 86, 97, 99, 131, 165, 166
Liquid phase epitaxy (LPE), 217, 224
Long-term
  damage, 65
  glycemic control, 65
  therapy, 71
Lower critical solution temperature (LCST),
  35, 43
Lubricants, 217
Lymphatic system, 142, 145
Lymphocytes, 87
Lyophilization, 104
Lysosome, 33

M

Macromolecules, 73, 75, 76, 84, 86, 135
Macrophages, 75, 82, 181
Magnetic
  field, 11, 39, 41
  hyperthermia, 41, 43
  particle imaging (MPI), 41, 43
  resonance imaging (MRI), 2, 38, 41, 184
Malfunctioning, 43
Mannose-like binding lectins (MLBLs), 123
Maximum concentration (C_{max}), 136, 146, 149
Mean residence time (MDT), 146, 155

Medical nanobots, 2, 7–10, 17, 19
  applications, 13
    artificial neurons, 19
    artificial phagocyte, 18
    atherosclerosis treatment, 19
    cancer therapy, 15
    cell repair and lysis, 19
    diabetes diagnosis and treatment, 16
    diagnosis and testing, 17
    respirocyte, 17
    surgery, 17
    targeted drug delivery, 15
Melt-dispersion technique, 168, 171
Metabolic
  disorders, 66
  system, 16
Metabolism, 10, 38, 41, 66, 79, 81, 83, 85,
  133
Metallization, 250, 252
Micelle, 15, 20, 33, 76, 77, 86, 131, 136,
  140, 142, 166
Microemulsion, 77, 79, 132, 137, 143, 168,
  171
Microfludization, 77
Micron, 7, 9, 10, 12, 13, 78, 132, 166, 168,
  170, 174
Microprocessor, 1, 20
Microspheres, 86, 131
Microvascular disorders, 66
Minimal
  solubility, 40
  toxicity cater, 31
Minoxidil, 33, 183
Mitochondrial dysfunction, 67
Molecular
  beam epitaxy (MBE), 217, 224
  diagnostics, 31, 38
  dynamics (MD), 227–229, 231, 232,
    234–237, 239–241
  mechanics (MM), 229, 231, 241
  medicine, 31, 38, 42
    gene therapy, 39
    metal nanoparticles (NPS), 38
    nucleic acid nanoparticles (NPS), 38
  plasticization, 101
  weight, 76, 140
Monocytes, 101
Monomer, 35, 75, 77
Mononuclear phagocytic processes, 75

Monte Carlo (MC), 228
Morbid cells, 40, 41
Mortality, 2, 7, 82
Mucin, 136, 145
Multidrug resistant (MDR), 31, 32, 37, 43
Multilamellar vesicles, 74
Myocardial infarction, 19, 66

N

Nano-arrays, 72
Nano-bioimaging, 32
Nano-bioprobing, 32
Nanobot, 1–4, 6–14, 16–20
  communication, 12
  components, 10
    capacitor, 11
    micro camera, 10
    payload, 11
    skeleton, 10
    swimming tail, 11
  elimination, 13
  monitors, 12
  properties, 9
  target site recognition, 11
Nanocarrier, 7, 82, 83, 97, 98, 110, 122
Nanocomposite, 58, 216–218, 224
Nanoemulsion, 1, 33, 36, 37, 77, 78, 129,
  131, 132, 135, 139, 140, 145, 155
Nanomaterials, 4, 5, 39, 40, 42, 43, 51, 53,
  54, 56, 59, 60, 72, 73, 79, 83, 224
  types, 53
    carbon-based materials, 53
    composites, 54
    dendrimers, 53
    metal-based materials, 53
Nanomedicine, 7, 32, 40, 55, 63, 71–73, 80,
  87, 88
  dendrimers, 75
  liposomes, 74
  nanoparticles (NPS) toxicity, 87
  niosomes, 73
  polymeric micelles, 76
    nanoemulsion, 77
    nanosuspension, 78
Nanometer, 1, 51–53, 71, 106, 132, 174
Nanoparticle (NP), 15, 16, 32–36, 38–40,
  42, 51, 54–57, 61, 72, 79, 81–88, 97, 98,
  106, 108, 111, 131, 141, 165, 166, 174,
  179, 181, 183–185, 251

Nanoscale, 7, 10, 52, 58, 216, 217
  drug-delivery vehicles, 33
Nanosomes, 1, 20
Nano-structured lipid carriers (NLCs), 131,
  155, 174
Nanostructures, 1–3, 19, 20, 38, 40, 53
Nanosuspensions, 78, 79, 86
Nanotech
  equivalent neuron, 19
  neurons, 19
Nanotechnology, 1, 2, 4–8, 13, 14, 19, 20,
  31, 32, 37, 40, 42, 43, 51–61, 63, 71–73,
  79, 82, 85, 86
  goals, 4
    chemical plants, 5
    electronics, 4
    healthcare system, 5
    manufacturing, 4
    pharmacy, 5
    sustainability, 5
    transportation, 5
  health impact, 54
    food system, 56
    health hazards, 56
    medical application, 54
    nanomedicine, 55
  natural impact, 57
    environmental application, 57
    environmental pollution, 58
  social and economic impact, 59
    economic growth, 60
    intellectual property issue, 59
    unemployment effect, 59
  theranostics, 40
    photothermal therapy, 40
    quantum dots, 41
    superparamagnetic iron oxide nanopar-
      ticles (SPIONS), 41, 42
National
  health institutes (NIH), 72
  technology initiative (NTI), 88
Negative zeta potential, 37
Nephropathy, 66
Neurodegenerative disorders, 7
Neuro-inflammation, 67
Neuronal
  cell membranes, 70
  conduction, 67
  pathways, 87

Neurons, 8, 17, 19, 70
Neuropathic pain (NeP), 63, 72, 80, 81
Neuropathy, 64–67
Neurotransmitters, 8, 70, 81
Next-generation
  antibiotics, 32
  pharmaceuticals, 31
Niosomes, 73, 74, 76, 86, 166
Nitrogen, 10, 56, 169
Non-covalent
  bonds, 98, 99
  interaction, 104
Non-immunogenicity, 37
Non-invasive techniques, 41, 130
Non-ionic surfactant, 73, 74
Non-little cell lung disease (NSCLC), 55
Non-renewable energy resources, 199
Noradrenaline, 70, 81
Novel pharmaceutics, 32
Nucleic acid, 38–40, 42, 43, 82, 84, 86, 184
Nucleotides, 39, 41, 82

O

Olfactory nerve, 57
Oligomers, 35, 101
Oligonucleotide, 33, 39, 184
  density, 39
  layers, 38
Oligosaccharide, 97
Omega-3 fatty acid, 37, 182
Opioids, 70
Optimal pharmacokinetic profile, 32
Oral
  administration, 78, 86, 123, 142, 152, 153
  bioavailability, 135, 136, 151, 183
  delivery, 37, 86, 130, 141, 147, 150
  insulin, 85, 86
  protein medications, 85
Organic
  fluorophores, 42
  solvent, 166, 170, 171
Orthostatic hypotension, 67
Ostwald ripening, 132
Ovalbumin (OVA), 106, 107, 119, 120, 124
Oxidation, 36, 84, 182, 215–217, 219–224
  resistance, 216, 219–222, 224
Oxidative
  damage, 84
  pressure, 57

stress, 36, 40, 41, 67, 69–71, 84, 85
Oxide scale
  growth, 215, 216
  spallation, 222
Oxygen, 10, 13, 14, 18, 43, 83, 85, 124, 216, 224
Ozone, 56, 58

P

Pancreatectomy, 4
Pancreatic
  exocrine disorders, 66
  β-cells, 65
Paracellular pathway, 36, 86
Paramagnetic solid lipid nanoparticles (pSLNs), 184
Pathogenesis, 67, 68, 71, 82, 83
PEGylation, 81, 86, 145
Pellets, 145–147, 150
Peptides, 33, 41, 43, 75, 76, 99, 135, 146, 172
Permeation, 86, 130, 146, 174, 177, 179
Perovskite solar cell, 209
Petroleum, 5, 199
pH, 33–35, 40, 86, 99, 103, 105, 118–122, 144, 146, 149, 176, 178
Pharmaceutical, 13, 32, 126, 154, 165, 166
  challenges, 63, 88
  designing, 31
  elution, 35
  industry, 15, 88
  sciences, 31, 32, 38, 43
Pharmacodynamics, 31
Pharmacokinetic, 72, 126, 151
  parameters, 8, 15, 130, 146
  performance, 181
  profile, 33, 79, 80, 81
  properties, 15
Phosphate buffer saline (PBS), 118, 120, 121
Phosphatidylcholine, 36
Phospholipid, 33, 74, 125, 147, 165
Photo electrochemical effect, 200
Photobleaching, 42
Photocurrent density, 202, 206–209
Photodynamic therapy, 40
Photoemission, 178
Photon correlation spectroscopy (PCS), 107, 143, 174
Photosensitivity, 35

Photothermal therapy, 40, 41, 43
Photovoltage, 202, 206, 207, 209
Physicochemical parameters, 32, 33, 42
Physiological
  clearance, 36
  degradation, 32, 43
  toxicity, 34, 41, 43
Piezoelectric substrate, 246, 248, 250
Piroxicam, 107, 108, 118, 120
Plasma, 13, 80, 81, 125, 134, 147, 149, 181
Plasmon resonance, 38
Plumbu (Pb), 209, 210
Poisson's ratio, 234
Pollutants, 60, 199
Poly methyl ether methacrylate (PMEM), 202
Poly(amidoamine) (PAMAM), 76, 124
Poly(lactic-co-glycolic acid) (PLGA), 34,
  39, 148
Polyacrylonitrile (PAN), 201, 202, 210
Polydispersity, 131, 132, 166
Polyethylene
  glycol (PEG), 36, 75, 81, 86, 125, 140,
    145, 202, 209, 210
  oxide complexes, 202
  terephthalate, 209
Polymer, 35, 39, 7, 35, 53, 73, 75, 76,
  78, 82, 86, 87, 99, 101, 141, 148, 186,
  199–204, 207, 250
Polymeric NPs, 34, 35, 36
Polymethyl
  ether methacrylate, 210
  methacrylate (PMMA), 202, 210
Polymorphism, 123
Polyol flux pathway, 67, 68
Polyplexes, 39
Polysaccharides, 115
Postsynaptic receptor blocking effects, 70
Potassium bromide pellet method, 112
Potent drugs, 72, 146
Powder x-ray diffraction (PXRD), 114, 115,
  176
Power plants, 215, 216, 221
Precipitation, 124, 144, 149, 168, 171
Prodrug technology, 86
Pro-inflammatory cytokine, 82, 87
Protected innovation law, 59
Protein
  drugs, 86, 147
  hormones, 99

kinase C activity, 67
molecules, 86
Proton, 34, 41
Pseudomonas aeruginosa, 172
Pseudo-ternary diagram, 133, 139

Q

QD-mediated Forster resonance energy
  transfer (QD-FRET), 42
Quantum
  confinement effect, 42
  dabs, 53
  dots, 20, 35, 38, 40, 42, 72
  speck, 53
Quasi solid-state polymer electrolytes, 202
Quercetin, 179–181
Quinidine, 70

R

Radial distribution function (RDF), 234, 241
Radio frequency (RF), 8, 17, 217, 224
Radiotherapy, 54
Ratiometric fluorescent approach, 36
Reactive oxygen species (ROS), 36, 43, 83,
  84, 85
Receptor molecule, 11
Red blood cell, 17, 18
Redox stress, 43
Reflux, 102, 103
Relative humidity, 248
Renewable energy sources, 215
Repulsion, 100, 175
Respirocyte, 14, 17, 18, 20
Retention effect, 35, 77
Reticuloendothelial system (RES), 8, 33, 99
Retinopathy, 66, 182
Retroviruses, 82
Ribonucleic acid (RNA), 34, 38, 82, 83
  interference (RNAi), 38
Robust candidature, 31, 32, 38

S

Scanning electron microscope (SEM), 104,
  106, 108, 109, 111, 175, 220, 222
Scavenger, 10, 71, 83
Second
  generation solar cells, 200
  line treatment, 81

Selective layer sintering (SLS), 217, 224
Self-assembly principle, 100
Self-double nano-emulsifying systems
    (SDEDDS), 144, 155
Self-emulsifying (SE), 129, 131, 133, 135,
    136, 139, 140, 143, 144, 147–150, 152,
    153
    drug delivery system (SEDDS), 129, 131,
        132, 135, 136, 139–142, 147, 148,
        152–154
    lipid types, 133
Self-micro-emulsifying drug delivery
    system (SMEDDS), 129, 132, 140–142,
    151, 152
Self-nano emulsifying drug delivery
    systems (SNEDDS), 129, 132, 133, 135,
    140–142, 144–148, 150, 151
Semiconductor, 4, 16, 41, 52, 53, 200
Serotonin and norepinephrine reuptake
    (SNRIs), 70
Serratiopeptidase (STP), 110, 121, 122, 125,
    126
Shear
    homogenization, 170
    modulus, 228, 234
    stress, 168
Shelf life, 33, 35, 124, 143
Short-circuit photocurrent density, 202
Single
    phase interfacial layer, 77
    walled carbon nanotubes, 227, 228, 241
    Walled Carbon Nanotubes, 227
    walled carbon nanotubes (SWCNTs),
        227–229, 233, 235, 236
Sodium chloride, 103, 178
Solar
    cells, 6, 199, 200, 209
    energy, 5, 199, 210
Solid lipid nanoparticles (SLNs), 131,
    165–184, 186
    administration routes, 172
        nasal route, 173
        ocular route, 173
        oral route, 172
        parenteral route, 172
        rectal route, 173
    applications, 179
        brain targeting, 179

cardiovascular diseases, 181
chemotherapy, 180
liver targeting, 181
ocular targeting, 182
uterus and ovary targeting, 182
biological properties, 166
blood components, 13
dispersions, 130
drug
    loading models, 167
    release mechanism, 168
evaluation, 174, 176
    general evaluation parameters, 174
limitations, 166
nasal evaluation, 178
    gamma scintigraphy imaging, 178
    histological studies, 178
    radiolabeling, 178
phase immobilization, 38
storage and stability, 178
technological properties, 166
topical evaluation, 177
    rheological studies, 177
    skin irritation studies, 177
    spredability, 177
Solidification, 52, 135, 141
Solvent
    dispersion methods, 75
    emulsification, 170
    evaporation, 77, 79
Sonication, 75, 102, 104, 126, 170
Sorbitol, 67, 71, 202
Spallation, 215, 216, 219, 222–224
Spherical nucleic acid (SNA), 38, 39
Spray drying technique, 147
Staphylococcus aureus, 172
Sterilization, 166, 173, 186
Stevens-Johnson syndrome, 81
Stone-Wales (SW), 227–230, 233–241
Streptozotocin, 78, 79, 84
Stretch-torsion interactions, 231
Sub-atomic
    assembling, 52
    engineering, 52
    nanotechnology, 52
Sucrose, 101, 104
Sugar coating quantification, 116
Super critical fluid, 170, 171

Supercapacitors, 202
Superparamagnetic metal oxide, 38
Supper critical anti-solvent (SAS), 168, 170
Surface
　acoustic wave (SAW), 245, 246, 248,
　　250–253
　additive bonding (SAB), 217
　coating, 41, 42
　ligand, 33, 36
　oligonucleotide, 39
　oxidation, 42
　shearing, 138
　tension, 13, 138, 140
　to-volume ration, 42
Surfactant, 77, 78, 130–133, 136–141, 144,
　165, 166, 168–172, 186

T

Targeted solid lipid nanoparticles (tSLNs),
　181
T-cell, 87, 154
Tersoff-Brenner potential, 228, 229, 236
Theranostics, 40, 41
Therapeutic, 7, 8, 31, 37, 43, 146
　action, 20
　activity, 126
　applications, 54, 72, 75
　effect, 39, 41, 80, 81, 99, 122, 136
　genes, 39
　index, 34, 72, 125
　molecules, 32, 33
　prototype, 32
　targets, 79
　　target to free radicals, 83
　　target to inflammation, 82
　task, 13
　temperature, 40
Thermal
　gradient, 40
　power plant, 221
　spraying, 215, 216
Thermo-analytical technique, 175
Thermolabile drugs, 135, 169
Thermoplastic granulation technique, 141
Three-dimensional
　geometry, 38
　structure, 101
Titanium dioxide, 53
Tobramycin, 173, 182

Tocopherol poly(ethylene glycol)succinate
　(TPGS), 37
Toluene, 251
Tomography, 38, 41, 55, 178
Topiramate, 70
Toxicity parameters, 15
Transmission electron microscopy (TEM),
　104, 106–111, 175, 220
Trehalose, 101, 104, 106, 108, 109, 114,
　115, 117, 118
Trial-error methods, 148
Triphenylmethane, 35
Trivial reaction, 34
Tumor
　cells, 15, 16, 34, 36, 40, 41, 180
　microenvironment, 34, 35, 36
　morbid cells, 33
　stroma fibroblasts, 41
　tissue aggregation, 76

U

Ulceration, 64
Ultra-low interfacial tension, 132
Ultrasonication, 170
Ultrasound, 12, 13, 17, 20
Unilamellar vesicles, 74
United States of America (USA), 52
Unsaturated fatty acids, 185
Unsaturation, 140, 143
Urey-Bradlay term, 231
Urinary retention, 82
UV spectrophotometer, 105

V

Vacancy defect, 84, 227–230, 233–241
Vaccines, 33, 39
Van der Waals force, 98–101
Vaporization, 17, 228
Vascular
　disorders, 66
　endothelial growth factor, 182
　system, 17
Venlafaxine, 70
Viscosity, 110, 144, 177, 202
Viscous body fluid, 12
Vitamin
　A, 136
　E, 37, 70
Volumetric liquid filling equipment, 135

W

Water
  filtration, 5
  in-oil, 37, 77, 144
  phase, 138
Wave frequencies, 9

X

X-ray, 114, 115
  diffraction (XRD), 114

Y

Young's modulus, 227–229, 234–240

Z

Zero order kinetics, 35
Zeta potential, 36, 118, 175
Zinc oxide, 253
Zipper effect, 35
Zoledronic acid, 80

For Product Safety Concerns and Information please contact our EU
representative GPSR@taylorandfrancis.com
Taylor & Francis Verlag GmbH, Kaufingerstraße 24, 80331 München, Germany